Computational Modelling in Behavioural Neuroscience

Classically, behavioural neuroscience theorizes about experimental evidence in a qualitative way. However, more recently there has been an increasing development of mathematical and computational models of experimental results, and in general these models are more clearly defined and more detailed than their qualitative counterparts. These new computational models can be set up so that they are consistent with both single-neuron and whole-system levels of operation, allowing physiological results to be meshed with behavioural data – thus closing the gap between neurophysiology and human behaviour.

There is considerable diversity between models with respect to the methodology of designing a model, the degree to which neurophysiological processes are taken into account and the way data (behavioural, electrophysiological, etc.) constrains a model. This book presents examples of this diversity and in doing so represents the state of the art in the field through a unique collection of papers from the world's leading researchers in the area of computational modelling in behavioural neuroscience.

Based on talks given at the third Behavioural Brain Sciences Symposium, held at the Behavioural Brain Sciences Centre, University of Birmingham, in May 2007, the book appeals to a broad audience, from postgraduate students beginning to work in the field to experienced experimenters interested in an overview.

Dietmar Heinke is a senior lecturer at the School of Psychology, University of Birmingham. His research interests include developing computational models for a broad range of psychological phenomena.

Eirini Mavritsaki is a research fellow at the University of Birmingham investigating the cognitive functions using models of spiking neurons.

Computational Modelling in Behavioural Neuroscience

Closing the gap between neurophysiology and behaviour

Edited by Dietmar Heinke and Eirini Mavritsaki

Psychology Press
Taylor & Francis Group

HOVE AND NEW YORK

First published 2009
by Psychology Press
27 Church Road, Hove, East Sussex BN3 2FA

Simultaneously published in the USA and Canada
by Psychology Press
711 Third Avenue, New York, NY 10017

*Psychology Press is an imprint of the Taylor & Francis Group,
an informa business*

First issued in paperback 2012

© 2009 Psychology Press

Typeset in Goudy by RefineCatch Limited, Bungay, Suffolk

Cover design by Anú Design

The publication has been produced with paper manufactured to strict
environmental standards and with pulp derived from sustainable
forests.

British Library Cataloguing in Publication Data
A catalogue record for this book is available from the British Library

Library of Congress Cataloging-in-Publication Data
Computational modelling in behavioural neuroscience : closing the gap
between neurophysiology and behaviour / edited by Dietmar Heinke
& Eirini Mavritsaki.
    p.; cm. – (Advances in behavioural brain science)
Includes bibliographical references.
1. Computational neuroscience. I. Heinke, Dietmar, 1965– II.
Mavritsaki, Eirini, 1975– III. Series.
[DNLM: 1. Neurosciences—methods. 2. Behavior—physiology. 3.
Models, Biological. WL 100 C738 2009]
QP357.5.C58 2009
612.80285—dc22   2008028187

ISBN: 978–1–84169–738–3 (hbk)
ISBN: 978–0–415–64685–7 (pbk)

Contents

Contributors

Harriet Allen, Behavioural Brain Science Centre, School of Psychology, University of Birmingham, Birmingham, B15 2TT, UK.

Andreas Backhaus, Behavioural Brain Science Centre, School of Psychology, University of Birmingham, Birmingham, B15 2TT, UK.

John A. Bullinaria, School of Computer Science, University of Birmingham, Birmingham, B15 2TT, UK.

Vassilis Cutsuridis, Department of Computing Science and Mathematics, University of Stirling, Stirling, FK9 4LA, UK.

Gustavo Deco, Institució Catalana de Recerca i Estudis Avançats (ICREA), Universitat Pompeu Fabra, Department of Technology, Computational Neuroscience, Passeig de Circumval-lació, 8, 08003 Barcelona, Spain.

Karl J. Friston, Wellcome Trust Centre for Neuroimaging, Institute of Neurology, University College London, London, WC1N 3BG, UK.

Bruce P. Graham, Department of Computing Science and Mathematics, University of Stirling, Stirling, FK9 4LA, UK.

Kevin N. Gurney, Adaptive Behaviour Research Group, Department of Psychology, University of Sheffield, Sheffield, S10 2TP, UK.

Dietmar Heinke, Behavioural Brain Science Centre, School of Psychology, University of Birmingham, Birmingham, B15 2TT, UK.

Glyn W. Humphreys, Behavioural Brain Science Centre, School of Psychology, University of Birmingham, Birmingham, B15 2TT, UK.

Stefan Kiebel, Wellcome Trust Centre for Neuroimaging, Institute of Neurology, University College London, London, WC1N 3BG, UK.

Ansgar Koene, Department of Psychology, University College London, London, WC1H 0AP, UK.

Martin Kreyling, Behavioural Brain Science Centre, School of Psychology, University of Birmingham, Birmingham, B15 2TT, UK.

Emilio Kropff, SISSA – International School of Advanced Studies, via Beirut 4, 34014, Trieste, Italy.

Eirini Mavritsaki, Behavioural Brain Science Centre, School of Psychology, University of Birmingham, Birmingham, B15 2TT, UK

Keith A. May, Department of Psychology, University College London, London, WC1H 0AP, UK

Michael C. Mozer, Department of Computer Science and Institute of Cognitive Science, University of Colorado, Boulder, CO 80309-0430, USA.

Edmund T. Rolls, Department of Experimental Psychology, University of Oxford, South Parks Road, Oxford, OX1 3UD, UK.

Aaron Sloman, School of Computer Science, University of Birmingham, Birmingham, B15 2TT, UK.

Klaas E. Stephan, Wellcome Trust Centre for Neuroimaging, Institute of Neurology, University College London, London, WC1N 3BG, UK.

Simon J. Thorpe, Centre de Recherche Cerveau & Cognition, Faculté de Médecine de Rangueil-Bât A3, 133, route de Narbonne, 31062 Toulouse Cedex9, France.

Thomas Trappenberg, Faculty of Computer Science, Dalhousie University, 6050 University Ave, Halifax, Nova Scotia, Canada, B3H 1W5.

Robert Ward, Wolfson Centre for Clinical and Cognitive Neuroscience, University of Bangor, Bangor, LL57 2AS, UK.

Ronnie Ward, Department of Computer Science, Texas A&M University, TX 77843, USA.

Matthew H. Wilder, Department of Computer Science and Institute of Cognitive Science, University of Colorado, Boulder, CO 80309-0430, USA.

Li Zhaoping, Department of Psychology, University College London, London, WC1H 0AP, UK.

Preface

Behavioural neuroscience is mainly concerned with understanding how neurophysiological processes, such as chemical processes in synapses or chemo-electrical processes in axonal spikes, ultimately lead to human behaviour. There are numerous empirical approaches to this issue – for example, experiments with stroke patients (neuropsychology), fMRI experiments, EEG experiments, and so forth. Classically, behavioural neuroscience theorises about such experimental evidence in a qualitative way. More recently, however, there has been an increasing development of mathematical and computational models of experimental results. In general, these models are more clearly defined and more detailed than their qualitative counterparts. The aim of this book is to give an overview of the state of the art of computational modelling in behavioural neuroscience.

These computational models consist of networks of processing units ("neurons"). The operations of such neurons are based on evidence from neurophysiological processes, while at the same time allowing whole-system behaviour to emerge from a network of these neurons. Such models can be set up so that they are consistent with both single-neuron and whole-system levels of operation, allowing physiological results to be meshed with behavioural data – that is, closing the gap between neurophysiology and human behaviour.

Apart from this very general definition of computational models in behavioural neuroscience, there is considerable diversity between models with respect to the methodology of designing a model, the degree to which neurophysiological processes are taken into account and the way data (behavioural, electrophysiological, etc.) constrain a model. In 15 chapters, this book presents examples of this diversity and, in doing this, covers the state of the art of computational modelling in behavioural neuroscience.

The book is formed from talks given at the third Behavioural Brain Sciences Symposium, held at the Behavioural Brain Sciences Centre, University of Birmingham, UK, in May–June 2007. We were delighted to host such a distinguished set of contributors, and our hope is that the meeting, as well as the chapters that emerged from it, will contribute both to progress in understanding how human behaviour relates to neurophysiological processes in general and, more specifically, to the advancement of computational modelling as part of this particular scientific quest.

In the following paragraphs we summarise the content of each chapter. The summaries are written for researchers working in the area of behavioural neuroscience who may not be not familiar with computational modelling. Thus, apart from stating the main implication of a chapter, we point out the human behaviour considered by each chapter and the particular neurophysiological processes taken into account by each model. We hope these introductions will make it easier for non-experts in computational modelling to read the chapters, allowing them to pick and chose chapters relevant for their own area of interest.

We have organised the chapters in an order reflecting differences in the way neurophysiological processes are taken into account in the models. The approaches to modelling can be roughly grouped into two types according to the way the "processing units" – the neurons in the systems – operate: "spiking neurons" and "rate-coded neurons". Generally, spiking neurons can be seen as taking more neurophysiological details into account than do rate-coded neurons. Even though, at first glance, it may be seen as preferable to use the more detailed approach, there are also many reasons why the rate-coded approach is more suitable, as discussed in chapter 15. The models presented in the first two chapters utilise spiking neuron models. In contrast, chapters 3–5 advocate that modellers move between the degrees of abstraction during the design process. Finally, the models in chapters 6–12 are based on rate-coded neuron models. Chapters 13 and 14 take more general approaches to understanding the brain, which do not easily fit into those three categories.

In chapter 1, Graham and Cutsuridis present a model of the microcircuitry in the CA1 region of the mammalian hippocampus, which covers a broad range of neurophysiological details including the properties of pyramidal cells, bistratified cells, basket cells, exo-axonic cells and oriens lacunosum-moleculare cells. The result is a multi-compartment model with AMPA, NMDA, GABA-A and GABA-B synapses. With this computer model, Graham and Cutsuridis support the hypothesis that the CA1 region coordinates both the storage of new information and the retrieval of old information, and the authors highlight the specific roles of the different neuron types in the solution of this problem.

In chapter 2, Thorpe summarises the implications of his "temporal coding hypothesis". Central to this hypothesis is that perceptual information is coded by the order in which neurons within a population fire ("order code"). He argues that such a single-spike-based code can explain the speed with which humans can process complex visual scenes. Interestingly, together with spike-time dependent plasticity (STDP), this approach can implement unsupervised learning of face features.

In chapter 3, Deco and Rolls present another example where processes on the cellular level, as modelled by integrate-and-fire neurons (one-compartment, point-like neurons), are important for the understanding of whole-system behaviour. The authors demonstrate that stochastic processes on the cellular level are intrinsically linked with probabilistic behaviour in decision-making, as expressed by Weber's law. In addition, they suggest that using rate-coded neurons (mean-field approach) to design the model first, and then employing a veridical transformation of the model to the spiking level, can be accomplished to simulate real neuronal behaviour.

In chapter 4, Humphreys, Mavritsaki, Allen, Heinke and Deco applied Deco and Rolls's mathematical framework to simulate data on human search over time as well as space, the spiking Search over Time and Space (sSoTS) model. In a first series of simulations, the authors show that behavioural deficits of stroke patients can be linked to lesions in the posterior parietal cortex and to reduced levels of neurotransmitters. In a second study, they demonstrate that sSoTS can be used to decompose complex neural circuits found in fMRI studies of human attention, separating circuits concerned with inhibition attention function from those concerned with top-down enhancement.

In chapter 5, Heinke, Mavritsaki, Backhaus and Kreyling present new versions of the selective attention for identification model (SAIM). In general, SAIM implements translation-invariant object identification in multiple-object scenes. In an approach analogous to Deco and Rolls's method, they demonstrate that the initial structure of SAIM, based on rate-coded neurons, can be used to guide the design of a new model utilising spiking neurons. In addition, they argue that a modelling approach for "bridging the gap" should not only include neurological processes, but should also be able to deal with ecologically valid stimuli. To show that SAIM can satisfy this requirement, the authors present an extension of SAIM that can select and recognise objects in naturalistic scenes.

In chapter 6, Gurney proposes a similar top-down approach to modelling, albeit within a Marrian framework (see also chapter 15). He illustrates this approach with a summary of his modelling work on the basal ganglia, which presents results from several levels of abstraction, ranging from conductance-based models of individual neurons, through spiking neuronal networks, to systems-level models with rate codes and embodied (robotic) models in a naturalistic environment.

The focus in the first two chapters is on spiking neurons, and the subsequent four chapters use a mix of models to close the gap between human behaviour and neurophysiology. In the next seven chapters, the modelling work is based on rate-coded neuron models.

In chapter 7, Zhaoping, May and Koene summarise simulation results from a physiologically plausible model of V1 supporting their "V1 saliency hypothesis". Central to this hypothesis is that processes in V1 can explain input-driven target detection ("odd-one out") in visual search tasks – for example, the detection of a red-vertical bar among red-horizontal and green-vertical bars. In addition, the authors present empirical support for V1 representing the saliency of stimuli. It should be noted that they consider their model as an alternative approach to the standard saliency-based approach (for an example, see chapter 13).

In chapter 8, Thomas Trappenberg shows that a simple dynamic neural field model can link a broad range of behavioural findings with physiological measurements in the brain – for example, a motion discrimination task. Trappenberg points out that because this model is so successful, it is safe to argue that it captures principal mechanisms that are important for "brain-style information processing".

In chapter 9, John Bullinaria presents a model for the origin of modularity in the brain by using a standard multi-layer perceptron. The structure of the network is modified by an evolutionary algorithm, and each generation is trained with a standard back-propagation algorithm to solve complex pattern-discrimination tasks. Within this framework, the author demonstrates that modularity emerges from a combination of behavioural requirements (e.g., pattern discrimination) and neurophysiological constraints (e.g., the size of the brain, the length of connections, etc.).

In chapter 10, Ward and Ward present results from their "research tool", termed "minimally cognitive agent". This tool models the performance of a visual agent (VA) as part of a perception–action cycle situated in a simple simulated environment. A VA consists of continuous-time recurrent neural network trained by a genetic algorithm to perform a specific task, such as catching two falling balls. In the first part of their chapter, the authors summarise their earlier results with the VA approach on selective attention and the implications of the results for the neurophysiology of selective attention (e.g., for the need for "reactive inhibition" in order to switch attention). In the second part of the chapter, they present new results of simulations with multiple VAs, leading to speculation that the organisation of the brain can be described as multiple VAs acting in concert.

In chapter 11, Kropff presents a new learning rule for an autoassociative memory which is optimal for storing correlated patterns. He argues that the storage of correlated patterns is behaviourally particularly relevant since semantic information must be represented in correlated patterns (e.g., as expressed in feature norms).

In chapter 12, Mozer and Wilder present a new theory to unify exogenous and endogenous attentional control. Central to their theory is the assumption that attentional control operates on two dimensions: the spatial scale of operation (local vs. global) and the degree of task focus (low vs. high). For instance, if the task is to search for a person with a red coat in street scenes, the attentional control setting is "local spatial scale" and "high degree of task focus". Mozer and Wilder integrate their attentional control system into a standard saliency-based computer model by means of associative mapping. Their results show that this approach can successfully simulate a broad range of data from visual search tasks.

In chapter 13, Friston, Stephan and Kiebel present an abstract, mathematical approach to understanding the brain. The authors argue that the brain's function and structure can be understood in terms of one simple principle, the minimisation of "free-energy". From this principled stance, they derive a broad range of conclusions concerning perception, action and learning. For instance, actions aim to minimise surprising exchanges with the environment, whereas perception serves to infer the causes of sensory inputs. As to the structure of the brain, the authors conclude from the free-energy principle that the brain's hierarchical structure can be interpreted as an attempt to represent causal hierarchies in the environment.

In chapter 14, Sloman combines the standpoints of philosophy and artificial intelligence with theoretical psychology and summarises several decades of research into the functions of vision in humans and other animals. This chapter focuses almost entirely on what needs to be explained, rather than presenting a computational model. However, from this survey, Sloman derives a list of requirements that future computational models will need to take into account.

The final chapter (chapter 15) is somewhat unusual. In this chapter, Dietmar Heinke presents the minutes of discussions at the symposium. In this way, the chapter gives an overview of the methodologies and approaches represented in this volume while (it is hoped) conveying some of the excitement of the discussions that took place.

Taken together, the contents of this book give an overview of the computational methods currently used in the field and how these methods intend to bridge the gap between behaviour and neurophysiology.

We are very grateful to several organisations and individuals who made both the workshop and this book possible. We thank the Biotechnology and Biological Sciences Research Council (BBSRC) for funding the workshop. We also thank the Experimental Psychology Society (EPS) for funding a satellite workshop ("The method of computational modelling: A practical introduction", http://www.comp-psych.bham.ac.uk/workshop.htm) to this research workshop, which allowed us to invite additional speakers (Howard Bowman and Thomas Trappenberg) who volunteered to present at both events. Thank you.

The running of the meeting was helped by Dietmar's PhD students, Christoph Boehme, Yuanyuan Zhao and Giles Anderson, and most especially by Elaine Fox and Jo Quarry. Thanks to you all.

Eirini Mavritsaki
Dietmar Heinke
March 2008

1 Dynamical information processing in the CA1 microcircuit of the hippocampus

Bruce P. Graham and
Vassilis Cutsuridis

A major challenge to understanding cortical function is the complexity found at both the single-cell and microcircuit levels. Here we outline what is known about the microcircuitry of the CA1 region of the mammalian hippocampus. We then explore the possible functional roles of the variety of neuronal types within this microcircuit during dynamic information processing. This is considered within the framework of CA1 acting as an associative storage device during encoding and retrieval of episodic memories.

1 Introduction

The local circuitry to be found in many parts of mammalian nervous systems consists of a complex architecture involving many different neuronal types connected in feedforward and feedback loops. Synaptic connections may be excitatory or inhibitory and target specific spatial locations on a neuron. In addition to synaptic input, a neuron and the microcircuit it is a part of are subject to diffuse neuromodulatory signals. Neural synaptic transmission and neuromodulation combine to provide a complex dynamics of neural activity and presumed information processing in a neuronal microcircuit.

Computational models of cognitive behaviour generally seek to provide a simple but cogent explanation of the functionality required to produce a particular behaviour. A model may be more or less interpretable in terms of the workings of a particular brain area, or set of connected areas. Often an artificial neural network (ANN) approach is used in which the simple computing units may correspond to populations of neurons rather than to individual biological neurons. The next level of biological detail is to use spiking neuron models where the identification with real neurons may be one-to-one. Such spiking models are of the integrate-and-fire type, or they may include explicit biophysical properties of a neuron in a compartmental model. Typically the neuronal types in such models are restricted to the principal excitatory cells, plus one or two sources of inhibition.

As we learn more about the details of real neural microcircuitry, it is clear that our current models lack the richness in spatial and temporal information processing that brain circuits possess. The challenge is to build models that include more of the known biological details – such as further cell types and more complex models of individual neurons – but remain simple enough that they are understandable and provide explanatory power for cognitive function. To explore the ways forward, here we outline what is known about a particular neuronal microcircuit: the CA1 region of the mammalian hippocampus. We then try to relate aspects of this microcircuit directly to the general cognitive function of the storage and recall of information in an associative memory.

2 The hippocampal CA1 microcircuit

For both historical and experimental reasons, the hippocampus is among the most widely studied of mammalian brain regions, yielding a wealth of data on network architecture, cell types, the anatomy and membrane properties of pyramidal cells and some interneurons, and synaptic plasticity (Andersen, Morris, Amaral, Bliss, & O'Keefe, 2007). Its basic functional role is hypothesized to be the formation of declarative, or episodic, memories (Andersen et al., 2007; Eichenbaum, Dudchenko, Wood, Shapiro, & Tanila, 1999; Wood, Dudchenko, & Eichenbaum, 1999). Various subsystems, such as dentate gyrus, CA3 and CA1, may be involved in the storage of information in context, such as location in a particular spatial environment (Andersen et al., 2007), with appropriate recoding of afferent information depending on familiarity or novelty (Treves & Rolls, 1994).

The mammalian hippocampus contains principal excitatory neurons (pyramidal cells in CA3 and CA1) and a large variety of inhibitory interneurons (Freund & Buzsaki, 1996; Somogyi & Klausberger, 2005). The circuitry they form exhibits different rhythmic states in different behavioural conditions. Multiple rhythms, such as theta (4–7 Hz) and gamma (30–100 Hz) oscillations, can coexist (Whittington & Traub, 2003). This dynamic complexity presumably corresponds to specific functional processing of information (Axmacher, Mormann, Fernandez, Elger, & Fell, 2006). Much work has been devoted to trying to understand the cellular and network properties that generate these rhythms (Buzsaki, 2002; Traub, Jefferys, & Whittington, 1999), but much is still to be been done to decipher the function of the detailed microcircuits. In particular, how is plasticity controlled so that it does not interfere with previously stored memories while appropriately assimilating familiar and new information? This is the fundamental question that we will address, concentrating on the operation of the CA1 area.

2.1 External inputs to CA1

The CA1 region is one of several stages of information processing in the hippocampus. Its major sources of input are from the CA3 region of the hippocampus and the entorhinal cortex. It sends excitatory output back to the entorhinal cortex, both directly and via the subiculum, and sends diverse outputs to a variety of other brain regions, such as the olfactory bulb. In addition, there are inhibitory projections from CA1 to the medial septum (MS) and back to CA3 (Sik, Ylinen, Penttonen, & Buzsaki, 1994). In turn, CA1 receives GABAergic inhibition and cholinergic neuromodulation from the MS (Freund & Antal, 1988; Frotscher & Lenrath, 1985).

CA1 also receives a variety of other neuromodulatory inputs, including dopaminergic and noradrenergic pathways. Much of this neuromodulation is directed to the distal apical dendrites of CA1 pyramidal cells, where it coincides with the entorhinal glutamatergic input (Otmakhova & Lisman, 2000).

2.2 Neuronal types and their connectivity

The basic hippocampal CA1 microcircuit is shown in Figure 1.1. The single excitatory cell type is the pyramidal cell (PC), which is the putative major

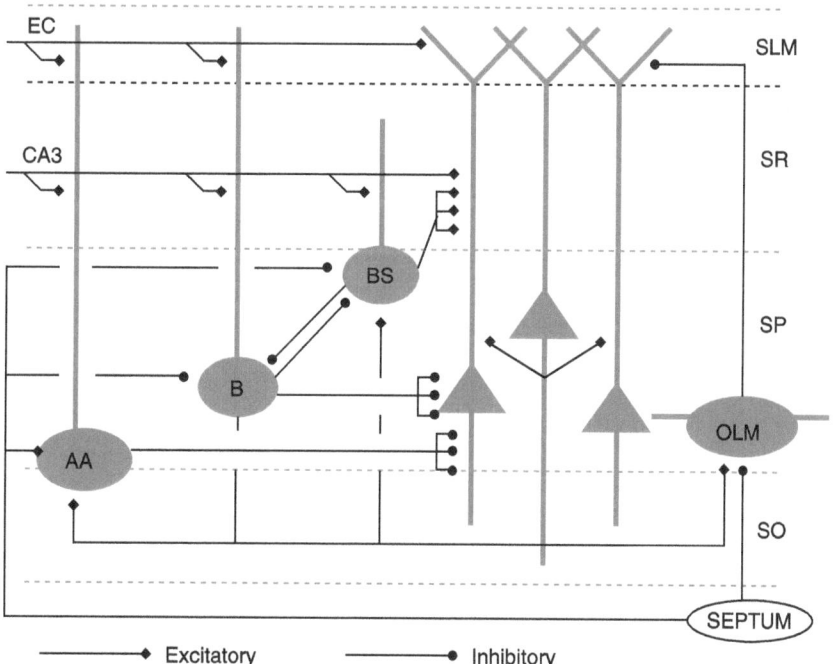

Figure 1.1 Hippocampal CA1 microcircuit showing major cell types and their connectivity. Large filled triangles: pyramidal cells. Large filled circles: CA1 inhibitory interneurons. EC: entorhinal cortex input; CA3: CA3 Schaffer collateral input; AA: axo-axonic cell; B: basket cell; BS: bistratified cell; OLM: oriens lacunosum-moleculare cell; SLM: stratum lacunosum moleculare; SR: stratum radiatum; SP: stratum pyramidale; SO: stratum oriens.

information processor for signals entering this brain region and is the major source of output from CA1. Pyramidal cells, here and elsewhere in the hippocampus and neocortex, have a large dendritic tree that is divided into apical and basal dendrites. These dendrites are the target for synaptic inputs that have distinct spatial segregation depending on the neuronal source.

Excitatory inputs from outside CA1 make connections on specific portions of the apical and basal dendrites of PCs (Ishizuka, Cowan, & Amaral, 1995). The Schaffer collateral input from pyramidal cells in the CA3 region of the hippocampus is exclusively to the proximal region of the apical dendrites constituting stratum radiatum (SR) and to the basal dendrites in stratum oriens (SO). Perforant path input from layer III of entorhinal cortex (EC) reaches the distal part of the apical dendritic tree in stratum lacunosum-moleculare (SL-M). Recurrent collaterals from other CA1 PCs synapse on the basal dendrites. Such collaterals are rather sparse in CA1, with only about 1% recurrent connectivity between pyramidal cells (Deuchars & Thomson, 1996). There are additional excitatory inputs from the thalamus to SL-M and from the amygdala to SO (Somogyi & Klausberger, 2005).

The pyramidal cells are surrounded by a variety of inhibitory interneurons (INs). These INs differ in morphology, pharmacology and connectivity (Freund & Buzsaki, 1996; Maccaferri & Lacaille, 2003; McBain & Fisahn, 2001; Somogyi & Klausberger, 2005). Though a complete catalogue of interneuronal types remains to be determined, at least 16 classes can be distinguished on anatomical, electrophysiological and pharmacological grounds (Somogyi & Klausberger, 2005). The most clear-cut types are basket cells, bistratified cells, axo-axonic (chandelier) cells and oriens lacunosum-moleculare (horizontal) cells. However, basket cells in particular consist of at least two subtypes: one that expresses parvalbumin and one that expresses cholecystokinin. Others include horizontal and radial trilaminar cells and INs that only synapse onto other INs (Freund & Buzsaki, 1996). A subclass of horizontal trilaminar cells (HTCs) sends axon collaterals out of the hippocampus to the medial septum (MS). There is also an inhibitory projection from CA1 to CA3. All these INs are inhibitory GABAergic cells.

Like excitatory afferents, different IN types target specific spatial regions on PCs (Megias, Emri, Freund, & Gulyas, 2001). They also receive excitatory input from particular pathways and may form synaptic (inhibitory) and gap junction (excitatory) connections with other INs (Gulyas, Megias, Emri, & Freund, 1999). In what follows we concentrate on four major classes of IN:

- **Basket cells (BCs)** receive feedforward excitation from CA3 and entorhinal PCs and feedback excitation from CA1 PCs. They form inhibitory connections on the perisomatic region of CA1 PCs, as well as with each other and with other IN classes. They also appear to form at least a partial syncitium through dendritic gap junctions with each other, ensuring high-frequency synchronization of their firing (Bartos, Vida, & Jonas, 2007).

- **Bistratified cells (BSCs)** are also driven by feedforward input, largely from CA3. They inhibit PCs in the same dendritic regions in SR and SO that are the site of CA3 input. They also inhibit other INs, including BCs.
- **Axo-axonic cells (AACs)** are driven in the same fashion as BCs, but they form synapses exclusively on the initial segment of PC axons.
- **Oriens lacunosum-moleculare (OLM) cells** are predominantly driven by CA1 PCs and provide feedback inhibition to the distal dendrites of PCs, corresponding to the site of entorhinal cortex input to these cells.

Recent data indicate that these cell types may be distinguished by their firing patterns in different brain states (Klausberger et al., 2003, 2004). The firing rate and timing of action potentials (APs) relative to the theta rhythm are distinct for the different cell types, arising from differences in network connectivity and intracellular properties. One factor here is differences in the short-term dynamics of the excitatory drive to these INs. Excitatory synapses onto BCs, BSCs and AACs are powerful and quickly depress in response to repeated stimulation (Ali, Deuchars, Pawelzik, & Thomson, 1998; Sun, Lyons, & Dobrunz, 2005). This results in these INs responding rapidly to the onset of excitatory drive and then adapting as the stimulus continues. In contrast, excitatory synapses onto OLM cells have low release probability and facilitate with repeated stimulation, resulting in OLM cells responding most strongly later in a stimulus rather than at the onset (Ali & Thomson, 1998; Losonczy, Zhang, Shigemoto, Somogyi, & Nusser, 2002). Thus inhibition onto CA1 PCs from OLM cells is delayed relative to these other inhibitory pathways. The difference in firing properties between IN types is a key indicator of their potential functional roles in different behavioural states.

2.3 Rhythm generation

Cellular activity shows distinct characteristics depending on the behavioural mode of an animal. This has been most extensively studied in rats. During exploration of the environment, the EEG recorded from CA1 exhibits a modulation in a frequency range of around 4–7 Hz, the so-called theta rhythm. At the same time, gamma-frequency (30–100 Hz) modulation of the EEG is also present. A typical pyramidal cell fires only one or two spikes per theta cycle and is not active in every cycle. Fast-spiking INs (BC, AAC, BSC) will fire multiple spikes at gamma frequency.

Microcircuit interneurons and external inputs are responsible for theta and gamma rhythm generation and modulation of PC synaptic plasticity. The network of BCs provides the robust gamma rhythm due to their fast-firing properties and mutual interconnections (Bartos et al., 2007). Inhibition from BCs onto PCs can synchronize PC firing (Cobb, Buhl, Halasy, Paulsen, & Somogyi, 1995).

Theta rhythm generation is highly complex and may take different forms in different *in vivo* and *in vitro* experimental preparations (Buzsaki, 2002). Recent modelling studies have demonstrated that slow inhibition provided by OLM cells coupled with fast inhibition from fast-spiking INs, such as BCs, can generate an intrinsic theta rhythm in CA1 (Orban, Kiss, & Erdi, 2006; Rotstein et al., 2005). The medial septum also oscillates at theta rhythm and provides rhythmic GABA-A inhibition, principally to interneurons in the hippocampus (Freund & Antal, 1988; Hasselmo & Fehlau, 2001). It also provides slower cholinergic modulation to multiple cellular targets (Frotscher & Lenrath, 1985; Hasselmo & Fehlau, 2001).

2.4 Synaptic plasticity

Experiments have revealed wide-ranging synaptic plasticity in the CA1 microcircuit. All excitatory inputs that have been studied, either onto PCs or onto INs, appear to be modifiable in response to patterns of pre- and post-synaptic activity (Bliss, Collingridge, & Morris, 2007). There is also some evidence for plasticity of inhibitory synapses onto pyramidal cells (Bliss et al., 2007).

The rules underpinning plasticity are largely Hebbian, in which correlated pre- and post-synaptic activity leads to a strengthening of the synaptic connection (long-term potentiation, LTP). Uncorrelated firing leads to a weakening of the synapse (long-term depression, LTD). The precise nature of the required correlations is still to be determined. There is evidence for spike-timing-dependent plasticity (STDP) at Schaffer collateral synapses onto PCs (Bi & Poo, 1998, 2001; Magee & Johnston, 1997). Plasticity may also depend purely on local dendritic activity rather than rely on spiking in the soma and axon (Golding, Staff, & Spruston, 2002; Holthoff, Kovalchuk, & Konnerth, 2006; Lisman & Spruston, 2005). This situation leads to the possibility of spatial specificity in learning, rather than just synapse specificity, in which activation of colocated synapses may increase the chances of all these synapses being modified (Mehta, 2004).

Not all plastic connections may be modified in a Hebbian fashion. Excitatory connections onto OLM INs appear to be subject to an anti-Hebbian learning rule in which presynaptic activity alone leads to LTP, whereas correlated pre- and post-synaptic activity results in LTD (Lamsa, Heeroma, Somogyi, Rusakov, & Kullmann, 2007).

3 Associative memory

The hippocampal regions CA3 and CA1 have been proposed to be auto- and heteroassociative memories, respectively (Treves & Rolls, 1994), for the storage of declarative information. Associative memory is one of the oldest ANN paradigms. It has been widely studied due to being plausibly a model of how certain brain regions, such as the hippocampus, may operate, but also due to the discovery of simple implementations that are analytically tractable (Amit, 1989; Hopfield, 1982; Willshaw, Buneman, & Longuet-Higgins, 1969).

The requirements for building a workable associative memory are rather simple. Memory patterns are encoded as the activity patterns across a network of computing units, or neurons. Patterns are stored in the memory by Hebbian modification of the connections between the computing units. A memory is recalled when an activity pattern that is a partial or noisy version of a stored pattern is instantiated in the network. Network activity then evolves to the complete stored pattern as appropriate units are recruited to the activity pattern, and noisy units are removed, by threshold-setting of unit activity. Memory capacity for accurate recall is strongly dependent on the form of patterns to be stored and the Hebbian learning rule employed.

Simple ANN models are amenable to mathematical analysis, leading to estimates of memory capacity (Amit, 1989) and the definition of optimal Hebbian learning rules (Dayan & Willshaw, 1991). Biologically plausible modifications to these simple models allow efficient memory storage in partially connected networks (Buckingham & Willshaw, 1993; Graham & Willshaw, 1995, 1997) with unreliable connections (Graham & Willshaw, 1999). Noise due to inputs to a neuron arriving over spatially extensive dendrites may not seriously reduce memory capacity and can be ameliorated by certain intracellular properties found in hippocampal pyramidal cell apical dendrites (Graham, 2001).

All of this work addresses the mechanics of pattern recall in networks containing a single (principal) neuron type. The mechanics of pattern storage and how it may be dynamically interleaved with recall are not considered. The cellular and network mechanisms underlying pattern specification, learning (storage) rules and threshold-setting during recall are not explicitly included. These mechanisms must be manifest in biological neural nets through the microcircuitry formed by the large variety of neuronal types.

3.1 Associative memory and the hippocampus

These considerations have led to the formulation of neural models of associative memory based on the architecture and operation of hippocampal areas CA3 and CA1 (Kunec, Hasselmo, & Kopell, 2005; Menschik & Finkel, 1998; Wallenstein & Hasselmo, 1997). These models include multiple cell types and their connectivity, with cells represented by biophysically based compartmental models of spiking neurons. The models seek to mimic the hippocampal activity seen in rats exploring a novel environment, absorbing and storing new spatial information (O'Keefe & Recce, 1993).

Theta and gamma rhythms are a feature of this activity. These models instantiate a working hypothesis that the theta rhythm, which is prominent during exploration, modulates episodes of storage of new information and recall of old information in its half-cycles (Hasselmo, Bodelon, & Wybl, 2002a; Hasselmo, Hay, Ilyn, & Gorchetchnikov, 2002b). During exploration an animal is likely to encounter both familiar and novel situations. Storage of new episodes with minimal interference from already encoded episodes takes place most efficiently if storage and recall are temporally separated in the encoding neural networks. Waxing and waning of GABA-mediated inhibition from the medial septum leads alternately to disinhibition and inhibition of PCs during a theta cycle, corresponding to ideal conditions for pattern recall and pattern storage, respectively. The higher-frequency gamma rhythms (30–100 Hz) constitute a basic clock cycle such that patterns of activity for storage and recall correspond to PCs that are active in a particular gamma cycle (Axmacher et al., 2006; Buzsaki & Chrobak, 1995; Lisman & Idiart, 1995).

Patterns of PC activity for storage are determined by the spatiotemporal correspondence of direct afferent input from the entorhinal cortex and indirect input via dentate gyrus onto CA3 PCs and via CA3 PC input onto CA1 PCs. Such patterns are stored autoassociatively in CA3 by Hebbian modification of recurrent connections between CA3 PCs, and heteroassociatively in CA1 by modification of CA3 input onto CA1 PCs (Hasselmo et al., 2002a).

Storage and recall dynamics are influenced by synaptic and intrinsic cellular properties and by alteration of these properties by neuromodulation with acetylcholine. Acetylcholine and GABA-B-mediated inhibition may serve to set appropriate conditions for pattern storage by reducing synaptic transmission while promoting plasticity on the modifiable pathways (Hasselmo, 1993; Hasselmo, Anderson, & Bower, 1992). Neuromodulation is slower than the theta rhythm and serves to generally bias the network towards storage (if, say, the animal is exploring a novel environment) or recall (if the environment is largely familiar). This bias may be controlled by inhibitory input to the medial septum from CA1, which is likely largest when CA1 PC cells are most active during recall, leading to a reduction in MS modulatory output back to CA1 (Hasselmo & Schnell, 1994; Hasselmo, Schnell, & Barkai, 1995).

4 Functionality of the microcircuit

Though these models are much closer to biological neural nets than ANN models, they still very much simplify the neuronal circuitry of the mammalian hippocampus. The role of inhibition has largely been confined to BCs acting to threshold PC activity during pattern recall (Sommer & Wennekers, 2001). Other ideas include the possibility that AACs provide the negative weights due to pattern storage required in some ANN models of associative memory (Menschik & Finkel, 1998).

The challenge remains to provide functional explanations that include more details of the known circuitry. Ideas concerning interneuronal network involvement in rhythm generation and control of PC networks are explored in Buzsaki and Chrobak (1995). Paulsen and Moser (1998) consider how GABAergic interneurons might provide the control structures necessary for phasing storage and recall in the hippocampus. Building on their ideas, we propose the following hypotheses concerning the functioning of the CA1 microcircuit, including a number of different neuronal types and their specific roles in storage and recall. We then present a model that instantiates these ideas (Figure 1.2).

4.1 Functional hypothesis

As described above, it has been suggested that the hippocampal theta rhythm (4–7 Hz) can contribute to memory formation by separating encoding (storage) and retrieval of memories into different functional half-cycles (Hasselmo et al., 2002a). Recent experimental data show that the activity of different neuronal types is modulated at specific phases relative to the theta rhythm (Klausberger et al., 2003). Given that PC firing is biased towards the recall phase (e.g., place cells firing when a rat is in a familiar location), then it follows from the experimental data that BCs and AACs fire in phase with the encoding (storage) cycle of the theta rhythm, whereas the PCs, BSCs, OLMs and GABAergic MS input to CA1 fire on the recall cycle (180° out of phase) (see also Kunec et al., 2005).

We propose (see also Paulsen & Moser, 1998) that during encoding (Figure 1.2A), when the MS input is minimal, the role of the BCs and AACs is to provide enough hyperpolarization for the prevention of PCs from firing, as their output is not relevant. During this phase a PC may receive input from EC in its distal dendrites and CA3 in its proximal dendrites. Those PCs that receive combined EC and CA3 inputs can show sufficient local activity (manifest as membrane depolarization and a rise in calcium level) in their proximal dendrites to lead to a strengthening of the active CA3-input synapses. This is aided by strong BC inhibition, which leads to activation of the hyperpolarization-activated, but depolarizing H-current, resulting in rebound excitation of PCs on each gamma cycle.

Experimental evidence (Leung, Roth, & Canning, 1995) has suggested that conduction latency of the EC-layer III input to CA1 lacunosum-moleculare (LM) dendrites is less than 9 ms (ranging between 5 and 8 ms), whereas the conduction latency of EC-layer II input to CA1 radiatum dendrites via the trisynaptic (via dentate gyrus and CA3) path is greater than 9 ms (ranging between 12 and 18 ms). Given that it is synchronous activity in EC layers II and III that carries the information to be stored in CA1, these different delays mean that forward pairing in time of the EC and CA3 inputs, as required by the encoding strategy, is impossible. A different mechanism is required to associate the two inputs. We suggest that the paired signal for learning is provided by a back-propagating action potential (BPAP) mediated by activation of the H channels due to strong hyperpolarization by the BCs and AACs on the PCs soma and axon. This BPAP is generated without full-blown action potential generation in the soma or axon (which is blocked by the BC and AAC input) and meets the incoming CA3 input at the PC

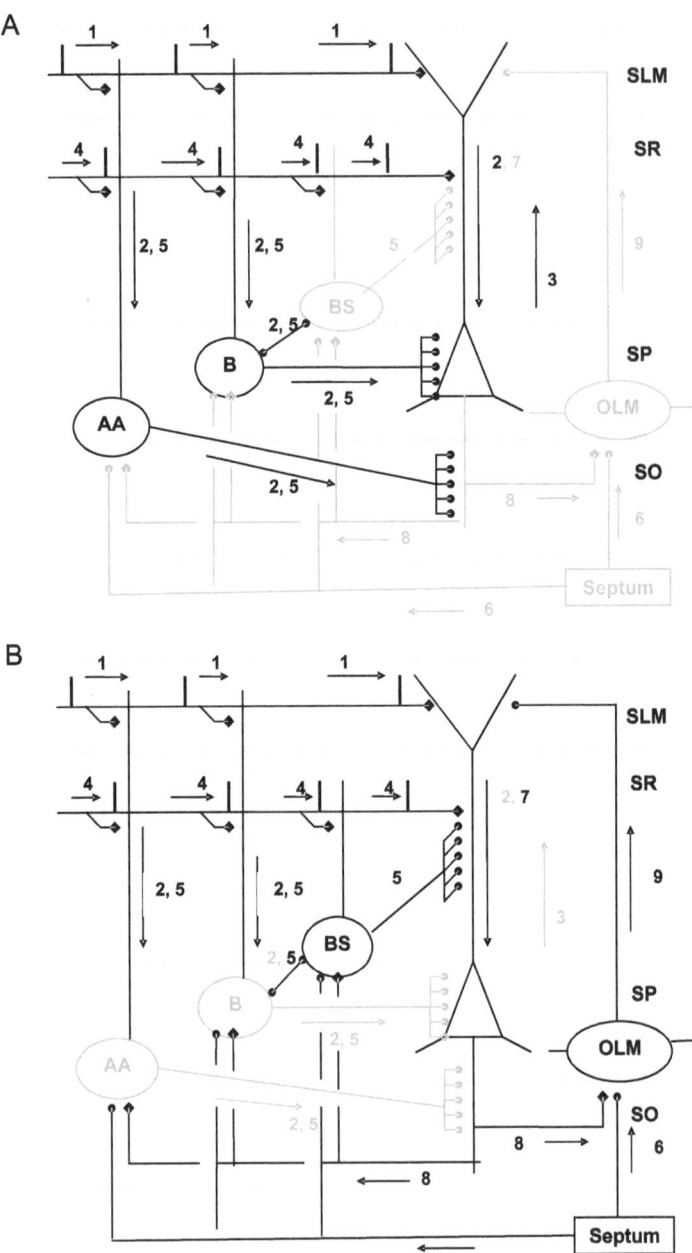

Figure 1.2 Active network pathways during (A) encoding cycle and (B) retrieval cycle. Only black solid-lined cells and pathways are active in each cycle. Numbers above and next to pathways indicate the temporal order of information processing during each cycle.

stratum radiatum medial dendrites to provide the underlying mechanism for associating the EC- and CA3-input patterns.

On the other hand, during retrieval (Figure 1.2B), when the BCs and AACs are silent due to a strong inhibitory input from the medial septum, the BSCs and OLM cells are active. The role of the BSCs is to provide a non-specific inhibitory signal to all PCs in the network that will raise the threshold enough to allow only the PCs that have learnt the EC–CA3-input association to fire (recall), whereas the role of the OLM cells is to inhibit the EC input to distal PC dendrites in order to prevent errors during retrieval. PC activity is due solely to strong CA3 input.

4.2 A computer model

To begin to explore these hypotheses, we are building a computer model of the CA1 microcircuit containing these major cell types. The initial small model consists of 100 PCs, 4 BCs, 2 BSCs, 2 AACs and 18 OLM cells.

- **Cellular morphology.** Moderately detailed compartmental models are used for the individual cells. The morphology and dimensions of the somatic, axonic and dendritic compartments of the model cells were adapted from Gulyas et al. (1999) and Megias et al. (2001). Compartments: PC, 15; B and AA, 17; BS, 13; OLM, 4. Cell structures and their firing properties are illustrated in Figure 1.3.
- **Cellular properties.** Each PC membrane contains a calcium pump and buffering mechanism, a calcium-activated mAHP potassium current, an LVA L-type Ca^{2+} current, an HVA L-type Ca^{2+} current, an MVA R-type Ca^{2+} current, an HVA T-type Ca^{2+} current, an H current, Hodgkin–Huxley-style sodium and delayed rectifier currents, a slow Ca^{2+}-dependent potassium current, a slow non-inactivating K^+ channel

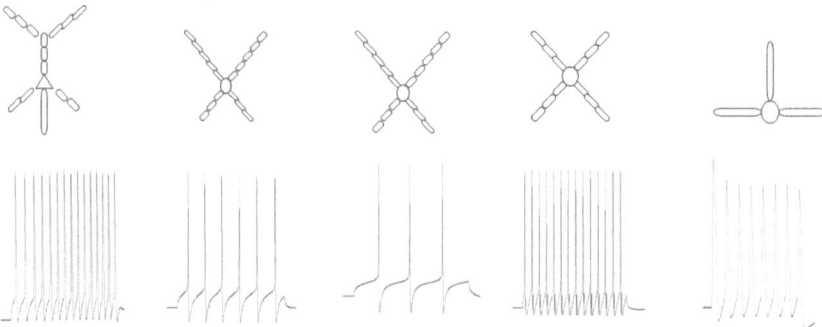

Figure 1.3 Compartmental structure models for the different cell types, plus their firing properties in response to depolarizing current injection (amplitude: 0.2 nA; duration: 200 ms). *From left to right:* pyramidal cell (PC), axo-axonic cell (AAC), basket cell (BC), bistratified cell (BSC), olm cell (OLM).

with HH-style kinetics and a K^+ A current (Poirazi, Brannon, & Mel, 2003a). Each BC, BSC and AAC contains a leak conductance, a sodium current, a fast delayed rectifier K^+ current, an A-type K^+ current, L- and N-type Ca^{2+} currents, a Ca^{2+}-dependent K^+ current and a Ca^{2+}- and voltage-dependent K^+ current (Santhakumar, Aradi, & Soltetz, 2005). Each OLM cell has a sodium (Na^+) current, a delayed rectifier K^+ current, an A-type K^+ current and an H current (Saraga, Wu, Zhang, & Skinner, 2003).

- **Synaptic properties.** AMPA, NMDA, GABA-A and GABA-B synapses are included. GABA-A are present in all strata, whereas GABA-B synapses are present in medium and distal SR and SLM dendrites. AMPA synapses are present in strata LM (EC connections) and radiatum (CA3 connections), whereas NMDA receptors are present only in stratum radiatum (CA3 connections).

- **Synaptic contacts.** AMPA only: all EC and CA1 PC recurrent connections; AMPA with NMDA: CA3 onto PCs. GABA-A synaptic contacts (Buhl, Halasy, & Somogyi, 1994): 8 by each AAC onto each PC axon; 9 by each BC onto each PC soma; 6 by each BSC onto each PC; 2 by each OLM cell with each PC cell.

- **Network connectivity.** Less than 1% recurrent connections between PCs. All-to-all connectivity for BCs and BSCs and between BCs and BSCs. No recurrent connections between AACs. All-to-all connectivity in PC–IN–PC loops for all types of IN.

- **Plasticity.** STDP learning rule at CA3–AMPA synapses on PCs (Song, Miller, & Abbott, 2000). Presynaptic spike times compared with timing of peak post-synaptic voltage amplitude due to a BPAP at the synapse. Synaptic strengthening (LTP due to an increase in AMPA conductance) occurs for a BPAP arriving just after the presynaptic spike (10-ms time window), whereas weakening (LTD) occurs if the BPAP arrives prior to the spike (similar 10-ms window.)

- **Inputs.** Excitatory inputs come from EC and CA3 Schaffer collaterals. PCs, BCs, AACs and BSCs receive CA3 input; PCs, BCs and AACs receive EC input. Initially, EC input arrives at PC apical LM dendrites between 0 and 9 ms (relative to the start of a theta cycle), whereas the CA3-input pattern arrives 9 ms later (Leung et al., 1995). Both EC and CA3 inputs are repeated to PC apical LM and medial radiatum dendrites, respectively, every 7 ms.

  Input for the medial septum provides GABA-A inhibition to all INs (strongest to BC and AAC). MS input is phasic at theta rhythm and is on for 70 ms during the retrieval phase, and off otherwise.

- **Storage and recall.** An experiment is conducted with the model in which a pattern of activity in CA1 is associated with a CA3 activity pattern. Initial CA3–CA1 synaptic conductances are set to random values, and so the pattern association takes place on top of this background synaptic noise. During the encoding (storage) phase, 20 randomly selected PCs exclusively receive EC input in the LM dendrites, creating

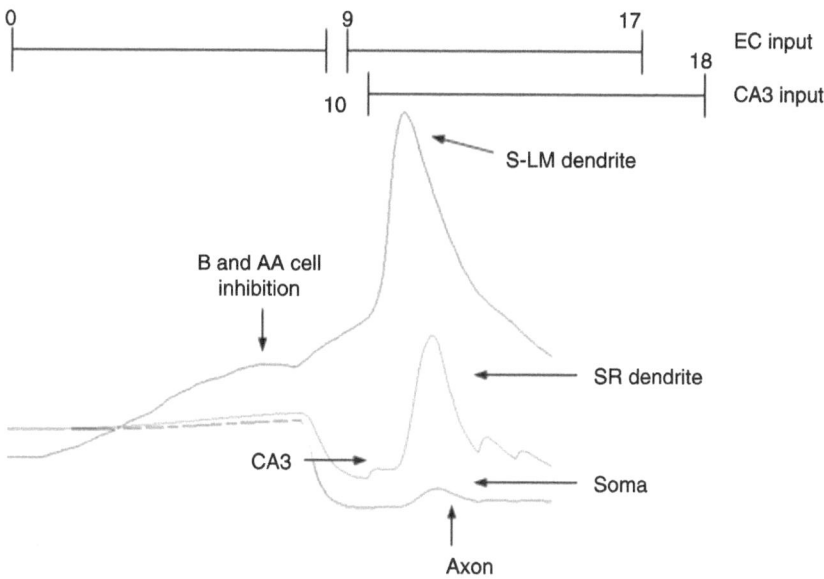

Figure 1.4 Post-synaptic signal of a CA1 pyramidal cell in response to EC and CA3 inputs. EC input is presented twice in two separate time intervals (0–8 ms and 9–17 ms). CA3 input is presented only once (10–18 ms). The inhibitory effects of the basket (B) cells and the axo-axonic (AA) cells on the pyramidal (P) cells are "seen" at about 6 ms. Due to the strong B and AA inhibition on the P soma and axon, an H-current-induced back-propagating action potential (BPAP) propagates back towards the SR dendrites of the P cell, where it coincides with the incoming CA3 and EC inputs. The SR dendrite of each P cell is the location where learning (storage) is taking place. Note that no action potential is generated in the soma or axon due to BC and AAC inhibition.

the CA1 activity pattern for storage. All PCs in the network are activated by the CA3 input in their medial radiatum dendrites. The STDP learning rule "teaches" the CA1 PCs to hetero-associate the H-current-activated BPAP with the incoming EC and CA3 inputs (Figure 1.4).

Cellular activity during a storage-and-recall cycle is shown in Figure 1.5. The pyramidal cell receives both EC and CA3 input during storage and thus becomes associated with the CA3 input. The PC is then active in response to CA3 input alone during the recall cycle.

5 Conclusions and further work

The hypotheses and model presented above are still very simple compared with what we know of the CA1 microcircuit and its putative role in different animal behaviours. More cell types and their connectivity could be included in the model. However, we still require further data on type-specific cell

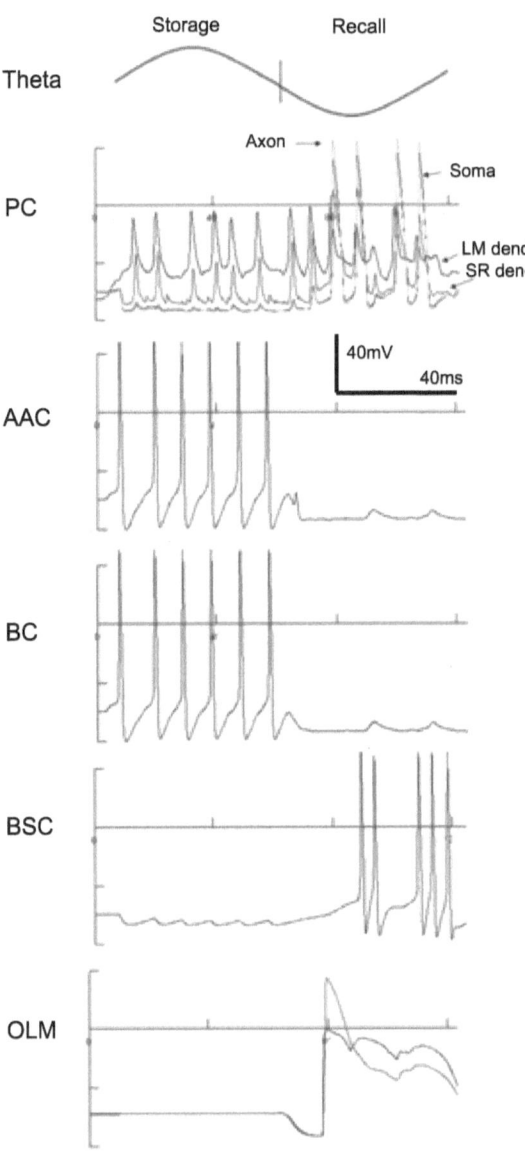

Figure 1.5 Firing responses of model cells during storage and recall of a theta cycle. From top to bottom: theta cycle oscillation, pyramical cell, axo-axonic cell, basket cell, bistratified cell and OLM cell.

properties and their *in vivo* firing patterns in particular behavioural states. We have chosen to concentrate on data related to environmental exploration in awake, behaving animals. Theories of hippocampal function also postulate how it interacts with neocortex in the formation of long-term memories (Morris, 2006; O'Reilly & Norman, 2002). In particular, there is evidence that information encoded during exploration is replayed in the hippocampus during sleep, possibly to drive memory consolidation in the neocortex (Ji & Wilson, 2007). A more complete model will propose the roles and activity dynamics of the different cell types in this behaviour too.

One aspect that we have not dealt with here is the complex membrane properties of neurons, particularly PCs that allow nonlinear integration of synaptic input. Detailed models of CA1 PCs have investigated the interaction of synaptic input with active membrane dynamics (Kali & Freund, 2005; Poirazi, Brannon, & Mel, 2003a, 2003b). Aspects of spatiotemporal cellular dynamics are lost in the reduced PC models used in large-scale network models. This can be redressed through new formulations of reduced models or through increased computing power that allows more complex cellular models to be used in networks.

Current models of specific brain circuits that include an aspect of learning usually only allow synaptic modification in one principal pathway. This is true here in that only the CA3 input to CA1 PCs is to modifiable synapses. In reality most, if not all, synaptic pathways are modifiable in the face of particular patterns of activity. For example, the entorhinal input to the distal dendrites of CA1 PCs is Hebbian-modifiable, and the post-synaptic signals in these dendrites are under specific inhibitory and neuromodulatory control (Remondes & Schuman, 2002). EC input can, in fact, appear largely inhibitory due to activation of feedforward interneurons and can result in a reduction of plasticity at CA3 synapses onto CA1 PCs (Remondes & Schuman, 2002). New models of CA1 function clearly need to take into account further aspects of this pathway (Pissadaki & Poirazi, 2007) – in particular, what learning may take place.

Also, the excitatory synapses on the inhibitory interneurons may be plastic, and hence the INs can be a part of smaller circuits within the global CA1 microcircuit capable of carrying out specific functionalities – for example, encoding Item A as opposed to Item B of a sequence of items A–B–A–B. Notably, OLM cells are active during slow-wave sleep but are silenced during sharp-wave ripples (SWRs), which are hypothesized to be recall episodes for consolidation of long-term memories in neocortex (Axmacher et al., 2006; Somogyi & Klausberger, 2005). In addition, the apparent learning rule at PC to OLM synapses leads to strengthening of these connections when PCs are active but OLM cells are silent (Lamsa et al., 2007). Thus it is likely that these synapses are being strengthened during SWRs, perhaps to reinforce their role during theta/gamma activity.

With any model, the great challenge is for the model to provide a consistent account of neural activity seen in different behavioural states and recorded in different experimental paradigms. Experimental data is often contradictory and difficult to combine due to reliance on very specific experimental protocols. *In vivo* data from animals in different behavioural states is clearly the most important to match but is usually insufficient in itself for the formulation of the model. For example, details of intracellular properties must be derived from wide-ranging *in vitro* experiments. Nonetheless, even given these limitations, models that (a) include more biological detail, (b) can match certain brain dynamics and (c) provide an instantiation of particular cognitive functions will definitely aid us in the quest of understanding how brains work.

6 References

Ali, A., Deuchars, J., Pawelzik, H., & Thomson, A. (1998). CA1 pyramidal to basket and bistratified cell EPSPs: Dual intracellular recordings in rat hippocampal slices. *Journal of Physiology, 507,* 201–217.

Ali, A., & Thomson, A. (1998). Facilitating pyramid to horizontal oriens-alveus interneurone inputs: Dual intracellular recordings in slices of rat hippocampus. *Journal of Physiology, 507,* 185–199.

Amit, D. J. (1989). *Modeling brain function: The world of attractor neural networks.* Cambridge: Cambridge University Press.

Andersen, P., Morris, R., Amaral, D., Bliss, T., & O'Keefe, J. (Eds.) (2007). *The hippocampus book.* Oxford: Oxford University Press.

Axmacher, N., Mormann, F., Fernandez, G., Elger, C., & Fell, J. (2006). Memory formation by neuronal synchronization. *Brain Research Reviews, 52,* 170–182.

Bartos, M., Vida, I., & Jonas, P. (2007). Synaptic mechanisms of synchronized gamma oscillations in inhibitory interneuron networks. *Nature Reviews Neuroscience, 8,* 45–56.

Bi, G.-q., & Poo, M.-m. (1998). Synaptic modifications in cultured hippocampal neurons: Dependence on spike timing, synaptic strength, and postsynaptic cell type. *Journal of Neuroscience, 18,* 10464–10472.

Bi, G.-q., & Poo, M.-m. (2001). Synaptic modification by correlated activity: Hebb's postulate revisited. *Annual Review of Neuroscience, 24,* 139–166.

Bliss, T., Collingridge, G., & Morris, R. (2007). Synaptic plasticity in the hippocampus. In P. Andersen, R. Morris, D. Amaral, T. Bliss, & J. O'Keefe (Eds.), *The hippocampus book* (pp. 343–474). Oxford: Oxford University Press.

Buckingham, J., & Willshaw, D. (1993). On setting unit thresholds in an incompletely connected associative net. *Network, 4,* 441–459.

Buhl, E., Halasy, K., & Somogyi, P. (1994). Diverse sources of hippocampal unitary inhibitory postsynaptic potentials and the number of synaptic release sites. *Nature, 368,* 823–828.

Buzsaki, G. (2002). Theta oscillations in the hippocampus. *Neuron, 33,* 325–340.

Buzsaki, G., & Chrobak, J. (1995). Temporal structure in spatially organized neuronal ensembles: A role for interneuronal networks. *Current Opinion in Neurobiology, 5,* 504–510.

Cobb, S., Buhl, E., Halasy, K., Paulsen, O., & Somogyi, P. (1995). Synchronization of neuronal activity in hippocampus by individual GABAergic interneurons. *Nature, 378,* 75–78.

Dayan, P., & Willshaw, D. (1991). Optimising synaptic learning rules in linear associative memories. *Biological Cybernetics, 65,* 253–265.

Deuchars, J., & Thomson, A. (1996). CA1 pyramid–pyramid connections in the rat hippocampus in vitro: Dual intracellular recordings with biocytin filling. *Neuroscience, 74,* 1009–1018.

Eichenbaum, H., Dudchenko, P., Wood, E., Shapiro, M., & Tanila, H. (1999). The hippocampus, memory, and place cells: Is it spatial memory or a memory space? *Neuron, 23,* 209–226.

Freund, T., & Antal, M. (1988). GABA-containing neurons in the septum control inhibitory interneurons in the hippocampus. *Hippocampus, 336,* 170–173.

Freund, T., & Buzsaki, G. (1996). Interneurons of the hippocampus. *Hippocampus, 6,* 347–470.

Frotscher, M., & Lenrath, C. (1985). Cholinergic innervation of the rat hippocampus as revealed by choline acetyltransferase immunocytochemistry: A combined light and electron microscopic study. *Journal of Comparative Neurology, 239,* 237–246.

Golding, N., Staff, N., & Spruston, N. (2002). Dendritic spikes as a mechanism for cooperative long-term potentiation. *Nature, 418,* 326–331.

Graham, B. (2001). Pattern recognition in a compartmental model of a CA1 pyramidal neuron. *Network, 12,* 473–492.

Graham, B., & Willshaw, D. (1995). Improving recall from an associative memory. *Biological Cybernetics, 72,* 337–346.

Graham, B., & Willshaw, D. (1997). Capacity and information efficiency of the associative net. *Network, 8,* 35–54.

Graham, B., & Willshaw, D. (1999). Probabilistic synaptic transmission in the associative net. *Neural Computation, 11,* 117–137.

Gulyas, A., Megias, M., Emri, Z., & Freund, T. (1999). Total number and ratio of excitatory and inhibitory synapses converging onto single interneurons of different types in the CA1 area of the rat hippocampus. *Journal of Neuroscience, 19,* 10082–10097.

Hasselmo, M. (1993). Acetylcholine and learning in a cortical associative memory. *Neural Computation, 5,* 32–44.

Hasselmo, M., Anderson, B., & Bower, J. (1992). Cholinergic modulation of cortical associative memory function. *Journal of Neurophysiology, 67,* 1230–1246.

Hasselmo, M., Bodelon, C., & Wyble, B. (2002a). A proposed function for hippocampal theta rhythm: separate phases of encoding and retrieval enhance reversal of prior learning. *Neural Computation, 14,* 793–817.

Hasselmo, M., & Fehlau, B. (2001). Differences in time course of ACh and GABA modulation of excitatory synaptic potentials in slices of rat hippocampus. *Journal of Neurophysiology, 86,* 1792–1802.

Hasselmo, M., Hay, J., Ilyn, M., & Gorchetchnikov, A. (2002b). Neuromodulation, theta rhythm and rat spatial navigation. *Neural Networks, 15,* 689–707.

Hasselmo, M., & Schnell, E. (1994). Laminar selectivity of the cholinergic suppression of synaptic transmission in rat hippocampal region CA1: Computational modeling and brain slice physiology. *Journal of Neuroscience, 14,* 3898–3914.

Hasselmo, M., Schnell, E., & Barkai, E. (1995). Dynamics of learning and recall at excitatory recurrent synapses and cholinergic modulation in rat hippocampal region CA3. *Journal of Neuroscience, 15,* 5249–5262.

Holthoff, K., Kovalchuk, Y., & Konnerth, A. (2006). Dendritic spikes and activity-dependent synaptic plasticity. *Cell and Tissue Research, 326,* 369–377.

Hopfield, J. (1982). Neural networks and physical systems with emergent collective computational abilities. *Proceedings of the National Academy of Science, 79,* 2554–2558.

Ishizuka, N., Cowan, W., & Amaral, D. (1995). A quantitative analysis of the dendritic organization of pyramidal cells in the rat hippocampus. *Journal of Comparative Neurology, 362,* 17–45.

Ji, D., & Wilson, M. (2007). Coordinated memory replay in the visual cortex and hippocampus during sleep. *Nature Neuroscience, 10,* 100–107.

Kali, S., & Freund, T. (2005). Distinct properties of two major excitatory inputs to hippocampal pyramidal cells: A computational study. *European Journal of Neuroscience, 22,* 2027–2048.

Klausberger, T., Magill, P., Maki, G., Marton, L., Roberts, J., Cobden, P., et al. (2003). Brain-state and cell-type-specific firing of hippocampal interneurons in vivo. *Nature, 421,* 844–848.

Klausberger, T., Marton, L., Baude, A., Roberts, J., Magill, P., & Somogyi, P. (2004). Spike timing of dendrite-targeting bistratified cells during hippocampal network oscillations in vivo. *Nature Neuroscience, 7,* 41–47.

Kunec, S., Hasselmo, M., & Kopell, N. (2005). Encoding and retrieval in the CA3 region of the hippocampus: A model of theta-phase separation. *Journal of Neurophysiology, 94,* 70–82.

Lamsa, K., Heeroma, J., Somogyi, P., Rusakov, D., & Kullmann, D. (2007). Anti-Hebbian long-term potentiation in the hippocampal feedback circuit. *Science, 315,* 1262–1266.

Leung, L., Roth, L., & Canning, K. (1995). Entorhinal inputs to hippocampal CA1 and dentate gyrus in the rat: A current-source-density study. *Journal of Neurophysiology, 73,* 2392–2403.

Lisman, J., & Idiart, M. (1995). Storage of 7±2 short-term memories in oscillatory subcycles. *Science, 267,* 1512–1514.

Lisman, J., & Spruston, N. (2005). Postsynaptic depolarization requirements for LTP and LTD: A critique of spike timing-dependent plasticity. *Nature Neuroscience, 8,* 839–841.

Losonczy, A., Zhang, I., Shigemoto, R., Somogyi, P., & Nusser, Z. (2002). Cell type dependence and variability in the short-term plasticity of EPSCs in identified mouse hippocampal interneurones. *Journal of Physiology, 542,* 193–210.

Maccaferri, G., & Lacaille, J.-C. (2003). Hippocampal interneuron classifications – making things as simple as possible, not simpler. *Trends in Neuroscience, 26,* 564–571.

Magee, J., & Johnston, D. (1997). A synaptically controlled, associative signal for Hebbian plasticity in hippocampal neurons. *Science, 275,* 209–213.

McBain, C., & Fisahn, A. (2001). Interneurons unbound. *Nature Reviews Neuroscience, 2,* 11–23.

Megias, M., Emri, Z., Freund, T., & Gulyas, A. (2001). Total number and distribution of inhibitory and excitatory synapses on hippocampal CA1 pyramidal cells. *Neuroscience, 102,* 527–540.

Mehta, M. (2004). Cooperative LTP can map memory sequences on dendritic branches. *Trends in Neuroscience, 27,* 69–72.

Menschik, E., & Finkel, L. (1998). Neuromodulatory control of hippocampal function: Towards a model of Alzheimer's disease. *Artificial Intelligence in Medicine, 13,* 99–121.

Morris, R. (2006). Elements of a neurobiological theory of hippocampal function:

The role of synaptic plasticity, synaptic tagging and schemas. *European Journal of Neuroscience, 23*, 2829–2846.

O'Keefe, J., & Recce, M. (1993). Phase relationship between hippocampal place units and the EEG theta rhythm. *Hippocampus, 3*, 317–330.

Orban, G., Kiss, T., & Erdi, P. (2006). Intrinsic and synaptic mechanisms determining the timing of neuron population activity during hippocampal theta oscillation. *Journal of Neurophysiology, 96*, 2889–2904.

O'Reilly, R., & Norman, K. (2002). Hippocampal and neocortical contributions to memory: Advances in the complementary learning systems framework. *Trends in Cognitive Sciences, 6*, 505–510.

Otmakhova, N., & Lisman, J. (2000). Dopamine, serotonin, and noradrenaline strongly inhibit the direct perforant path-CA1 synaptic input, but have little effect on the Schaffer collateral input. *Annals of the New York Academy of Sciences, 911*, 462–464.

Paulsen, O., & Moser, E. (1998). A model of hippocampal memory encoding and retrieval: GABAergic control of synaptic plasticity. *Trends in Neuroscience, 21*, 273–279.

Pissadaki, E., & Poirazi, P. (2007). Modulation of excitability in CA1 pyramidal neurons via the interplay of entorhinal cortex and CA3 inputs. *Neurocomputing, 70*, 1735–1740.

Poirazi, P., Brannon, T., & Mel, B. (2003a). Arithmetic of subthreshold synaptic summation in a model CA1 pyramidal cell. *Neuron, 37*, 977–987.

Poirazi, P., Brannon, T., & Mel, B. (2003b). Pyramidal neuron as a two-layer neural network. *Neuron, 37*, 989–999.

Remondes, M., & Schuman, E. (2002). Direct cortical input modulates plasticity and spiking in CA1 pyramidal neurons. *Nature, 416*, 736–740.

Rotstein, H., Pervouchine, D., Acker, C., Gillies, M., White, J., Buhl, E., et al. (2005). Slow and fast inhibition and an H-current interact to create a theta rhythm in a model of CA1 interneuron network. *Journal of Neurophysiology, 94*, 1509–1518.

Santhakumar, V., Aradi, I., & Soltetz, I. (2005). Role of mossy fiber sprouting and mossy cell loss in hyperexcitability: A network model of the dentate gyrus incorporating cell types axonal typography. *Journal of Neurophysiology, 93*, 437–453.

Saraga, F., Wu, C., Zhang, L., & Skinner, F. (2003). Active dendrites and spike propagation in multicompartmental models of oriens-lacunosum/moleculare hippocampal interneurons. *Journal of Physiology, 552*, 673–689.

Sik, A., Ylinen, A., Penttonen, M., & Buzsaki, G. (1994). Inhibitory CA1–CA3–Hilar region feedback in the hippocampus. *Science, 265*, 1722–1724.

Sommer, F., & Wennekers, T. (2001). Associative memory in networks of spiking neurons. *Neural Networks, 14*, 825–834.

Somogyi, P., & Klausberger, T. (2005). Defined types of cortical interneurone structure space and spike timing in the hippocampus. *Journal of Physiology, 562*(1), 9–26.

Song, S., Miller, K., & Abbott, L. (2000). Competitive "Hebbian" learning through spike-timing-dependent synaptic plasticity. *Nature Neuroscience, 3*, 919–926.

Sun, H., Lyons, S., & Dobrunz, L. (2005). Mechanisms of target-cell specific short-term plasticity at Schaffer collateral synapses onto interneurones versus pyramidal cells in juvenile rats. *Journal of Physiology, 568*, 815–840.

Traub, R., Jefferys, J., & Whittington, M. (1999). *Fast oscillations in cortical circuits*. Cambridge, MA: MIT Press.

Treves, A., & Rolls, E. (1994). Computational analysis of the role of the hippocampus in memory. *Hippocampus, 4*, 374–391.

Wallenstein, G., & Hasselmo, M. (1997). GABAergic modulation of hippocampal population activity: Sequence learning, place field development, and the phase precession effect. *Journal of Neurophysiology, 78*, 393–408.

Whittington, M., & Traub, R. (2003). Inhibitory interneurons and network oscillations in vitro. *Trends in Neuroscience, 26*, 676–682.

Willshaw, D., Buneman, O., & Longuet-Higgins, H. (1969). Non-holographic associative memory. *Nature, 222*, 960–962.

Wood, E., Dudchenko, P., & Eichenbaum, H. (1999). The global record of memory in hippocampal neuronal activity. *Nature, 397*, 613–616.

2 Why connectionist models need spikes

Simon J. Thorpe

Many connectionist and neural network models of cognition work on the assumption that information is transmitted between the processing nodes (neurons) using continuous activation levels. Translated into neurophysiological terms, this effectively considers that neurons use a rate-based coding scheme, often using Poisson-like activation. I will review evidence that this form of coding is incompatible with the speed with which complex natural scenes can be processed. An alternative coding scheme uses the fact that the most strongly activated neurons will tend to fire first, and, as a consequence, information can be extracted from the order in which neurons within a population fire. I will illustrate the power of such an approach by showing that spike-time dependent plasticity, when coupled with temporally structured firing, automatically leads to high weights being concentrated on the earliest firing inputs. We have recently shown that such a mechanism will naturally result in neurons that respond selectively and rapidly to frequently occurring input patterns.

1 Introduction: closing the gap between neurophysiology and behaviour

There is an enormous amount of information available concerning the behavioural capacities of organisms, including ourselves. Indeed, much of experimental psychology can be thought of as an attempt to describe the performance of human subjects and to understand what limits our ability to perform particular tasks. At the same time, we have more and more details concerning the neural hardware that is presumably used to generate the behaviour. However, it is clear that there is a tendency for these two types of knowledge to remain relatively independent, and there are no doubt many researchers in psychology who really are not particularly interested in the details of the hardware. Likewise, there are whole armies of scientists working at the cellular and molecular levels who are quite happy to continue accumulating detailed information without worrying particularly about what impact, if any, those details have for understanding higher level phenomena such as perception, memory and consciousness. Nevertheless, I would argue that true understanding requires being able to bridge the two levels and gain insights into those aspects of neurophysiology that are critical for behaviour. This is clearly one of the main aims of the current meeting.

For me, the re-emergence of brain-style modelling in the 1980s with the development of connectionism and parallel distributed processing is a good illustration of the sort of work that has made a serious attempt to make this bridge (McClelland & Rumelhart, 1986; Rumelhart & McClelland, 1986). In many ways, there is clear evidence that the approach was and has continued to be successful. There has been a steady flow of studies over the last two decades that have used models based loosely on neural network architectures to increase our understanding of a wide range of phenomena, ranging from category learning, visual search, language, speech, development to cognitive disorders (for a good sample of the range, see Dawson, 2005; Ellis & Humphreys, 1999; Houghton, 2005; Quinlan, 1991). A typical connectionist model will involve large numbers of processing nodes that are linked to each other via connections whose weights typically vary between −1 and +1, or between 0 and +1. Each node has an activation level, which again can typically vary continuously between a minimum and a maximum value, and at each processing step the activation value of each node is recalculated on the basis of some form of weighted sum of all the nodes' inputs, sometimes with some residual activation from the preceding step. All of this sounds reasonable to someone interested in neural implementation, because the processing nodes of a connectionist model are strongly reminiscent of neurons, and the connections are roughly equivalent to the synapses in a real neural network. It should be noted that many researchers working in connectionist modelling go out of their way to stress that the nodes are not necessarily the equivalent of individual neurons: they could, for instance, correspond to groups of neurons. In this way, many researchers who have worked with "localist" connectionist models, in which individual nodes can correspond to concepts such as words, stress that they are not necessarily arguing for "grandmother cell" coding (Thorpe, 2002). Interestingly, the recent reports of individual neurons in the human temporal lobe responding selectively to individuals (the famous "Jennifer Aniston" cell: Quian Quiroga, Reddy, Kreiman, Koch, & Fried, 2005; Waydo, Kraskov, Quian Quiroga, Fried, & Koch, 2006) mean that the whole issue of just how close connectionist nodes might be to individual neurons is currently a hot subject.

However, it is not this link between connectionist models and the neural underpinnings that I wish to address in this chapter. Instead, I will concentrate on another key feature of biological neural networks that has rarely made its way into the current crop of connectionist models but has, I feel, been unjustly neglected – namely, the fact that in real brains, neurons send information using spikes. I will begin by caricaturing what I believe to be to be the generally accepted view that dominates not just connectionist models, but also much of experimental neuroscience – namely, the idea that neurons transmit information using rate coding and, more specifically, rate coding based on a Poisson process. Then I will go on to consider some of the arguments for believing that rate coding is not the whole story and that, more specifically, it is vital to take into account the information that can be encoded in the relative timing of spikes across populations of neurons. In the final sections, I will discuss some recent work on the link between Spike-Time Dependent Plasticity and temporal coding, work that suggests that temporal coding using patterns of spikes across populations of neurons may well be a key to understanding how the brain operates.

2 Traditional rate-based coding

Ever since the very first recordings from individual nerve fibres by Lord Adrian in the 1920s (Adrian & Matthews, 1927), it has been known that when the intensity of a sensory stimulus is increased, there is often an associated increase in the rate of firing in the sensory neuron. These changes in firing rate can be summarized by plotting a peri-stimulus time histogram (PSTH) of the number of spikes per time bin around the time of stimulus presentation. By summing the responses of the neuron across many repeated presentations of the stimulus, one can obtain a curve that represents how the probability of firing a spike varies over time. For many experimental studies, this firing-rate curve is considered to be a full enough description of the neuron's response, at least for all practical purposes. This view is reinforced by the widely held view that spike generation by neurons is essentially a Poisson process – that is, a mechanism in which for a given level of activation, the probability of emitting a spike at any particular moment is effectively constant – a bit like throwing a dice repeatedly at each time step. Of course, neurophysiologists know that the Poisson-process assumption is not strictly true, because neurons have refractory periods that prevent them firing immediately after having generated a spike. As a consequence, when the firing rate is very high, spiking becomes automatically more regular than would be expected for a true Poisson process. Thus, for a neuron that emits spikes that themselves have a duration of 1 ms, it is clearly impossible to fire above 1 kHz, but the relative refractory period will in fact make firing more regular even at substantially lower rates. Despite these obvious departures from true Poisson behaviour, it is clear that for many researchers, the basic truth of the Poisson rate code hypothesis seems well established.

However, it is very important to distinguish between two very different views of why neuronal firing looks roughly like a Poisson process. If we record the activity of a single cortical neuron for a long period and note that the emission of spikes obeys the statistics of a Poisson process, one could interpret this by imagining the neuron to be essentially a deterministic system embedded in a network composed of a very large number of neurons whose activity is unknown to the experimenter. If one was able to record simultaneously from all the thousands of afferents that contact the neuron, one might suppose that the firing of the neuron, despite being basically a Poisson process, was actually perfectly predictable on the basis of the pattern of afferent activity, together with a detailed knowledge of the geometry of the cell's dendritic arborization, the various ionic currents involved and other mechanisms involved in spike generation. Clearly, such an experimental test is currently impossible for obvious technical reasons. Nevertheless, sophisticated simulation studies give every reason to believe that one could, in principle, succeed in such an approach.

The other way of viewing the activity of the neuron is to imagine that the neuron is performing some sort of weighted sum of its inputs (as in the typical connectionist model), producing some sort of internal activation value, and then generating a Poisson process at the output to encode that activation in the form of a spike train. In other words, we would have a view of the neuron in which the production of spikes really does include a truly random process at the level of spike generation.

When pushed, most researchers who have used this approach will admit that it is not a realistic view of neural processing. However, I firmly believe that the underlying belief that spike initiation is intrinsically noisy is one that has permeated much of the research that aims to understand the links between brain function and neural activity. There are numerous examples where researchers have attempted to model processing using neurons that effectively include a random number generator at the output stage. This is even the case for modelling sensory receptor cells, where it is clear that the only input that is involved in activating the neuron is the receptor potential that is directly produced by the stimulus. In the absence of massive inputs from thousands of other inputs, it is literally impossible to invent any plausible mechanism that could generate a Poisson process, and yet scientists will often insist on adding such a random Poisson process just before the output. There are reasons for this. One is that Poisson processes have mathematical properties that are well understood and it is possible to perform sophisticated analysis of such systems. Nevertheless, when it comes to initiating spikes in sensory fibres that are not being bombarded by thousands of uncontrolled inputs, there seems to be little justification for not using a model such as a leaky integrate-and-fire mechanism – that is, a model that does not need to "play dice with spike initiation".

But the implications of this deeply entrenched idea that neurons are intrinsically noisy are far reaching and have had a profoundly limiting effect on the development of models of brain function. Essentially, by assuming that spike initiation is effectively a noisy process that can be modelled by Poisson rate statistics, the natural conclusion has been that we can model neural networks with neurons that have continuous output functions. If neurons generate spikes according to a Poisson process, it is clear that we can fully describe the neurons' output as a time-varying process that is summarized by the instantaneous firing rate. And this instantaneous firing rate is effectively a continuous analogue value that can itself be represented by a floating-point number. In such a context it is no surprise that the vast majority of connectionist models have used this basic formalism – namely, the idea that the brain can be thought of as a large network of nodes each possessing an activation level that can (for modelling purposes) be represented by a floating-point number, often in the range 0.0–1.0, where 1.0 corresponds to the neuron's (or the node's) maximum level of activation. Once this idea has been accepted, it becomes perfectly natural to suppose that we can have an arbitrarily high level of precision for this activation level.

This view is summarized in Figure 2.1, in which we see that although the neuron is actually receiving inputs in the form of patterns of spikes, these patterns effectively get converted into floating-point numbers (corresponding to instantaneous firing rate) at each synaptic connection. The neuron then performs some (potentially very complicated) non-linear weighting function involving the firing rates of the inputs and calculates a new floating-point number corresponding to its own internal activation level, which is then translated into a spike sequence using a Poisson-like rate process.

As we will see in the next section, there are a number of very severe problems with such a view, related in part to the fact that encoding anything with a Poisson process is really extremely inefficient. However, for the time being I would just like to stress the fact that this cartoon version of neural computation basically assumes that spikes are not actually important – they are just nature's way of sending analogue values. One might imagine that if natural selection had discovered a way of making axons out of copper wire with good insulation, our brains might well have ended up sending signals as analogue voltages rather than spikes. In later sections we will look at some of the ways in which coding with spikes transforms the sorts of computation that can be performed by the brain. But first, we will look at some of the most serious limitations of Poisson-based rate coding.

3 The limits of Poisson rate coding

The main problem with using a Poisson rate code stems from its inherent unreliability. Some time ago, Jacques Gautrais looked at how accurately one can estimate the underlying firing rate of a neuron using a Poisson rate coding scheme as a function of the amount of time available for measurement (Gautrais & Thorpe, 1998). Suppose that the neuron is firing with a mean firing rate of 1 spike every 10 ms (i.e., an observed frequency of 100 Hz), and that we have just 10 ms to make a judgement. What can we say about the actual firing rate, given that one spike is received in 10 ms? It turns out that the best we can do is say that there is a probability of 90% that the actual rate lies in the range 5–474 Hz(!). Of course, by listening to the neuron for longer, we can increase the precision of our estimate, but we would need to listen for several seconds to get a measure accurate to within 10%. Many people would argue that the obvious solution would be not to use a single neuron but, rather, a population of redundant cells. With enough

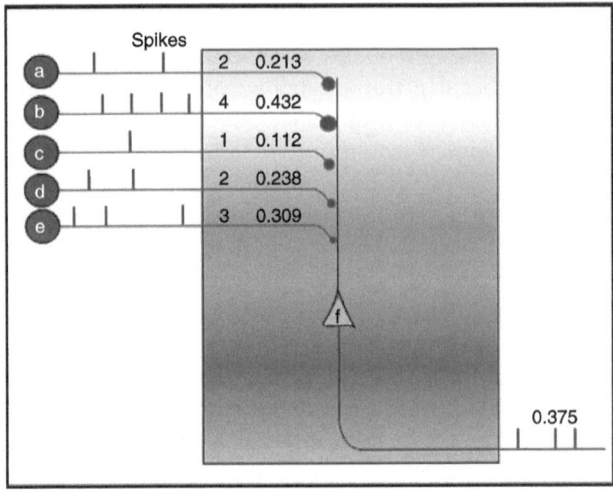

Figure 2.1 A caricature of the commonly held view that spikes in afferent fibres can be thought of as representing analogue floating-point numbers, thanks to rate coding. The neuron then performs some sort of non-linear weighted sum of the inputs and then generates an output activation value that is then converted into a spike train using a Poisson process to generate spikes with the appropriate density.

neurons, one can in principle code an analogue value with arbitrary precision. Unfortunately, the calculations show that with 30 redundant neurons, and a temporal window of 10 ms, we would still only be able to get the 90% confidence interval down to 72–135 Hz – that is, roughly 100 ± 30 Hz. Indeed, to get the confidence level to ±10% would require no less than 281 redundant and independent neurons. Remember that here we are talking about trying to estimate the firing rate of populations of neurons firing very rapidly (at around 100 Hz). The problems only get worse if we wanted to estimate the mean firing rate of a population of cells firing at only 40 Hz.

One could argue that maybe the brain does not really care about absolute firing rates. Instead, it might be that a more important question would be to ask which of two or more groups of neurons is firing fastest. This is essentially the classic winner-takes-all problem that is a standard question in many neural networks. Here again, the calculations show that using Poisson rate coding would be catastrophic (Gautrais & Thorpe, 1998). Suppose we have two populations, A and B. One is firing at 100 Hz and the other at 75 Hz, and we would like to determine which is which as quickly as possible. To keep the error rate under 5% using a window of 10 ms would require 76 redundant and independent neurons in each group. With a smaller difference in rate (e.g., 100 Hz vs. 90 Hz), the problems just get worse.

It would be reasonable to ask why I have been putting so much importance on using an observation window that is only 10 ms long. The reason lies in an observation that I originally made in a chapter in a book published in 1989 (Thorpe & Imbert, 1989). It was known that face-selective neurons in the primate inferotemporal (IT) cortex can respond selectively to flashed visual stimuli with onset latencies of around 100 ms (Perrett, Rolls, & Caan, 1982), and subsequent studies have demonstrated that this face selectivity can be fully present right at the very beginning of the neuronal response (Oram & Perrett, 1992). Given that the total number of anatomical layers between the photoreceptors of the retina and the neurons in IT cortex is roughly 10, it follows that, on average, each processing stage has only about 10 ms in which to make a response before the next stage has to respond. In fact, the constraints are probably even more serious because the roughly 100 ms needed to activate neurons in the IT cortex must also include the conduction delays caused by the fact that transmission of spikes along axons is a relatively slow process. While certain peripheral nerves can conduct action potentials at up to 100 ms^{-1}, cortico-cortical connections are probably limited to just $1-2 \text{ ms}^{-1}$ (Nowak & Bullier, 1997). With several tens of millimetres to cover between V1 and IT cortex, these incompressible conduction delays will probably constitute a sizable proportion of the time taken to initiate firing at later stages.

These severe timing constraints mean that we really do need to look at the question of how information can be transmitted efficiently using spiking neurons. We will look at some of the alternatives in the next section, but for now it is worth noting that more and more scientists are realizing that even if neural activity can look roughly Poisson in nature, this does not mean that we need to think of neurons as having a built-in random Poisson process attached to the spike initiation process. Many studies have shown how purely deterministic mechanisms can lead to activity patterns that look close to Poisson (Deneve, 2008; Gerstner & Werner, 2002) without the need to posit specific randomness in the spike generation process. And on the other side, the last few years have seen a large number of experimental studies that have demonstrated that the timing of spike emission can be remarkably precise (Rieke, Warland, De Ruyter, & Bialek, 1997; VanRullen, Guyonneau, & Thorpe, 2005). In addition, there has been a recent study looking at the detailed statistics of spike train emission by neurons in monkey temporal cortex that has explicitly excluded the Poisson hypothesis, at least for the initial part of the neural response (Amarasingham, Chen, Geman, Harrison, & Sheinberg, 2006). While the study fell short of describing precisely what sort of model was appropriate, it was clear that the reliability of the initial transient was too high to be explained by a simple Poisson model.

Thus, to conclude this initial section, it seems clear that while the idea that the neurons in neural network (or the nodes in a connectionist model) can be considered as having a continuously varying activation level that can be summarized by a single analogue value is appealing because of its simplicity, there is considerably more that needs to be taken into consideration if we are truly to link cognitive capacities with their neurophysiological underpinnings.

4 Computing with spikes: the temporal coding option

We have seen that an idea that permeates much thinking about computation in the brain is that, essentially, the spikes that neurons emit are just the translation via a Poisson process of the neurons' level of activation. If so, then clearly the precise timing of spikes is of no real importance, and it is reasonable to continue with connectionist models whose nodes have continuous-valued outputs. However, once we accept that spike emission is not limited by a built-in random number generator, then a whole range of new and interesting options becomes available. One idea that has received a great deal of attention over the last two decades is the idea that synchrony of firing between neurons could be important for a range of issues that includes the binding of features (Singer, 1999; Singer & Gray, 1995). The basic idea is that when neurons within a population produce spikes that are closely aligned in time, they are more likely to be able to trigger spikes in post-synaptic neurons because of temporal summation. This basic idea has led to a large field of research in which scientists have looked at the degree of synchronization between different cortical areas and found that such measures show very interesting variations that can be related to higher order functions such as conscious perception and attention (Rodriguez, George, Lachaux, Martinerie, Renault, & Varela, 1999; Varela, Lachaux, Rodriguez, & Martinerie, 2001). There has also been a lot of work on the possible roles of oscillatory activity, particularly in the gamma frequency range (30 Hz and up) that has been linked with a wide range of cognitive functions (Tallon-Baudry, 2004; Tallon-Baudry, Bertrand, & Fischer, 2001). Reviewing this large area of research would be beyond the scope of the present chapter. Instead, I would like to concentrate on another type of processing that involves spike timing, one that depends not so much on the degree of synchronization between neurons but, rather, on the fact that neurons will actually tend to fire at different times, dependent on how strongly activated they are.

The idea that the timing of spike emission depends on the strength of the stimulation follows very naturally from the simplest integrate-and-fire models (Burkitt, 2006a, 2006b). In its simplest form we can think of a neuron as a capacitor into which we inject current. The neuron has a threshold at which it emits a spike, and the time taken to reach the threshold depends on the strength of the input. Variations of onset latency with stimulus strength have been reported for many systems (VanRullen et al., 2005), but it is particularly well established in the auditory system (Heil, 2004). Indeed, it is well known that the reason why a sound located on the left can be localized is because the intensity difference between the two ears leads to a difference in arrival times for spikes. Spikes coming from the left ear will have shorter latencies. However, the generalization of the idea that the order in which neurons fire can be extended to other modalities has been difficult. I originally proposed in a short 4-page report (Thorpe, 1990), which was reprinted in a collection of papers aimed at physicists (Thorpe, 1994), that relative spike arrival times could be used as a general principle for sensory coding. The basic idea is illustrated in Figure 2.2. But it has taken a long time for the experimental support for the idea to arrive. One of the very first studies to test explicitly the idea of order coding was a study on human mechanoreceptors that showed that tactile stimuli can be encoded efficiently just on the basis of the order in which afferents fire, without the need to encode the firing rates of individual afferents (Johansson & Birznieks, 2004). Very recently, a new study of population responses in the salamander retina has shown convincingly that the order of firing of retinal ganglion cells is a remarkably robust means for encoding visual stimuli (Gollisch & Meister, 2008). They also show that using the order of firing, rather than absolute latencies or the spike count, has a number of advantages, including the fact that the code is very robust to changes in the contrast of the image: as contrast changes, the number of spikes emitted can vary a lot, but the relative order in which cells fire changes little.

Although the experimental confirmation is very recent, earlier simulation studies had already shown that the order of firing was potentially a very

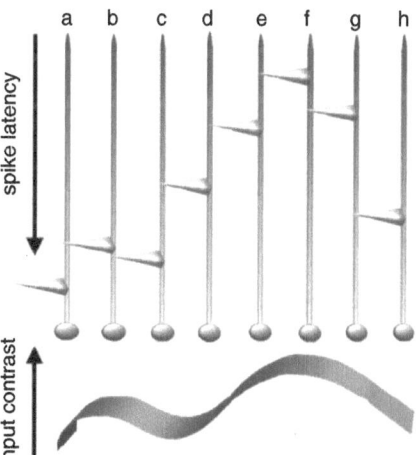

Figure 2.2 The basic principle underlying coding by order of firing. If we consider a typical integrate-and-fire neuron with a threshold, it is clear that the time taken to reach the threshold will depend on the effectiveness of the input. As the intensity of the stimulation increases, the latency of the spike will tend to decrease. With a population of cells, this means that a pattern of input will be translated into a wave of spikes in which the order of firing can be used as a code. With 8 neurons, there are 8! (more than 40,000) different orders that can be generated (VanRullen & Thorpe, 2002).

efficient way to encode information. For example, in the theoretical work performed by Gautrais that was mentioned earlier (Gautrais & Thorpe, 1998), it was demonstrated that while using spike counts was essentially a very inefficient way of determining which set of neurons was the most active, the problem becomes almost trivial when one looks at the order in which cells fire. Almost by definition, the first neurons to fire will correspond to the ones with the strongest activation. This therefore provides a very simple and robust solution to solving the classic winner-take-all (WTA) problem – the first neuron to fire in a population is likely to be the most strongly activated one. If we add lateral inhibition such that, once one neuron has fired, inhibition prevents any others from firing, we have a solution for solving the WTA problem with an arbitrarily large number of inputs, a solution that can even be implemented in hardware. If the inhibitory circuits only become active once a certain number of neurons within the population have fired, then we have a way of controlling the proportion of neurons active within the population, a problem known as the k-WTA problem. Suppose that we have 100 different neurons in a cortical column and the inhibitory feedback circuits are adjusted so that the inhibition kicks in as soon as 10 of the neurons have fired; this gives us a simple way of ensuring that, irrespective of the strength of the incoming stimulus, the proportion of active cells can be fixed.

Another study that explored the advantages associated with using the temporal ordering of spikes across a population of cells was a paper published in 2001 that looked at the question of how the retina might send information efficiently (VanRullen & Thorpe, 2001). We used a very simple "toy" retina containing ganglion cells with ON- and OFF-centre receptive fields at a range of scales and asked the question of whether it would be possible for the brain to effectively reconstruct the image simply on the basis of the order in which the ganglion cells fired. We used the idea that in response to a flashed stimulus, the order of firing will naturally reflect how effective the stimulus was in activating the cell. The reconstruction process used a strategy in which every time an afferent fired, its receptive field was added to the reconstructed image, but with a weighting factor that depended on the order in which the neurons fired: the very first neurons to fire were given a maximal weighting of 100%, but neurons firing later were less and less effective. Specifically, we used a desensitization function that was derived experimentally, on the basis of the mean contrast associated with neurons firing with different ranks. We found that this curve is really very steep, since by the time 1% of the cells has fired, the average energy had dropped to just a few percent of the original value. In other words, the remaining 99% of cells contribute only a relatively small percentage of the information. More to the point, we noticed that by the time 1% of the neurons had fired, the reconstructed image was clearly identifiable, and it was perfectly possible to decide whether (for example) the image contained an animal, even on the basis of this very preliminary analysis.

5 Brain mechanisms for using spike-order-based information

Given the success of such results, it is natural to ask whether information based on the ordering of spikes could indeed be used by later stages of the visual system to allow processing. We have explored this idea over a number of years during which we have developed a software simulation system called SpikeNet that attempts to implement some of these key ideas. Some of the earliest work that we did was on face processing, and we were able to show that it is possible to construct a purely feed-forward architecture in which each neuron is only allowed to emit one spike (thus preventing the use of a true rate code) but which is nevertheless capable of detecting the presence of faces in natural images (VanRullen, Gautrais, Delorme, & Thorpe, 1998). Later work demonstrated that a similar architecture was capable of identifying individual faces, despite substantial variations in contrast and luminance and the addition of noise to the images (Delorme & Thorpe, 2001). These encouraging results led to the creation of a spin-off company in 1999 called SpikeNet Technology that has, over the last few years, continued to develop the same basic concept (see http://www.spikenet-technology.com). Although the current technology has diverged somewhat relative to the primate visual system, the underlying ideas are still intact, in that the system continues to be based entirely on a purely feed-forward processing architecture in which each neuron only gets to fire at most once per stimulus presentation. Some of the basic ideas are illustrated in Figure 2.3 where we have an initial stimulus profile that is encoded in the order of firing of a set of afferents (a–e). The afferents each have synaptic weight connections to a second set of neurons (f–j). Normally, these cortical neurons should receive the same amount of excitation irrespective of the order in which the inputs fire. However, by adding a feed-forward shunting inhibition circuit (FF in Figure 2.3), we can progressively reduce the sensitivity of the cortical neurons as a function of the wave of spikes – much as we did in the simulations of retinal transmission. As a result, the amount of activation in a given cortical neuron will be maximal when the order of firing matches the order of the weights. Ideally, the synapses with the largest weights should be activated first, when the shunting inhibition is minimal, and those with the smallest weights activated at the end.

Another computational trick that works well in the case of temporally structured spiking activity is to add a feedback-inhibition circuit (FB in Figure 2.3) that performs the k-WTA operation mentioned earlier. By fixing the threshold for initiating the feedback-inhibition circuit, we can control the percentage of neurons that are allowed to fire at each stage of the system. Obviously, the case illustrated in Figure 2.3 is extremely simplified, with just 5 afferents and 5 neurons at each cortical stage. However, it is clear that when the numbers are scaled up, with thousands of afferents and with hundreds of units at each stage, then the computational possibilities become extremely

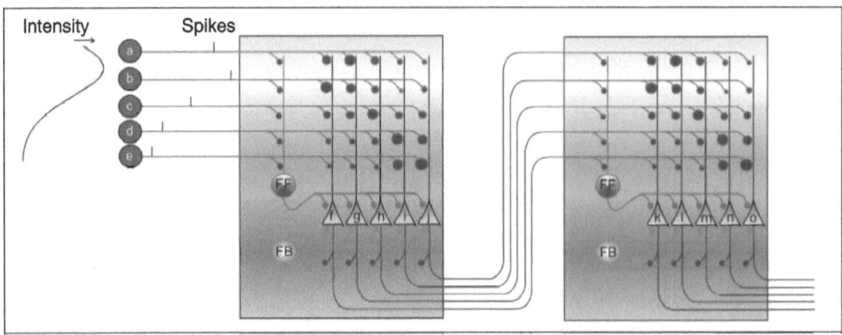

Figure 2.3 Some simple cortical circuits that can be used to process spike order information. Neurons a–e respond to an intensity profile in the input by generating an ordered wave of spikes, with the shortest latency spikes corresponding to the places where the intensity is highest. Neurons f–j are cortical neurons that receive inputs from the afferents that have different weights. FF is a feed-forward inhibitory circuit that progressively desensitizes the cortical neurons as spikes arrive from the periphery. As a result, activation in the cortical neurons is maximal when the order of arrival of the spikes matches the order of synaptic weights: the highest weight synapses need to be activated first. In this case, neuron f will respond first because its weights match the input. FB is a feedback-inhibition circuit that can prevent more than a fixed number of cortical cells firing, thus effectively solving the classic k-WTA problem.

rich, especially if the same sort of architecture is cascaded through multiple layers.

The latest version of SpikeNet's recognition engine is effectively based on the same sorts of ideas – namely, generating a wave of spikes in response to each new image, in which the most excited neurons fire first, and controlling the proportion of neurons firing at each stage. To learn a particular input pattern, we fix the weights of the connections as a function of the order of firing, with the highest weights corresponding to the earliest firing inputs. For each visual pattern that we want to recognize, we generate an array of units at the top end of the system, with effectively one unit for each point in the visual field. When the ordering of spikes in the earlier processing layers matches the weight set of the neurons in the recognition layer, the neurons go over threshold and generate spikes. This spiking activity in the recognition layers can then be used to localize previously learned visual forms within the image. Despite the system's simplicity (a pure feed-forward architecture, and activity strictly limited to no more than one spike per neuron), performance turns out to be impressive. When run on a simple desktop processor, it is possible to locate and identify arbitrary visual forms in full-format video sequences in real time. Each model (corresponding to an array of units in the recognition layer) will respond to a range of variations in the input image, including, for example, changes in orientation of roughly ±10°, size changes of roughly ±10% and large variations in contrast (Thorpe, Guyonneau, Guilbaud, Allegraud, & VanRullen, 2004).

The recognition capacities of the SpikeNet system are impressive and demonstrate the potential of processing based on asynchronous spike propagation. Since no units ever fire more than once, this effectively rules out processing based on rate-based coding. Nevertheless, there are a number of features of SpikeNet that do not fit easily with biology. For example, we use a weight-sharing scheme in which all the neurons in the recognition layer have the same set of weights, thus allowing the system to achieve position invariance by simply duplicating the same recognition mechanism at each point in the image. This is unlikely to be a viable strategy in real biological vision systems, at least in cases where very large numbers of different visual forms need to be recognized. If we wanted to recognize 100,000 different visual objects at any point within an image and if we had to have 100×100 separate mechanisms to achieve position invariance, this would require a billion neurons. It is more likely that the primate visual system builds up selectivity more gradually, by progressively increasing receptive field size from layer to layer. The evidence suggests that the neurons at the top of the visual hierarchy are in fact only relatively poorly selective to the position of the object within the visual field (Rousselet, Thorpe, & Fabre-Thorpe, 2004), although there may be more position dependence than had been thought previously.

Another non-biological feature is the way the learning mechanism works. Essentially, the weight sets are fixed instantaneously on the basis of the order in which the afferents fire – with high weights on the early firing inputs, and low or zero weights on those that fire later. However, while this sort of instantaneous learning is relatively straightforward to do in a computer simulation, it is much less obvious how the same method could be achieved with real neurons. Imagine a neuron with 1000 afferents that are activated by an incoming stimulus. We would like to fix the weights with high values for the afferents that fired first, but since synapses do not have a convenient time stamp to say when they last fired, it would be very difficult for the cell to work out which inputs fired first. However, as we shall see in the next section, a biologically plausible learning rule called *spike-time dependent plasticity* is capable of achieving just such a computational trick.

6 Spike-time dependent plasticity and temporal coding

Spike-time dependent plasticity (STDP) refers to a family of learning mechanisms in which the strength of a synapse depends on the relative timing of firing in the incoming afferent and on the timing of a spike in the post-synaptic cell (Dan & Poo, 2006). In the standard version, when an afferent spike precedes a post-synaptic spike, the synaptic strength is increased, whereas it is decreased when the afferent spike occurs afterwards. There have been many studies of the possible computational implications of such plasticity rules (see, e.g., Song, Miller, & Abbott, 2000) but in this section I would like to look at the particular relevance of STDP in the context of temporally coded information. It appears that there may be a very interesting link between STDP and the sort of fast processing that we have been interested in for a number of years.

Consider the case of a neuron receiving activity from a large number of afferents. Let us suppose that the afferents are being activated by some external stimulus and that this generates a wave of spikes that converge upon the post-synaptic neuron. If the synaptic weights of the afferents are all initially low and random, the neuron will take a relatively long time to depolarize enough to reach threshold. However, once it reaches threshold and fires a spike, the STDP mechanism will be activated and the synaptic strength of the afferents activated just before the post-synaptic spike will be strengthened. As a consequence, if the same pattern is presented a second time, the neuron will tend to fire slightly early. The natural consequence of this process is that if the same pattern is presented over and over again, the response latency of the post-synaptic neuron will become shorter and shorter. We have looked at this phenomenon in detail (Guyonneau, VanRullen, & Thorpe, 2005) and found that this tendency to concentrate high synaptic weights on the earliest firing inputs is remarkably robust. It is possible to add a considerable amount of spike-time jitter to the firing of individual afferents or to add additional random background firing without preventing the neuron from finishing with the weights concentrated on the earliest firing inputs. Even more remarkably, we were able to set up a simulation in which the firing rate of the afferents was negatively correlated with the latency of the first spike. But even this situation, in which the earliest firing inputs emit only one spike, whereas those starting to fire later could fire at very high rates, did not prevent the convergence occurring. It is as if the presence of the STDP mechanism almost forces the neurons to ignore the firing rates of the inputs – only the timing of the spikes is critical in the end.

In other simulations (Guyonneau, VanRullen, & Thorpe, 2004) we looked at what happens when, instead of having just one neuron listening to a set of afferents, there are numerous neurons in parallel, all receiving the same sets of afferents, but with weight sets that are initially set to different random values. Interestingly, we can here use the WTA trick mentioned earlier in which the first neuron to fire inhibits all the others from firing. In such a situation, when the input patterns are being continuously varied, different neurons in the output layer will become "specialized" for particular types of input using what amounts to an unsupervised learning procedure. With input patterns corresponding to 15 × 15 pixel fragments of natural images, Rudy Guyonneau found that if he starts with a large number of initially unspecified neurons (1024), by the time the network has stabilized, less than 300 of them have become selective. Furthermore, the receptive fields of these cells look very similar to the orientation-selective simple cells characteristic of primary visual cortex (Guyonneau, 2006) (see Figure 2.4).

If the outputs of these simple-like receptive fields are then fed into a second layer of neurons equipped with STDP and lateral inhibition to limit the number of cells that can fire in response to each input, Guyonneau found that the cells in the next layer developed receptive field properties that were somewhat more complicated and included neurons selective to curves and corners (see Figure 2.5). It would appear that simply providing this sort of hierarchically organized neural architecture with natural images as inputs leads to the formation of receptive field properties that are reasonably well matched to the types of receptive field organization seen in the real visual system. For the moment, these sorts of simulations have been relatively limited in scale (the inputs were, after all, limited to 15 × 15 pixel image patches). But it seems conceivable that the same approach could in principle be used to guide the development of a truly hierarchical processing architecture.

In a more recent extension of this work, Timothée Masquelier has taken a richer architecture, largely inspired by the biologically inspired feed-forward

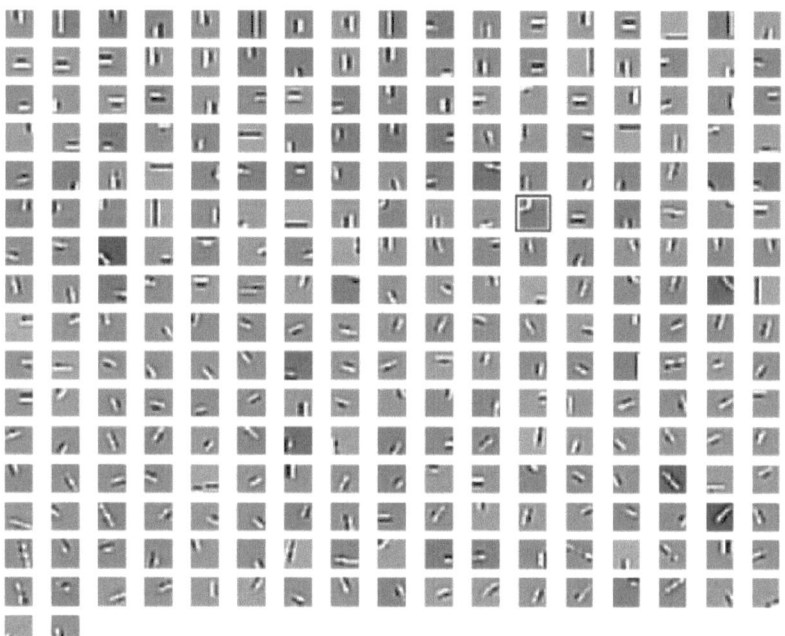

Figure 2.4 Receptive fields formed using an STDP-based rule in a cortical layer composed of 1024 neurons using 15 × 15 pixel fragments from natural images as inputs. The inputs were ON- and OFF-centre receptive fields similar to retinal ganglion cells. Note that only 274 of the cells became specialized, and all but one (highlighted with a square) had oriented receptive fields similar to cortical simple cells (Guyonneau, 2006).

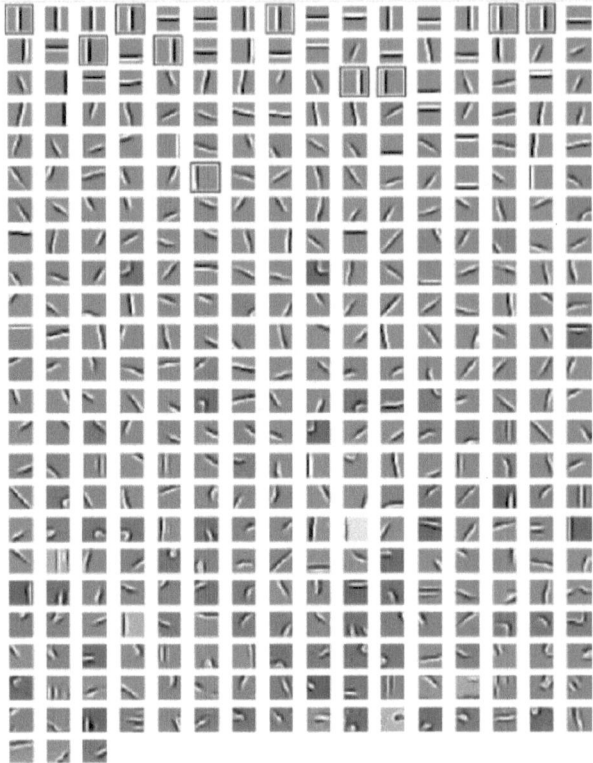

Figure 2.5 Receptive fields formed using an STDP-based rule in the second cortical layer using 15 × 15 pixel fragments from natural images as inputs. The inputs to the second layer were oriented Gabor-like patches, similar to cortical simple cells. Note the presence of elongated receptive fields (top of figure) but also a substantial number of neurons with more complex receptive fields selective to, for example, corners and curves (Guyonneau, 2006).

architectures used by Tomaso Poggio's group at MIT (Serre, Kreiman, Kouh, Cadieu, Knoblich, & Poggio, 2007; Serre, Oliva, & Poggio, 2007), and looked at what happens when an STDP-based learning mechanism at the top end of the hierarchy is coupled with the temporal wave of spikes. When such an architecture was stimulated with a set of images from the CalTech Face database, neurons at the top end of the system were found to progressively become tuned to parts of the face (Masquelier & Thorpe, 2007). For example, Figure 2.6 shows how after about 200 image presentations, the three classes of neurons in the system have become selective to the upper part of the face (top square (A)), the lower part of the face (bottom square (C)) and the nose and right eye (smaller square in middle (B)). It is important to realize that this development of selectivity emerged by a completely unsupervised

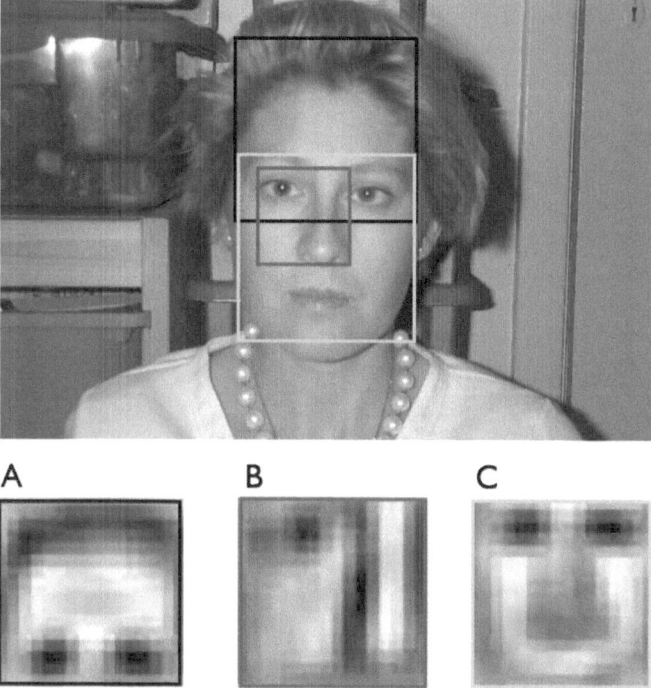

Figure 2.6 STDP-based unsupervised learning of face features. The system has three classes of neurons that initially have random selectivity. The figure shows that after roughly 200 presentations of images randomly selected from the CalTech Face database, the three different types of neurons have become selective to three parts of the face. (For details see Masquelier & Thorpe, 2007.)

process, since at no point was the system told anything about faces. The selectivity developed simply because the set of training images contained faces. In other experiments, the training set contained photographs of motorcycles, and in this case the neurons became selective to parts of motorcycles.

The combination of STDP and temporal coding thus seems to provide a powerful way of permitting the development of selective neuronal responses as a function of the environment. Depending on the nature of the visual environment to which the system is exposed, we can expect that receptive field properties will appear that provide a useful vocabulary to describe the world. Once that vocabulary has emerged, one can suppose that supervised learning mechanisms will also be involved in the process of grouping these higher level features into useful behavioural categories. However, it may be that this process does not take place directly in the visual pathways. There is increasing evidence that the categorization of visual stimuli may actually occur in structures such as prefrontal cortex, rather than in areas like inferotemporal cortex (Freedman, Riesenhuber, Poggio, & Miller, 2001; Thorpe & Fabre-Thorpe, 2001).

7 Beyond the feed-forward wave

Much of the material that we have looked at in this chapter has concerned the rather specific situation where a flashed visual stimulus is processed very rapidly. Clearly, the temporal constraints imposed by the fact that the brain has to make very rapid decisions in order to determine whether a briefly presented image contains an animal are very severe (Bacon-Mace, Mace, Fabre-Thorpe, & Thorpe, 2005; Kirchner & Thorpe, 2006), and one might suppose that the processing strategies required in such cases may differ from those used more generally. However, in a recent study (Masquelier, Guyonneau, & Thorpe, 2008) we have looked at the potential role of STDP in a more generic case, where there is no obvious start to the processing stream.

In that study, we considered the case of a single neuron receiving inputs from a large number (2000) of afferents, each of which was generating spikes according to a time-varying Poisson process that was completely independent for each input (see Figure 2.7). There was thus absolutely no structure in the afferent activity pattern. However, we then chose to select a random subset of the afferents (typically half of them, although it could be much less), and then copy–paste a 50-ms-long section of their activity pattern at random irregular intervals (see the darker tinted spikes in the figure). The problem of finding the repeating pattern is a computationally difficult one and would normally require a lot of intensive computation. However, we found that if the neuron that was "listening" to the afferents is equipped with STDP, the neuron almost invariably starts to respond selectively each time the repeating pattern occurs. Furthermore, with more extensive training, the neuron gradually responds faster and faster to the repeating pattern so that in the end it fires systematically within a few milliseconds of the start of the pattern.

The implications of this observation seem to be quite profound. Essentially, it means that any neuron equipped with STDP is potentially capable of detecting the presence of any repeating pattern of activity in its afferents, even when that pattern has absolutely nothing special about it in terms of firing rate. Indeed, in the simulations we performed we effectively did everything possible to hide the pattern. Nevertheless, STDP allows the pattern to be detected, and this was possible even if we added considerable amounts of jitter to the pattern, or added additional background activity, or even deliberately dropped a significant proportion of the spikes from repetition to repetition.

Remember as well that all this was done with only one neuron listening to the activity patterns in the 2000 afferents. In more recent tests we have looked at the case where several neurons are simultaneously listening to the

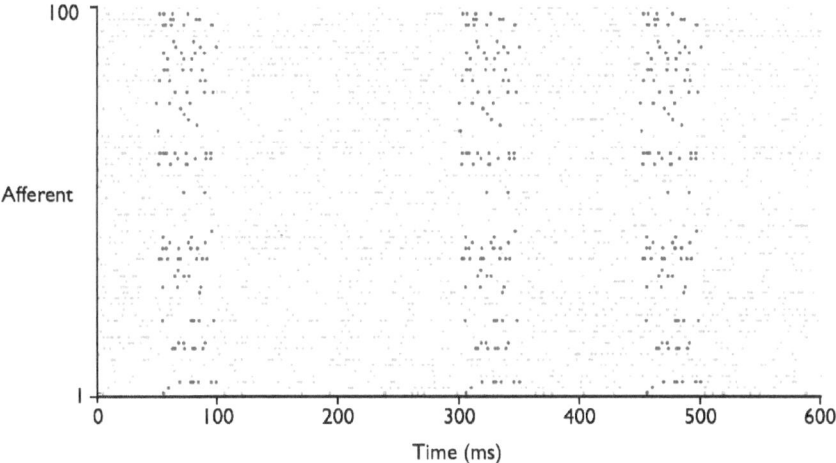

Figure 2.7 Spatio-temporal spike pattern. Here we show in the darker tint a repeating 50-ms-long pattern that concerns 50 afferents among 100 (Masquelier et al., 2008).

same set of afferents. As before, lateral inhibition is used to prevent more than one neuron firing at the same time (Masquelier, Guyonneau, & Thorpe, in press). Interestingly, when the afferent activity contains several different repeating patterns, each different neuron will tune into a different pattern. Furthermore, when the afferent pattern is relatively long, one can obtain the situation where more than one neuron can become tuned to the same repeating pattern but respond at different times. Thus during a 50-ms-long pattern, one neuron might respond within 5 ms of the start of the stimulus, another after 20 ms and yet another at about 35 ms (the relative spacing between the spikes depends essentially on the duration of the inhibition caused by the spiking of the first neuron to fire).

So far, we have not really been able to scale up this sort of simulation study beyond the case of a few neurons. And in all the examples we have looked at so far, we have restricted ourselves to simple feed-forward networks where the neurons receive information in parallel from a large number of independent afferents. However, it seems likely that the sorts of behaviour that would be obtained with a large number of interconnected neurons could be very interesting. Imagine a network composed of a billion neurons, each with roughly 10^4 connections and where each neuron is effectively looking for any form of repeating pattern among its afferents. In such a situation, we would not only have the potential for developing fast feed-forward architectures capable of detecting particular patterns in the input. We would also have a system that could use any information capable of allowing the system to predict its own behaviour. Could this be one of the secrets of intelligence?

8 Concluding remarks

In this review chapter, we started by considering the "conventional" connectionist modelling framework, in which processing nodes send analogue activation levels. This sort of approach has been very successful in a wide range of areas. However, there is an increasingly wide gap between this form of modelling – which essentially takes for granted the idea that the only form of coding used by neurons is rate coding – and the increasingly large range of neurophysiological studies that have demonstrated that the details of spike timing can be very significant. We have seen that part of the reason for this reluctance to take into account the more specific details of spike timing lies in the (false) belief that spike emission is intrinsically noisy – that in some way, a neuron converts a nice, clean internal activation level into spike trains using a random Poisson process at the output. It is important to realize just how misguided this view is. While it is true that Poisson statistics give a reasonably good description of what neuronal activity looks like, this does not mean that the spiking activity of neurons has to be considered random in any way. It is perfectly conceivable that to all extents and purposes, spike emission in the brain could be completely deterministic. The fact that we cannot predict when a neuron will next generate a spike may be entirely explained by the fact that no one has ever recorded from all the 10,000 afferents converging simultaneously onto a given neuron. And, of course, no one will be able to do that any time soon, despite the remarkable progress that has been made over the last few years in monitoring the activity of large numbers of cells simultaneously (Ohki, Chung, Ch'ng, Kara, & Reid, 2005; Ohki & Reid, 2007). Once one accepts that the timing of spikes can be important, then a whole range of new possibilities is opened. In the review, we have concentrated on just one particular aspect – namely, the possibility that information could be encoded in the relative ordering of spikes across a population of neurons, and the possibility that STDP could allow neurons downstream to learn about these patterns. But there are plenty of other options to be explored. What seems to be clear is that if we are really to bridge the gap between behaviour and brain mechanisms, it would be a good thing to move beyond the conventional connectionist modelling paradigm to one in which spikes are important.

One final remark concerns the question of the possible role of noise in brain function. There are many researchers (including some of the contributors to the third Behavioural Brain Sciences Symposium) who believe that a brain without noise would, quite possibly, simply not be able to function – that noise is needed for some types of computation. This may be true, but I am not sure that the evidence for such a view is convincing. There is a substantial literature on stochastic resonance showing that adding noise to neurons can improve their ability to transmit information (Stemmler, 1996) and increase the system's sensitivity to weak stimuli (Longtin, 1993). However, I would note that while there may well be a need for the modeller to add noise to the model to make it work well, in the case of the real brain there may no such need. We can probably all agree that for a typical cortical neuron, only a small percentage of the neuron's inputs will be directly dependent on the incoming stimulus and therefore under experimenter control. The other 90% of synapses will effectively be random and therefore appear to be noise. Similarly, for many modelling studies, it may well be perfectly justified to model the input activity as essentially Poisson in nature. However, I feel that it is essential to realize that this is only a useful technique for modelling. The real error occurs when people imagine that real neurons actually do have a random number generator at the spike initiation process. In this chapter, I have tried to show why this false belief has actually slowed progress in bridging the gap between behaviour and neurophysiology by making it hard for many scientists to take seriously the idea that spike timing really is important.

9 References

Adrian, E. D., & Matthews, R. (1927). The action of light on the eye: Part I. The discharge of impulses in the optic nerve and its relation to electric changes in the retina. *Journal of Physiology, 63*, 378–414.

Amarasingham, A., Chen, T. L., Geman, S., Harrison, M. T., & Sheinberg, D. L. (2006). Spike count reliability and the Poisson hypothesis. *Journal of Neuroscience, 26*(3), 801–809.

Bacon-Mace, N., Mace, M. J., Fabre-Thorpe, M., & Thorpe, S. J. (2005). The time course of visual processing: Backward masking and natural scene categorisation. *Vision Research, 45*(11), 1459–1469.

Burkitt, A. N. (2006a). A review of the integrate-and-fire neuron model: I. Homogeneous synaptic input. *Biological Cybernetics, 95*(1), 1–19.

Burkitt, A. N. (2006b). A review of the integrate-and-fire neuron model: II. Inhomogeneous synaptic input and network properties. *Biological Cybernetics, 95*(2), 97–112.

Dan, Y., & Poo, M. M. (2006). Spike timing-dependent plasticity: From synapse to perception. *Physiological Reviews, 86*(3), 1033–1048.

Dawson, M. R. W. (2005). *Connectionism: A hands-on approach*. Oxford, UK: Blackwell.

Delorme, A., & Thorpe, S. J. (2001). Face identification using one spike per neuron: Resistance to image degradations. *Neural Networks, 14*(6–7), 795–803.

Deneve, S. (2008). Bayesian spiking neurons I: Inference. *Neural Computation, 20*(1), 91–117.

Ellis, R., & Humphreys, G. (1999). *Connectionist psychology: A text with readings*. Hove, UK: Psychology Press.

Freedman, D. J., Riesenhuber, M., Poggio, T., & Miller, E. K. (2001). Categorical representation of visual stimuli in the primate prefrontal cortex. *Science, 291*(5502), 312–316.

Gautrais, J., & Thorpe, S. (1998). Rate coding versus temporal order coding: A theoretical approach. *Biosystems, 48*(1–3), 57–65.

Gerstner, W., & Werner, M. (2002). *Spiking neuron models.* Cambridge, UK: Cambridge University Press.

Gollisch, T., & Meister, M. (2008). Rapid neural coding in the retina with relative spike latencies. *Science, 319,* 1108–1111.

Guyonneau, R. (2006). *Codage par latence et STDP. Des stratégies temporelles pour expliquer le traitement visuel rapide.* Unpublished thesis, Université Paul Sabatier, Toulouse, France.

Guyonneau, R., VanRullen, R., & Thorpe, S. J. (2004). Temporal codes and sparse representations: A key to understanding rapid processing in the visual system. *Journal of Physiology, Paris, 98*(4–6), 487–497.

Guyonneau, R., VanRullen, R., & Thorpe, S. J. (2005). Neurons tune to the earliest spikes through STDP. *Neural Computation, 17*(4), 859–879.

Heil, P. (2004). First-spike latency of auditory neurons revisited. *Current Opinion in Neurobiology, 14*(4), 461–467.

Houghton, G. (2005). *Connectionist models in cognitive psychology.* Hove, UK: Psychology Press.

Johansson, R. S., & Birznieks, I. (2004). First spikes in ensembles of human tactile afferents code complex spatial fingertip events. *Nature Neuroscience, 7*(2), 170–177.

Kirchner, H., & Thorpe, S. J. (2006). Ultra-rapid object detection with saccadic eye movements: Visual processing speed revisited. *Vision Research, 46*(11), 1762–1776.

Longtin, A. (1993). Stochastic resonance in neuron models. *Journal of Statistical Physics, 70,* 309–327.

Masquelier, T., Guyonneau, R., & Thorpe, S. J. (2008). Spike timing dependent plasticity finds the start of repeating patterns in continuous spike trains. *PLoS ONE, 3*(1), e1377.

Masquelier, T., Guyonneau, R., & Thorpe, S. J. (in press). Competitive STDP-based spike pattern learning. *Neural Computation.*

Masquelier, T., & Thorpe, S. J. (2007). Unsupervised learning of visual features through spike timing dependent plasticity. *PLoS Computational Biology, 3*(2), e31.

McClelland, J. L., & Rumelhart, D. E. (1986). *Parallel distributed processing: Explorations in the microstructure of cognition: Vol. 2. Psychological and biological models* (8th ed.). Cambridge, MA: MIT Press.

Nowak, L. G., & Bullier, J. (1997). The timing of information transfer in the visual system. In K. S. Rockland, J. H. Kaas, & A. Peters (Eds.), *Extrastriate visual cortex in primates* (Vol. 12, pp. 205–241). New York: Plenum Press.

Ohki, K., Chung, S., Ch'ng, Y. H., Kara, P., & Reid, R. C. (2005). Functional imaging with cellular resolution reveals precise micro-architecture in visual cortex. *Nature, 433*(7026), 597–603.

Ohki, K., & Reid, R. C. (2007). Specificity and randomness in the visual cortex. *Current Opinion in Neurobiology, 17*(4), 401–407.

Oram, M. W., & Perrett, D. I. (1992). Time course of neural responses discriminating different views of the face and head. *Journal of Neurophysiology, 68*(1), 70–84.

Perrett, D. I., Rolls, E. T., & Caan, W. (1982). Visual neurones responsive to faces in the monkey temporal cortex. *Experimental Brain Research, 47*(3), 329–342.

Quian Quiroga, R., Reddy, L., Kreiman, G., Koch, C., & Fried, I. (2005). Invariant visual representation by single neurons in the human brain. *Nature, 435*(7045), 1102–1107.

Quinlan, P. (1991). *Connectionism and psychology.* New York: Harvester Wheatsheaf.

Rieke, F., Warland, D., De Ruyter, R., & Bialek, W. (1997). *Spikes: Exploring the neural code*. Cambridge, MA: MIT Press.

Rodriguez, E., George, N., Lachaux, J. P., Martinerie, J., Renault, B., & Varela, F. J. (1999). Perception's shadow: Long-distance synchronization of human brain activity. *Nature, 397*(6718), 430–433.

Rousselet, G. A., Thorpe, S. J., & Fabre-Thorpe, M. (2004). How parallel is visual processing in the ventral pathway? *Trends in Cognitive Sciences, 8*(8), 363–370.

Rumelhart, D. E., & McClelland, J. L. (1986). *Parallel distributed processing: Explorations in the microstructure of cognition: Vol. 1. Foundations*. Cambridge, MA: MIT Press.

Serre, T., Kreiman, G., Kouh, M., Cadieu, C., Knoblich, U., & Poggio, T. (2007). A quantitative theory of immediate visual recognition. *Progress in Brain Research, 165C*, 33–56.

Serre, T., Oliva, A., & Poggio, T. (2007). A feedforward architecture accounts for rapid categorization. *Proceedings of the National Academy of Sciences, USA, 104*(15), 6424–6429.

Singer, W. (1999). Neuronal synchrony: A versatile code for the definition of relations? *Neuron, 24*(1), 49–65, 111–125.

Singer, W., & Gray, C. M. (1995). Visual feature integration and the temporal correlation hypothesis. *Annual Review of Neuroscience, 18*, 555–586.

Song, S., Miller, K. D., & Abbott, L. F. (2000). Competitive Hebbian learning through spike-timing-dependent synaptic plasticity. *Nature Neuroscience, 3*(9), 919–926.

Stemmler, M. (1996). A single spike suffices: The simplest form of stochastic resonance in model neurons. *Network: Computation in Neural Systems, 7*(4), 687–716.

Tallon-Baudry, C. (2004). Attention and awareness in synchrony. *Trends in Cognitive Sciences, 8*(12), 523–525.

Tallon-Baudry, C., Bertrand, O., & Fischer, C. (2001). Oscillatory synchrony between human extrastriate areas during visual short-term memory maintenance. *Journal of Neuroscience, 21*(20), RC177.

Thorpe, S., & Imbert, M. (1989). Biological constraints on connectionist modelling. In R. Pfeifer, Z. Schreter, F. Fogelman-Soulié, & L. Steels (Eds.), *Connectionism in perspective* (pp. 63–92). Amsterdam: Elsevier.

Thorpe, S. J. (1990). Spike arrival times: A highly efficient coding scheme for neural networks. In R. Eckmiller, G. Hartmann, & G. Hauske (Eds.), *Parallel processing in neural systems and computers* (pp. 91–94). Amsterdam: Elsevier.

Thorpe, S. J. (1994). Spike arrival times: A highly efficient coding scheme for neural networks. In H. Gutfreund & G. Toulouse (Eds.), *Biology and computation: A physicist's choice*. Singapore: World Scientific.

Thorpe, S. J. (2002). Localized versus distributed representations. In M. A. Arbib (Ed.), *The handbook of brain theory and neural networks* (2nd ed., pp. 643–645). Cambridge, MA: MIT Press.

Thorpe, S. J., & Fabre-Thorpe, M. (2001). Neuroscience: Seeking categories in the brain. *Science, 291*(5502), 260–263.

Thorpe, S. J., Guyonneau, R., Guilbaud, N., Allegraud, J. M., & VanRullen, R. (2004). SpikeNet: Real-time visual processing with one spike per neuron. *Neurocomputing, 58–60*, 857–864.

VanRullen, R., Gautrais, J., Delorme, A., & Thorpe, S. (1998). Face processing using one spike per neurone. *Biosystems, 48*(1–3), 229–239.

VanRullen, R., Guyonneau, R., & Thorpe, S. J. (2005). Spike times make sense. *Trends in Neurosciences, 28*(1), 1–4.

VanRullen, R., & Thorpe, S. J. (2001). Rate coding versus temporal order coding: What the retinal ganglion cells tell the visual cortex. *Neural Computation, 13*(6), 1255–1283.

VanRullen, R., & Thorpe, S. J. (2002). Surfing a spike wave down the ventral stream. *Vision Research, 42*(23), 2593–2615.

Varela, F., Lachaux, J. P., Rodriguez, E., & Martinerie, J. (2001). The brainweb: Phase synchronization and large-scale integration. *Nature Reviews Neuroscience, 2*(4), 229–239.

Waydo, S., Kraskov, A., Quian Quiroga, R., Fried, I., & Koch, C. (2006). Sparse representation in the human medial temporal lobe. *Journal of Neuroscience, 26*(40), 10232–10234.

3 Stochastic neurodynamical computation of brain functions

Gustavo Deco and Edmund T. Rolls

Neuronal correlates of brain functions are typically characterized by a high degree of variability in firing activity that manifests itself both within and between trials. These fluctuations are themselves of functional relevance and reflect computational dynamics that build on probabilistic transitions between attractor states. This suggests that probabilistic behavior is intrinsically linked with the stochasticity at cellular levels. In this chapter, we review the theoretical framework of stochastic neurodynamics that allows us to investigate the roles of noise and neurodynamics in the computation of probabilistic behavior. This is important for neuroscience, because it provides a theoretical framework that goes beyond the traditional noiseless neurodynamical analyses, and for neurophysiology, because transitions between states and not just averages across states are important to analyze.

1 Introduction

The challenge to unravel the primary mechanisms underlying brain functions demands explicit description of the computation performed by the neuronal and synaptic substrate (Rolls & Deco, 2002). Computational neuroscience aims to understand that computation by the construction and simulation of microscopic models based on local networks with large numbers of neurons and synapses that lead to the desired global behavior of the whole system. A suitable level of description of the microscopic level is captured by the spiking and synaptic dynamics of one-compartment, point-like models of neurons, such as integrate-and-fire models (Brunel & Amit, 1997). These dynamics allow the use of realistic biophysical constants (e.g., conductances, delays, etc.) in a thorough study of the actual time scales and firing rates involved in the evolution of the neural activity underlying cognitive processes, for comparison with experimental data.

Nevertheless, an in-depth analytical study of these detailed microscopic models is not possible, and therefore a reduction of the hypothesized models is necessary in order to establish a systematic relation between structure (parameters), dynamics, and functional behavior (i.e., to solve the "inverse" problem). A simulation only of phenomena of a complex system, such as the brain, is in general not useful, because there are usually no explicit underlying first principles.

In order to overcome this problem, statistical physics methods have been introduced, to analytically reduce the system. In particular, *mean-field* techniques (Brunel & Amit, 1997; Brunel & Wang, 2001) are used for simplifying the detailed spiking dynamics at least for the stationary conditions – that is, for periods after the dynamical transients. The stationary dynamics of a population of neurons can be described by a reduced rate-equation describing the asymptotic stationary states of the associated average population rate. The set of stationary, self-reproducing rates for the different populations in the network can be found by solving a set of coupled self-consistency equations. This reduced dynamics allows a thorough analysis of the parameter space enabling a selection of the parameter region that shows the emergent behavior of interest (e.g., short-term memory, attention). After that, with this set of parameters, one can perform the simulations described by the full integrate-and-fire scheme, which includes the effects of fluctuations and is able to describe the computation performed at the transients. In particular, we stress the fact that fluctuations (noise) are not just a concomitant of neural processing, but can play an important and unique role in cortical function. We show how computations can be performed through stochastic dynamical effects, including the role of noise in enabling probabilistic jumping across barriers in the energy landscape describing the flow of the dynamics in attractor networks. We show how these effects help to understand decision-making in the brain, and to establish a link between neuronal firing variability and probabilistic behavior.

2 Theory: from spiking neurons to reduced rate-models

The computation underlying brain functions emerges from the collective dynamics of spiking neuronal networks. A spiking neuron transforms a large set of incoming input spike trains, coming from different neurons, into an output spike train. Thus, at the microscopic level, neuronal circuits of the brain encode and process information by spatiotemporal spike patterns. We assume that the transient (non-stationary) dynamics of spiking neurons is properly captured by one-compartment, point-like models of neurons, such as the leaky integrate-and-fire (LIF) model (Tuckwell, 1988). In the LIF model, each neuron i can be fully described in terms of a single internal variable – namely, the depolarization $V_i(t)$ of the neural membrane. The basic circuit of a leaky LIF model consists of a capacitor C in parallel with a resistor R driven by a synaptic current (excitatory or inhibitory post-synaptic potential; EPSP or IPSP, respectively). When the voltage across the capacitor reaches a threshold θ, the circuit is shunted and a δ-pulse (spike) is generated and transmitted to other neurons. The sub-threshold membrane potential of each neuron evolves according to a simple RC circuit, with a time constant $\tau = RC$ given by the following equation:

$$\tau \frac{dV_i(t)}{dt} = -[V_i(t) - V_L] + \tau \sum_{j=1}^{N} J_{ij} \sum_{k} d(t - t_j^{(k)}), \qquad (1)$$

where V_L is the leak potential of the cell in the absence of external afferent inputs, and where the total synaptic current flow into cell i is given by the sum of the contributions of δ-spikes produced at presynaptic neurons, being J_{ij} (the efficacy of synapse j) and $t^{(k)}$ (the emission time of the kth spike from the jth presynaptic neuron).

In the brain, local neuronal networks are composed of a large number of neurons that are massively interconnected. The set of coupled differential equations (Equation 1) above describe the underlying dynamics of such networks. Direct simulations of these equations yield a complex spatiotemporal pattern, covering the individual trajectory of the internal state of each neuron in the network. This type of direct simulation is computationally expensive, making it very difficult to analyze how the underlying connectivity relates to various dynamics. One way to overcome these difficulties is by adopting the population-density approach, using the Fokker–Planck formalism (Knight, 2000; Knight, Manin, & Sirovich, 1996; Omurtag, Knight, & Sirovich, 2000). In this approach, individual integrate-and-fire neurons are grouped together into populations of statistically similar neurons. A statistical description of each population is given by a probability density function that expresses the distribution of neuronal states (i.e., membrane potential) over the population. In general, neurons with the same state $V(t)$ at a given time t have a different history because of random fluctuations in the input current $I(t)$. The main source of randomness is from fluctuations in the currents. The key assumption in the population-density approach is that the afferent input currents impinging on neurons in one population are uncorrelated. Thus, neurons sharing the same state $V(t)$ in a population are indistinguishable. The population density $p(u,t)$ expresses the fraction of neurons at time t that have a membrane potential $V(t)$ in the interval $[u, u + du]$. The evolution of the population density is given by the Chapman-Kolmogorov equation

$$p(u, t + dt) = \int_{-\infty}^{+\infty} p(u - \varepsilon, t) \rho(\varepsilon \mid u - \varepsilon) d\varepsilon, \qquad (2)$$

where $\rho(\varepsilon \mid u) = prob\{V(t + dt) = u + \varepsilon \mid V(t) = u\}$ is the conditional probability that generates an infinitesimal change $\varepsilon = V(t + dt) - V(t)$ in the infinitesimal interval dt. The temporal evolution of the population density can be reduced to a simpler differential equation by the mean-field approximation. In this approximation, the currents impinging on each neuron in a population have the same statistics, because, as we mentioned above, the history of these currents is uncorrelated. The mean-field approximation entails replacing the time-averaged discharge rate of individual cells with a common time-dependent population activity (ensemble average). This assumes ergodicity for all neurons in the population. The mean-field technique allows us to discard the index denoting the identity of any single neuron. The resulting differential equation describing the temporal evolution of the population density is called the Fokker–Planck equation, and reads:

$$\frac{\partial p(u,t)}{\partial t} = \frac{1}{2\tau}\sigma^2(t)\frac{\partial^2 p(u,t)}{\partial u^2} + \frac{\partial}{\partial u}\left[\left(\frac{u - V_L - \mu(t)}{\tau}\right)p(u,t)\right] \tag{3}$$

In the particular case that the drift is linear and the diffusion coefficient $\sigma^2(t)$ is given by a constant, the Fokker–Planck equation describes a well-known stochastic process called the Ornstein–Uhlenbeck process (Risken, 1996). The Ornstein–Uhlenbeck process describes the temporal evolution of the membrane potential $V(t)$ when the input afferent current is $\mu(t) + \sigma\sqrt{\tau}w(t)$, with $w(t)$ a white-noise process. This can be interpreted, by means of the central limit theorem, as the case in which the sum of many Poisson processes becomes a normal random variable with mean $\mu(t)$ and variance σ^2.

The non-stationary solutions of the Fokker–Planck equation (Equation 3) describe the dynamical behavior of the network. However, these simulations, like the direct simulation of the original network of spiking neurons (Equation 1), are computationally expensive and their results probabilistic, which makes them unsuitable for systematic explorations of parameter space. On the other hand, the stationary solutions of the Fokker–Planck equation (Equation 3) represent the stationary solutions of the original integrate-and-fire neuronal system. The stationary solution of the Fokker–Planck equation satisfying specific boundary conditions (see Lansky, Sacerdote & Tomassetti, 1995; Ricciardi & Sacerdote, 1979) yields the *population transfer function of Ricciardi* (ϕ):

$$v = \left[t_{ref} + \tau\sqrt{\pi}\int_{(V_{reset} - V_L - \mu/\sigma)}^{(\theta - V_L - \mu/\sigma)} e^{x^2}\{1 + erf(x)\}dx\right]^{-1} = \phi(\mu,\sigma), \tag{4}$$

where $erf(x) = 2/\sqrt{\pi}\int_0^x e^{-y^2}dy$. In Equation (4), t_{ref} is the refractory time and V_{reset} the resetting potential.

The population transfer function gives the average population rate as a function of the average input current. For more than one population, the network is partitioned into populations of neurons whose input currents share the same statistical properties and fire spikes independently at the same rate. The set of stationary, self-reproducing rates v_i for different populations

i in the network can be found by solving a set of coupled self-consistency equations, given by:

$$v_i = \phi(\mu_i, \sigma_i). \tag{5}$$

This reduced system of equations allows a thorough investigation of parameters. In particular, one can construct bifurcation diagrams to understand the non-linear mechanisms underlying equilibrium dynamics and in this way solve the "inverse problem" – that is, to select the parameters that generate the attractors (steady states) that are consistent with the experimental evidence. This is the crucial role of the mean-field approximation: to simplify analyses through the stationary solutions of the Fokker–Planck equation for a population density under the diffusion approximation (Ornstein–Uhlenbeck process) in a self-consistent form. After that, one can then perform full non-stationary simulations using these parameters in the integrate-and-fire scheme to generate *true dynamics*. The mean-field approach ensures that these dynamics will converge to a stationary attractor that is consistent with the steady-state dynamics we require (Brunel & Wang, 2001; Del Giudice, Fusi, & Mattia, 2003). An extended mean-field framework that is consistent with the integrate-and-fire and synaptic equations described above – that is, that considers both the fast and slow glutamatergic excitatory synaptic dynamics (AMPA and NMDA) and the dynamics of GABA inhibitory synapses – can be found in Brunel and Wang (2001).

The mean-field approach has been applied to model single neuronal responses, fMRI activation patterns, psychophysical measurements, and the effects of pharmacological agents and of local cortical lesions (Corchs & Deco, 2002, 2004; Deco & Lee, 2002; Deco, Pollatos, & Zihl, 2002; Deco & Rolls, 2002, 2003, 2004, 2005a; Deco, Rolls, & Horwitz, 2004; Rolls & Deco, 2002; Szabo, Almeida, Deco & Stetter, 2004). The next section reviews one of these examples, in the context of decision-making.

3 The computational correlates of decision-making

Decision-making is a key brain function of intelligent behavior. A number of neurophysiological experiments on decision-making provide evidence on the neural mechanisms underlying perceptual comparison, by characterizing the neuronal correlates of behavior (Romo, Hernandez, & Zainos, 2004; Romo, Hernandez, Zainos, Lemus, & Brody, 2002; Romo, Hernandez, Zainos & Salinas, 2003). In particular, Hernandez, Zainos, and Romo (2002), Romo and Salinas (2001, 2003), and Romo et al. (2003, 2004) have studied the neural mechanisms underlying perceptual comparison by measuring single-neuron responses in monkeys trained to compare two mechanical vibrations applied sequentially to the tip of a finger. The subjects have to report which of the two stimuli has the higher frequency. They found neurons in the ventral premotor cortex (VPC) whose firing rate depended only on the difference between the two applied frequencies, the sign of that difference being the determining factor for correct task performance (Romo et al., 2004). These neurons reflect the implementation of the perceptual comparison process and may underlie the process of decision-making.

Figure 3.1 shows a biophysically realistic computational model for a probabilistic decision-making network that compares two mechanical vibrations applied sequentially (f1 and f2). The attractor neuronal network model implements a dynamical competition between neurons. The model enables a formal description of the transients (non-stationary) and probabilistic character of behavior (performance) by the explicit use, at the microscopic level, of spiking and synaptic dynamics of one-compartment integrate-and-fire (IF) neuron models. The network contains excitatory pyramidal cells and inhibitory interneurons. The excitatory recurrent post-synaptic currents (EPSCs) are mediated by AMPA (fast) and NMDA-glutamate (slow) receptors, whereas external EPSCs imposed on the network are driven by AMPA receptors only. Inhibitory post-synaptic currents (IPSCs) to both excitatory and inhibitory neurons are mediated by GABA receptors. Neurons are clustered into populations. There are two subtypes of excitatory population – namely, specific and non-specific. Specific populations encode the result of the comparison process in the two-interval vibrotactile discrimination task – that is, if f1 > f2 or f1 < f2 (see Deco & Rolls, 2006).

The attractors of the network of IF neurons can be studied exhaustively by using the associated reduced mean-field equations. The set of stationary, self-reproducing rates v_i, for the different populations i, can be found by solving a set of coupled self-consistency equations. This enables *a posteriori*

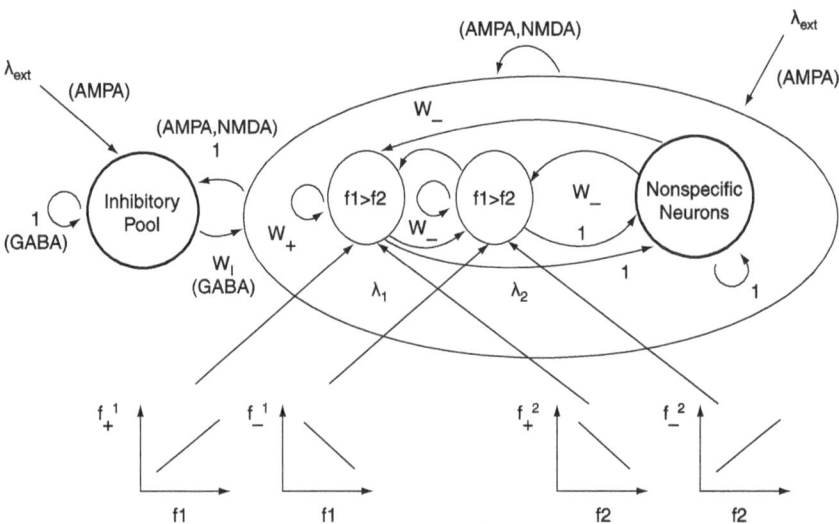

Figure 3.1 Decision-making neuronal network. (After Deco & Rolls, 2006, with permission from Wiley-Blackwell.)

selection of parameter regions that contain desired behaviors. In the present case, the essential requirement is that, for the stationary conditions, different attractors are stable. The attractors of interest for our task correspond to the activation (high spiking rates) or inactivation (low spiking rates) of the neurons in the specific populations f1 > f2 and f1 < f2. The activation of the specific population f1 > f2 (f1 < f2) and the simultaneous lack of activation of the complementary population f1 < f2 (f1 > f2), corresponds to encoding the "single-state" categorical decision f1 > f2 (f1 < f2) that produces the motor response that reports the decision. The lack of activation of both specific populations (the *spontaneous state*) would correspond to an encoding state that cannot lead to a behavioral decision – that is, there is no answer, or a motor response is generated randomly. The same happens if both specific populations are activated to the same degree (the *pair state*). Because responses in animals are probabilistic in nature, the operating point of the network should be such that both categorical decisions (i.e., both states) are stable. In addition, we have also shown that the model predicts behavior consistent with Weber's law if, and only if, the spontaneous state is also a stable state – that is, when the dynamical operating point of the network is in a regime of multistability. In this way, Weber's law provides evidence on the operating point of the network.

Figure 3.2 shows numerical simulations corresponding to the response of VPC neurons during the comparison period (to be contrasted with the

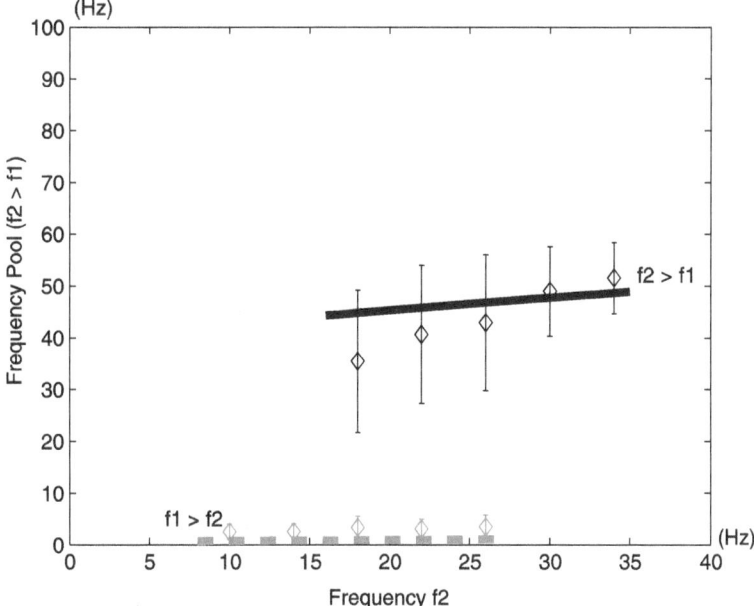

Figure 3.2 Responses of a neuron of the population f2 > f1 during the comparison period of f1 and f2 stimulation.

experimental results shown in Fig. 2 of Romo et al., 2004). Figure 3.2 shows the average firing rate as a function of f1 and f2, obtained with the spiking simulations (the diamond-points correspond to the average values over 200 trials, and the error bars to the standard deviation). The lines correspond to the mean-field calculations: the black line indicates f1 < f2 (f2 = f1 + 8 Hz) and the gray line f1 > f2 (f2 = f1 − 8 Hz). The average firing rate of the population f1 < f2 depends only on the sign of f2 − f1 and not on the magnitude of the difference, |f2 − f1|, confirming again that Weber's law cannot be encoded in the firing rate, but only in the probability with which that firing rate can be reached (which depends on the sign and magnitude of the difference between f1 and f2).

The comparison of two stimuli to determine which is the more intense becomes more difficult as they become more similar. The "difference-threshold" is the amount of change needed for us to recognize that a change has occurred. Weber's law (enunciated by Ernst Heinrich Weber [1795–1878]) states that the ratio of the difference-threshold to the background intensity is a constant.

Figure 3.3 shows the probability of correct discrimination as a function of the difference between the two presented vibrotactile frequencies to be compared. We assume that f1 > f2 by a Δ-value – that is, f1 = f2 + Δ (labeled the "Delta frequency (f1 − f2)" in Figure 3.3). Each diamond-point in Figure 3.3 corresponds to the result calculated by averaging 200 trials of the full spiking simulations. The lines were calculated by fitting the points with a logarithmic function. A correct classification occurs when, during the 500-ms comparison period, the network evolves to a "single-state" attractor that shows a high level of spiking activity (larger than 10 Hz) for the population f1 > f2, and simultaneously a low level of spiking activity for the population f1 < f2 (at the level of the spontaneous activity). One can observe from the different panels corresponding to different base vibrotactile frequencies, f2, that for reaching a threshold of correct classification of, for example, 85% (horizontal dashed line in Figure 3.3), the difference between f1 and f2 must become larger as f2 increases. The second panel from the top of Figure 3.3 shows a good fit between the actual neuronal data (Romo & Salinas, 2003) for the f2 = 20 Hz condition (indicated by *) and the results obtained with the model. Figure 3.4 plots the critical discrimination Δ-value corresponding to an 85%-correct performance level (the "difference-threshold") as a function

---

Caption to Figure 3.3 opposite
Probability of correct discrimination (±SD) as a function of the difference between the two presented vibrotactile frequencies to be compared. In the simulations, we assume that f1 > f2 by a Δ-value (labeled "Delta frequency (f1 − f2)") – that is, f1 = f2 + Δ. The horizontal dashed line represents the threshold of correct classification for a performance of 85% correct discrimination. The second panel down includes actual neuronal data described by Romo and Salinas (2003) for the f2 = 20 Hz condition (indicated by *). (After Deco & Rolls, 2006, with permission from Wiley-Blackwell.)

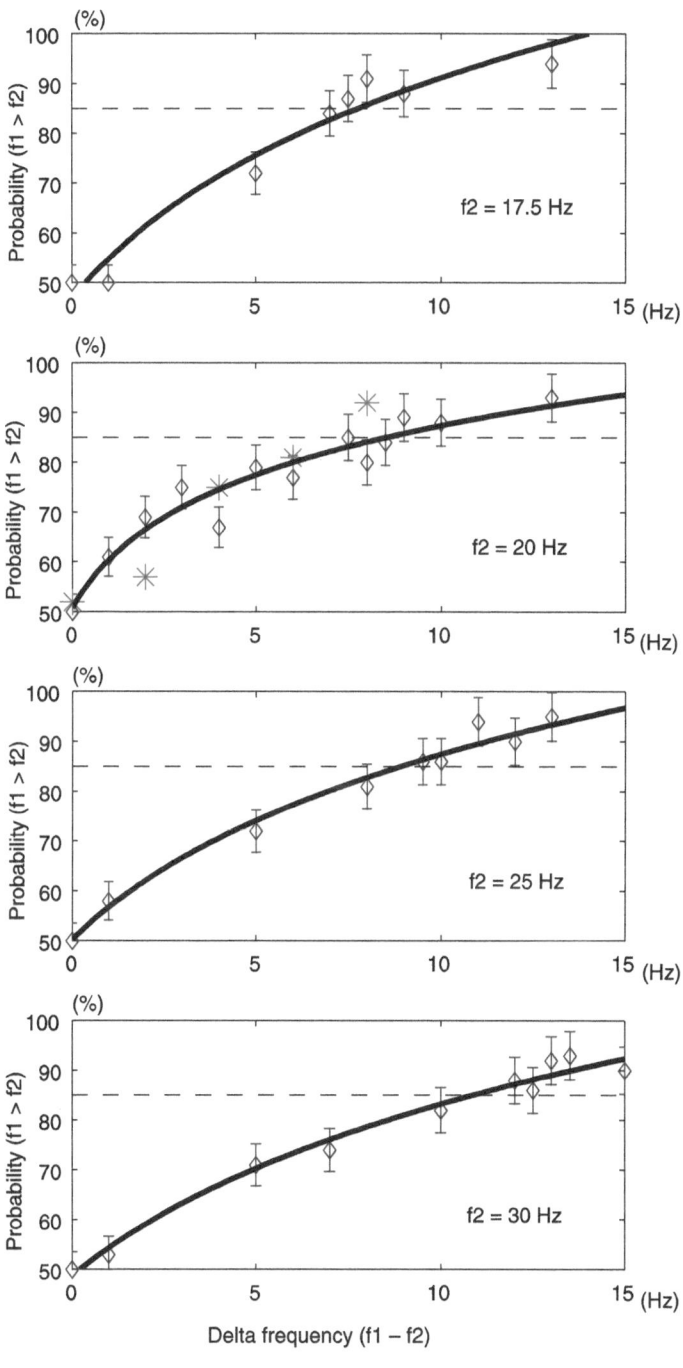

Figure 3.3

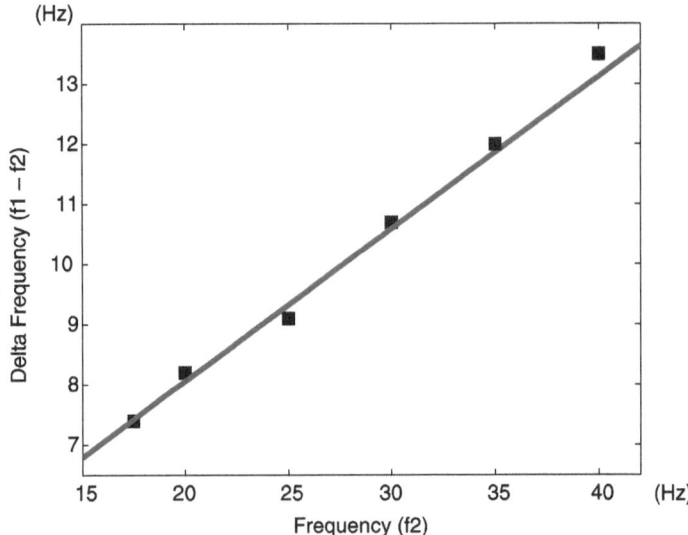

Figure 3.4 Weber's law for the vibrotactile discrimination task. The critical discrimination Δ-value ("difference threshold") is shown corresponding to an
85%-correct performance level as a function of the base frequency, f2. The
"difference-threshold" increases linearly as a function of the base frequency.
(After Deco & Rolls, 2006, with permission from Wiley-Blackwell.)

of the base frequency, f2. The "difference-threshold" increases linearly as a
function of the base frequency. This corresponds to Weber's law for the
vibrotactile discrimination task.

The analysis shown in Figures 3.3 and 3.4 suggests that Weber's law, and
consequently the ability to discriminate two stimuli, is encoded in the probability of performing a transition to the correct final attractor, and, indeed,
once in the attractor the neurons have a high firing rate that reflects the
binary decision, and not the Δ-value (Deco & Rolls, 2006).

We confirmed this prediction in behavioral tests of vibrotactile discrimination in humans and proposed a computational explanation of perceptual
discrimination which accounts naturally for the emergence of Weber's law
(Deco, Scarano, & Soto-Faraco, 2007). We concluded that the neurodynamical mechanisms and computational principles underlying the decision-
making processes in this perceptual discrimination task are consistent with a
fluctuation-driven scenario in a multistable regime (for more details, see
Deco et al., 2007).

4 Conclusions

We conclude with some key properties of this biased attractor model of
decision-making.

The decisions are taken probabilistically because of the finite-size noise due to spiking activity in the integrate-and-fire dynamical network, with the probability that a particular decision is made depending on the biasing inputs provided by the sensory stimuli f1 and f2.

The statistical fluctuations in the network are due to the finite-size noise, which approximates to the square root of the firing rate/number of neurons in the population (see Mattia & DelGiudice, 2002), as shown by Deco and Rolls (2006).

This is the first time of which we are aware when the implementation of a psychophysical law is not the firing rate of the neurons, nor the spike timing, nor single-neuron based, but instead is based on the synaptic connectivity of the network and on statistical fluctuations due to the spiking activity in the network.

The way in which the system settles (i.e., the probability of reaching one attractor state or the other from the initial spontaneous state, and the time it takes) depends on factors that include the distortion of the attractor land-scapes produced by the biasing inputs, which will influence both the shapes and the depths of the attractor basins and the finite-size noise effects.

An interesting aspect of the model is that the recurrent connectivity, and the relatively long time constant of the NMDA receptors (Wang, 2002), together enable the attractor network to accumulate evidence over a long time period of several hundred milliseconds. This is an important aspect of the functionality of attractor networks.

Although the model described here is effectively a single attractor net-work, we note that the network need not be localized to one brain region. Long-range connections between cortical areas enable networks in different brain areas to interact in the way needed to implement a single attractor network. The requirement is that the synapses between the neurons in any one pool be set up by Hebb-like associative synaptic modification, and this is likely to be a property of connectivity between areas as well as within areas (Rolls & Deco, 2002; Rolls & Treves, 1998). In this sense, the decision could be thought of as distributed across different brain areas.

The current model of decision-making is part of a unified approach to attention, reward-reversal, and sequence learning, in which biasing inputs influence the operation of attractor networks that operate using biased com-petition (Deco & Rolls, 2005b; Rolls, 2005, 2008; Rolls & Deco, 2002). The same approach is now seen to be useful in understanding decision-making and its relation to Weber's law, in understanding the details of the neuronal responses that are recorded during decision-making, and in understanding many other aspects of decision-making, attention, and memory recall (Loh, Rolls, & Deco, 2007; Rolls, 2008).

5 References

Brunel, N., & Amit, D. (1997). Model of global spontaneous activity and local structured delay activity during delay periods in the cerebral cortex. *Cerebral Cortex, 7,* 237–252.

Brunel, N., & Wang, X. (2001). Effects of neuromodulation in a cortical networks model of object working memory dominated by recurrent inhibition. *Journal of Computational Neuroscience, 11,* 63–85.

Corchs, S., & Deco, G. (2002). Large-scale neural model for visual attention: Integration of experimental single cell and fMRI data. *Cerebral Cortex, 12,* 339–348.

Corchs, S., & Deco, G. (2004). Feature-based attention in human visual cortex: Simulation of fMRI data. *NeuroImage, 21,* 36–45.

Deco, G., & Lee, T. S. (2002). A unified model of spatial and object attention based on inter-cortical biased competition. *Neurocomputing, 44–46,* 775–781.

Deco, G., Pollatos, O., & Zihl, J. (2002). The time course of selective visual attention: Theory and experiments. *Vision Research, 42,* 2925–2945.

Deco, G., & Rolls, E. T. (2002). Object-based visual neglect: A computational hypothesis. *European Journal of Neuroscience, 16,* 1994–2000.

Deco, G., & Rolls, E. T. (2003). Attention and working memory: A dynamical model of neuronal activity in the prefrontal cortex. *European Journal of Neuroscience, 18,* 2374–2390.

Deco, G., & Rolls, E. T. (2004). A neurodynamical cortical model of visual attention and invariant object recognition. *Vision Research, 44,* 621–644.

Deco, G., & Rolls, E. T. (2005a). Neurodynamics of biased competition and cooperation for attention: A model with spiking neurons. *Journal of Neurophysiology, 94,* 295–313.

Deco, G., & Rolls, E. T. (2005b). Attention, short term memory, and action selection: A unifying theory. *Progress in Neurobiology, 76,* 236–256.

Deco, G., & Rolls, E. T. (2006). A neurophysiological model of decision-making and Weber's law. *European Journal of Neuroscience, 24,* 901–916.

Deco, G., Rolls, E. T., & Horwitz, B. (2004). "What" and "where" in visual working memory: A computational neurodynamical perspective for integrating fMRI and single-neuron data. *Journal of Cognitive Neuroscience, 16,* 683–701.

Deco, G., Scarano, L., & Soto-Faraco, S. (2007). Weber's law in decision-making: Integrating behavioral data in humans with a neurophysiological model. *Journal of Neuroscience, 27*(42), 11192–11200.

Del Giudice, P., Fusi, S., & Mattia, M. (2003). Modeling the formation of working memory with networks of integrate-and-fire neurons connected by plastic synapses. *Journal of Physiology, Paris, 97,* 659–681.

Hernandez, A., Zainos, A., & Romo, R. (2002). Temporal evolution of a decision-making process in medial premotor cortex. *Neuron, 33,* 959–972.

Knight, B. (2000). Dynamics of encoding in neuron populations: Some general mathematical features. *Neural Computation, 12*(3), 473–518.

Knight, B., Manin, D., & Sirovich, L. (1996). Dynamical models of interacting neuron populations, In E. C. Gerf (Ed.), *Symposium on robotics and cybernetics: Computational engineering in systems applications.* Lille, France: Cite Scientifique.

Lansky, P., Sacerdote, L., & Tomassetti, F. (1995). On the comparison of Feller and Ornstein–Uhlenbeck models for neural activity. *Biological Cybernetics, 73,* 457–465.

Loh, M., Rolls, E. T., & Deco, G. (2007). A dynamical systems hypothesis of schizophrenia. *PLoS Computational Biology, 3,* e228.

Mattia, M., & DelGiudice, P. (2002). Attention and working memory: A dynamical model of neuronal activity in the prefrontal cortex. *Physical Review E, 66,* 51917–51919.

Omurtag, A., Knight, B., & Sirovich, L. (2000). On the simulation of large populations of neurons. *Journal of Computational Neuroscience, 8,* 51–53.

Ricciardi, L., & Sacerdote, L. (1979). The Ornstein–Uhlenbeck process as a model for neuronal activity. I. Mean and variance of the firing time. *Biological Cybernetics, 35,* 1–9.

Risken, H. (1996). *The Fokker–Planck equation: Methods of solution and applications.* Berlin: Springer-Verlag.

Rolls, E. T. (2005). *Emotion explained.* Oxford, UK: Oxford University Press.

Rolls, E. T. (2008). *Memory, attention, and decision-making: A unifying computational neuroscience approach.* Oxford, UK: Oxford University Press.

Rolls, E. T., & Deco, G. (2002). *Computational neuroscience of vision.* Oxford, UK: Oxford University Press.

Rolls, E. T., & Treves, A. (1998). *Neural networks and brain function.* Oxford, UK: Oxford University Press.

Romo, R., Hernandez, A., & Zainos, A. (2004). Neuronal correlates of a perceptual decision in ventral premotor cortex. *Neuron, 41,* 165–173.

Romo, R., Hernandez, A., Zainos, A., Lemus, L., & Brody, C. (2002). Neural correlates of decision-making in secondary somatosensory cortex. *Nature Neuroscience, 5,* 1217–1225.

Romo, R., Hernandez, A., Zainos, A., & Salinas, E. (2003). Correlated neuronal discharges that increase coding efficiency during perceptual discrimination. *Neuron, 38,* 649–657.

Romo, R., & Salinas, E. (2001). Touch and go: Decision-making mechanisms in somatosensation. *Annual Review of Neuroscience, 24,* 107–137.

Romo, R., & Salinas, E. (2003). Flutter discrimination: Neural codes, perception, memory and decision making. *Nature Reviews Neuroscience, 4,* 203–218.

Szabo, M., Almeida, R., Deco, G., & Stetter, M. (2004). Cooperation and biased competition model can explain attentional filtering in the prefrontal cortex. *European Journal of Neuroscience, 19,* 1969–1977.

Tuckwell, H. (1988). *Introduction to theoretical neurobiology.* Cambridge, UK: Cambridge University Press.

Wang, X. J. (2002). Probabilistic decision making by slow reverberation in cortical circuits. *Neuron, 36,* 955–968.

4 Modelling visual search in biologically plausible neural networks

Whole-system behaviour, neuropsychological breakdown and BOLD signal activation

*Glyn W. Humphreys, Eirini Mavritsaki,
Harriet Allen, Dietmar Heinke, and
Gustavo Deco*

We discuss attempts to model aspects of human visual selection in a bio-logically plausible neural network – the spiking Search over Time and Space model (sSoTS). We show how, by incorporating physiologically realistic parameters into the model, aspects of human search over time as well as space can be captured. These parameters are also useful for simulating neuropsychological deficits in selection found in patients with lesions to posterior parietal cortex. Finally, we show how the model can be used to help decompose neural networks revealed by fMRI to be involved in spatial selection over time, providing a model-based analysis linking neuroanatomy to cognitive function. We discuss the utility of developing biologically plaus-ible networks for high- as well as well as low-level cognitive operations.

1 Introduction

A great deal of neuroscientific evidence has accumulated over the past 20 years concerning the functional and neural processes underlying human visual attention. For example, there is a general consensus that visual atten-tion is contingent on a network of neural circuits in frontal and parietal cortex, which control both voluntary and reflex orienting to visual signals (Corbetta & Shulman, 2002). The neural regions that allocate attention to the spatial locations of stimuli also overlap to some degree with the areas involved in responding to targets on the basis of their temporal properties (Coull, 1998; Kanwisher & Wocjiulik, 2000), and both spatial and temporal selection can be impaired after damage to posterior parietal cortex (e.g., Humphreys, Olivers, & Yoon, 2006b; Olivers & Humphreys, 2004). The cor-tical processes involved in spatial and temporal selection are themselves modulated by additional cortical and sub-cortical structures that regulate neurotransmitter systems to determine the efficiency of attentional selection. Posner and Petersen (1990), for example, proposed an influential account in which the norepinephrine system, implicated in arousal, was proposed to modulate the neural and functional processes involved in stimulus selection. Posner and Petersen proposed that a critical level of activity was required in the arousal/vigilance system, based on norepinephrine (NE) function, in order to maintain attention during a task and that reductions in vigilance could in turn lead to deficits in selection. This account gathers support from studies of patients with unilateral neglect, where impaired arousal may combine with spatially selective deficits in attention to generate impairments in selecting stimuli on the contralesional side of space (Husain & Rorden, 2003; Robertson, 1994). Indeed, some aspects of spatial neglect can be reduced when (non-spatial) phasic arousal is temporarily increased in patients (Robertson & Manly, 1999; Robertson, Mattingley, Rorden, & Driver, 1998). To the extent that right rather than left parietal and frontal regions are critical for arousal (particularly dorso-lateral prefrontal cortex, DLPFC; Coull, 1998; Paus, Jech, Thompson, Comeau, Peters, & Evans, 1997; Posner & Petersen, 1990), then this can provide one account of why there is a general dominance in the incidence of neglect after right-hemisphere lesions (Bisiach, 1999).

Our growing knowledge about the functional and neural mechanisms of selective attention has been incorporated into increasingly sophisticated computational models of human visual selection (e.g., see Deco & Rolls, 2005; Heinke & Humphreys, 2003; Itti & Koch, 2000; Mozer & Sitton, 1998). The importance of these models is that they can generate a system-level account of performance, providing a means of examining how interactions between different components in a complex network give rise to coherent behaviour. This can be particularly important for understanding neurological syndromes such as unilateral neglect. In such syndromes there can be a variety of dissociations between contrasting symptoms. Relating the behaviour of patients back to an explicit model provides one way in which a complex set of symptoms can be understood and related to one another. The models can also be used in conjunction with empirical tests of human performance, providing a critical cycle between prediction, test, and further refinement of the model (see Heinke & Humphreys, 2005). One example of this comes from work in our laboratory on the development of the selective attention for identification model (SAIM; Heinke & Humphreys, 2003). SAIM is a connectionist model that has been successful in simulating a broad range of experimental evidence on both normal attentional behaviour and attentional disorders, while also generating new empirical predictions that have been tested in psychophysical experiments (see Heinke et al., chapter 5, this volume). The model uses a spatial window to select visual information for object recognition, using both bottom-up and top-down constraints (e.g., effects of stored knowledge of objects; Heinke & Humphreys, 2003). When the effects of brain lesions were simulated (e.g., by adding noise to the operations of "neurons" in critical regions), then impairments such as spatial neglect were apparent. In addition, SAIM generated predictions about how top-down information interacted with the strength of bottom-up signals for attention, which have been tested empirically (Heinke, Humphreys, & Tweed, 2006).

Models such as SAIM provide a good account of how the combined action of cognitive sub-systems generates overall behaviour. However, such models have their limitations as frameworks for understanding information processing in real biological systems. For example, the models often use simplified activation functions, modified by single parameters that influence, for example, the gain in excitation with signal strength (e.g., Servan-Schreiber, Printz, & Cohen, 1990), and consequently the models may not approximate the more complex modulations generated through different neuron-transmitter systems in the human brain. Also the processing units in models such as SAIM typically mimic the behaviour of many hundreds of neurons rather than individual neurons, and it can be difficult to compare temporal processing within the model directly with real-time analyses in a biological system. Depending on what the research question is, these limitations may be important. We have been interested in exploring human search over time as well as space (e.g., for a review, see Watson, Humphreys, & Olivers, 2003), where we wish specifically to model data on the time course of selection. Here it may be useful to have a model that can relate more directly to the time course of activation at a neural level. Similarly we have been interested in simulating non-spatial as well as spatial aspects of neurological disorders such as unilateral neglect, and for this it may be advantageous to have models that can reflect neurotransmitter functions in the brain (see above). To address these issues, we have developed a neural network model of selection – the spiking Search over Time and Space model (sSoTS) – which incorporates both biologically plausible temporal processing parameters and modulatory effects of neurotransmitters on activity, to gain a finer-grained analysis of normal and disordered human search. In the present chapter we review our work using sSoTS and discuss the utility of taking steps towards incorporating biological reality into models.

2 The sSoTS model

In contrast to attempts to simulate human visual attention within a connectionist framework (Heinke & Humphreys, 2003; Itti & Koch, 2000; Mozer & Sitton, 1998), there have been relatively few attempts to develop more detailed neural accounts. Deco and colleagues (Deco & Rolls, 2005; Deco & Zihl, 2001) have simulated human attention using models based on "integrate-and-fire" neurons, which utilise biologically plausible activation functions and output in terms of neuronal spikes (rather than, e.g., a continuous value, as in many connectionist systems). These authors showed that a model with parallel processing of the input through to high levels was able to simulate classic "attentional" (serial) aspects of human search (e.g., contrasting search when targets are defined by simple features with search when targets are defined by conjunctions of features), providing an existence proof that a model incorporating details of neuronal activation functions could capture aspects of human visual attention. The Deco and Zihl (2001) model, which forms the starting point for our own simulations (below), used a "mean-field approximation" to neuronal function, where the actual fluctuating induced local field u_i for each neuron i was replaced by its average, and so the model does not capture dynamic operations at the level of individual neurons as, for example, levels of neurotransmitter vary. Recently we elaborated this account at the level of individual neurons to simulate human spatial and temporal exploration of the visual field. This model, the sSoTS model (Mavritsaki, Heinke, Humphreys, & Deco, 2006, 2007), uses a system of spiking neurons modulated by NMDA, AMPA, and GABA transmitters along with an I_{AHP} current, as originally presented by Brunel and Wang (2001; see also Deco & Rolls, 2005). The architecture of the model is illustrated in the top panel of Figure 4.1.[1] The sSoTS model uses two layers of feature maps to encode the characteristics of visual stimuli (their colour and their shape) and a layer in which units respond to the presence of any feature at a given location (the "location map"). Activity in the location map may provide an index of saliency (Itti & Koch, 2000). This is the skeleton architecture needed to explore the contrast between feature and conjunction search, which provides one of the basic tenets of research on human visual attention (see Treisman & Gelade, 1980). Each pool of units (for each feature and for the location map) is composed of 80:20 excitatory to inhibitory neurons, following the ratio of inter-neurons (inhibitory) and pyramidal cells (excitatory) in the brain (Abeles, 1991). Each layer also contains an inhibitory pool that provides global inhibition in the system and a non-specific pool that simulates the neurons in the given brain areas that are not involved in the particular brain function being modelled (for discussion, see Deco & Rolls, 2005). The spiking neurons in the system receive inputs from external neurons not included in the network, corresponding to changing states in the external world. When no external signal is present, input from these neurons provides a level of "background noise". Bottom-up input generates competition between neurons in the feature pools for each represented location, with activation in the location map feeding back to the feature pools to help resolve competition in favour of the location receiving most evidence for some item being present. In addition to these bottom-up and feedback waves of activation, search tasks are simulated by providing a further external

1 Although we have labelled the shape maps as A and H maps (Figure 4.1), this is done purely for the ease of matching the results of the model with psychological data on single-feature, conjunction and preview search involving blue and green As and Hs (e.g., Watson & Humphreys, 1997). The maps may more accurately be thought of as representing simple local properties of shape, such as edges at particular orientations, which may also distinguish letter shapes in search.

a.

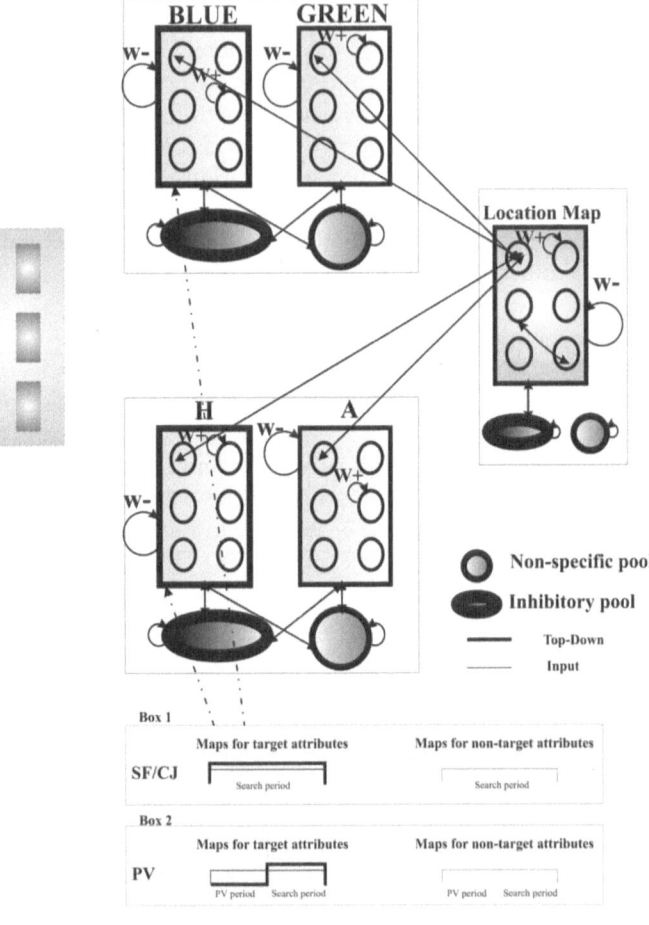

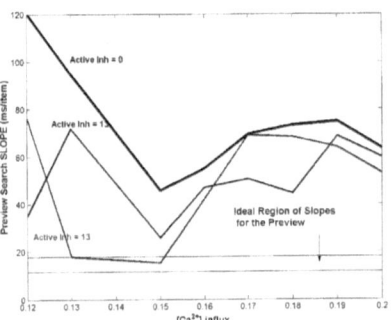

b.

Figure 4.1

source of top-down input into feature maps corresponding to properties of the expected targets (e.g., if a red item is expected, the red feature pool receives top-down activation). This top-down signal gives a competitive bias favouring a target over distractor stimuli not containing the target's features, as suggested by verbally expressed models of human attention (e.g., the biased competition model; Desimone & Duncan, 1995). Over time, the model converges in an approximate winner-take-all fashion upon a target, with reaction times based on the real-time operation of the neurons. Search time in the model is determined by the degree of overlap between the features of the target and those of distractors, with search times lengthening as competition increases, giving rise to differences in search efficiency matching those of human observers (Mavritsaki et al., 2006, 2007; see Duncan & Humphreys, 1989).

The sSoSTs model also successfully simulates data on human search over time as well as space. In preview search (Watson & Humphreys, 1997), one set of distractors precedes the other distractors and the target. Provided that the interval between the initial items and the search display is over 450 ms or so, the first set of distractors has little impact on performance (Humphreys, Kyllinsbæk, Watson, Olivers, Law, & Paulson, 2004). The sSoTS model mimics this time course due to the contribution of two processes: (1) a spike frequency-adaptation mechanism generated from a (biologically plausible) slow $[Ca^{2+}]$-activated K^+ current, which reduces the probability of spiking after an input has activated a neuron for a prolonged time period (Madison & Nicoll, 1984) (Figure 4.2) and (2) a top-down inhibitory input that forms an active bias against known distractors (Figure 4.1b). The slow action of frequency-adaptation simulates the time course of preview search. The top-down inhibitory bias matches substantial data from human psychophysical studies (Humphreys, Olivers, & Braithwaite, 2006a; Watson et al., 2003), while our explorations of the parameter space for sSoTS further indicate that it is a necessary component in order to approximate the behavioural data on preview search (Mavritsaki et al., 2006) (see Figure 4.1b).

Activation within the model in "single-feature" and "conjunction" search tasks, classically examined in the psychology literature on visual search (Treisman & Gelade, 1980), is shown in Figure 4.3a. Activation at the target's position in the location map rises more sharply and reaches a higher level in the single-feature relative to the conjunction condition, leading to faster

---

Caption to Figure 4.1 opposite
(a) The architecture of the sSoTS model. Boxes 1 and 2 at the bottom show the time periods for the displays (single-feature, SF; conjunction, CJ; and preview, PV). The dotted lines indicate the top-down inhibition applied to the features and locations of distractors during the preview period. (b) Illustration of search through the parameter space in sSoTS, relating frequency adaptation (x-axis) to the slope of preview search (y-axis), with three levels of top-down inhibition (three lines). The fit to human data occurs within the area highlighted by the dotted rectangle. These parameters are then set for subsequent simulations.

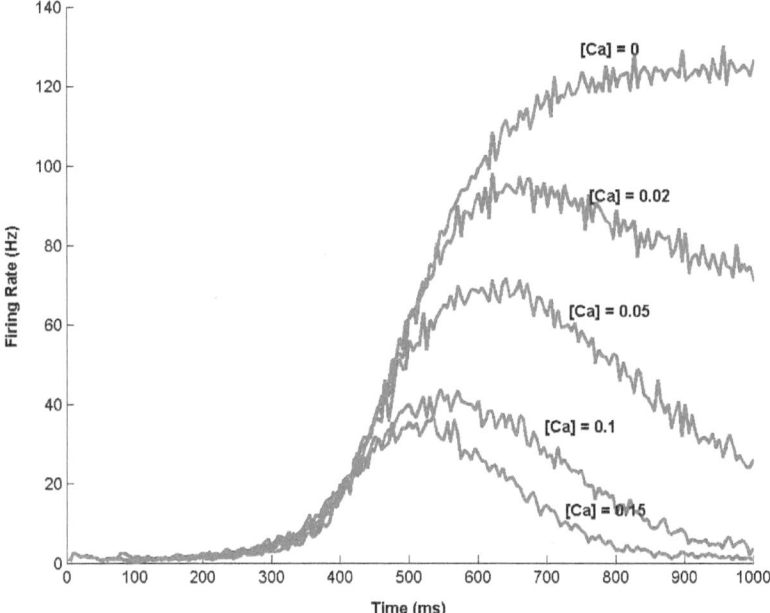

Figure 4.2 Illustration of variations in the neuronal firing rate as a function of the Ca²⁺ parameter.

reaction times for single-feature search (matching the psychological data; see Figure 4.3b). Activation during preview search, where half of the distractors (in a particular colour) are presented before the remaining distractors and the target, is shown in Figure 4.4a. Here it can be seen that there is initial activation from the preview items followed by frequency adaptation so that the new target has relatively little competition for selection when it is subsequently presented. Within sSoTS, preview search is as efficient (in terms of the slope of the search function) as when only the second set of items is presented in the search display (the single-feature baseline condition), while it is more efficient than in a conjunction baseline, when all the items (the preview and the search displays) are presented together (Figure 4.4b).

Human preview search selectively decreases in efficiency (relative to the single-feature and conjunction baselines) when participants are given a secondary task during the preview period (Humphreys, Watson & Jolicoeur, 2002; Watson & Humphreys, 1997). This result is consistent with preview search dependent on top-down, resource-demanding processes that help bias search away from the old (previewed) items. In sSoTS the result can be simulated by reducing top-down inhibition from the old items, which we suggest arises when there is competition for the resources that would normally induce the inhibitory bias. Data are shown in Figure 4.4b.

a.

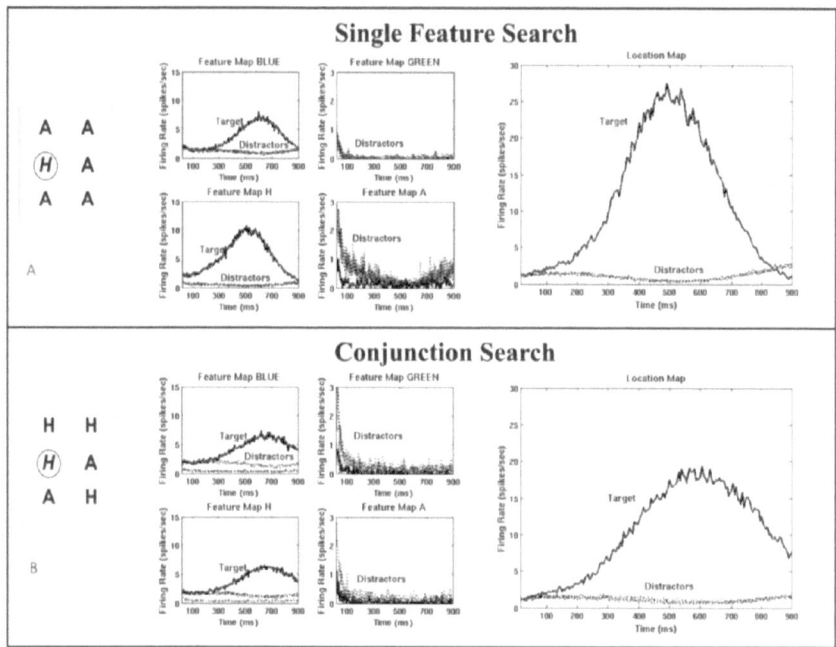

b.

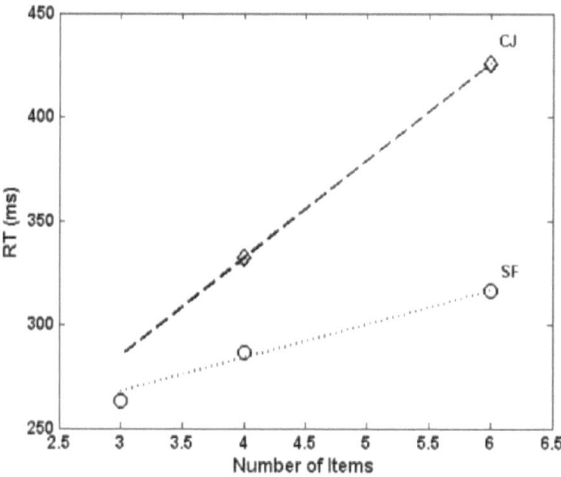

Figure 4.3 (a) Activation in the feature and location maps in a "single-feature" search
task (target differs from the distractors in terms of a single feature) and in a
"conjunction" search tasks (target shares at least one feature with each
distractor and differs only in the combination of its features). (b) Search
times for single-feature and conjunction targets in sSoTS.

a.

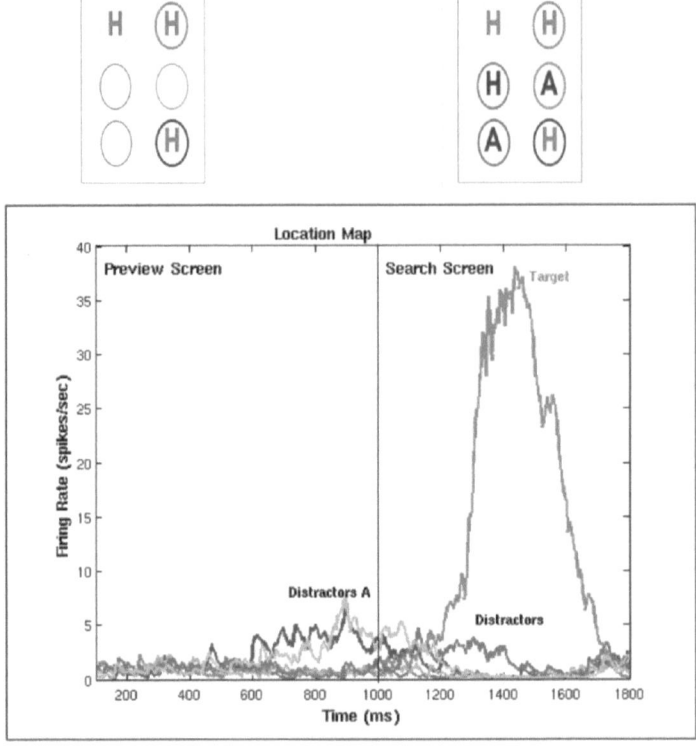

Figure 4.4 (a) Activation profile for units in the location map in the preview condition. The display on the top left indicates which of the 6 field locations
were occupied by a distractor during the preview period. The subsequent
(*right*) display indicates which locations are occupied by old distractors and
new search items, in the search display. The different shades of grey represent contrasting locations. *Opposite* (b) Search times for previewed targets
compared against single-feature (*left*) and conjunction (*right*) baseline conditions. For comparisons with the single-feature condition, display sizes
reflect the number of items presented in the new search display in the
preview condition. For comparisons with the conjunction condition, display sizes reflect the number of items finally present in the field (for discussion, see Humphreys et al., 2002).

The work on preview search provides an example of how the timing of
human visual selection can be captured within a biologically plausible neural
network model that assumes realistic temporal parameters for its activation
functions. We have also assessed whether neuropsychological problems in
selection emerge when the model is damaged.

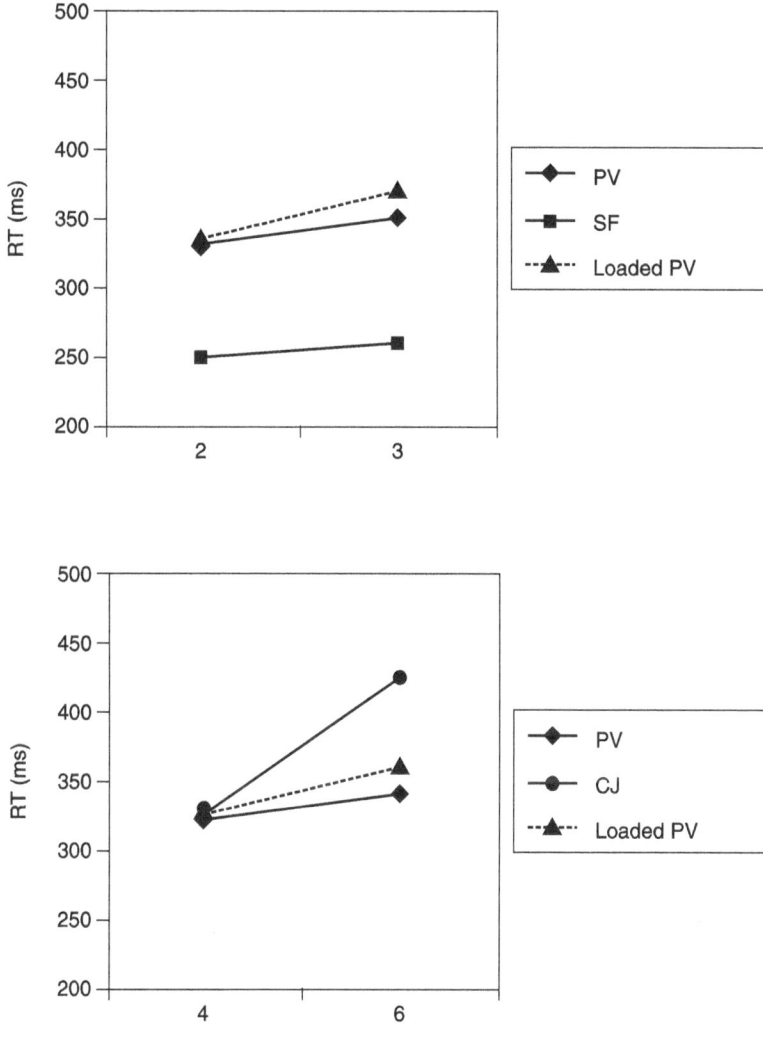

b

Figure 4.4 continued

3 Lesioning sSoTS

The feature maps in sSoTS can be conceptualised as representing neurons in ventral visual cortex, which respond to local properties of colour and shape (see Cowey, 1985). In contrast, units in the location map may correspond to neurons in posterior parietal cortex (PPC) that code spatial location and/or saliency in a manner that is independent of the driving feature values (see Gottlieb, Kusunoki & Goldberg, 1998). Lesions to the PPC, then, may be simulated by reducing the numbers of neurons present in pools representing particular positions in the location map (Humphreys, Mavritsaki, Heinke, & Deco, in press). When neurons are reduced for the positions on one side of the location map, sSoTS shows a spatial bias in selection. In addition, the detrimental effects of the lesion are stronger on preview and conjunction search than on single-feature search (Figure 4.5). These results arise because the lesion makes targets falling at impaired locations vulnerable to increased competition from distractors falling at unimpaired locations, and the effects of the increased competition are greater in preview and conjunction search, when the target must be defined by activation in more than one feature map and distractors have features shared with targets. In addition, damage to

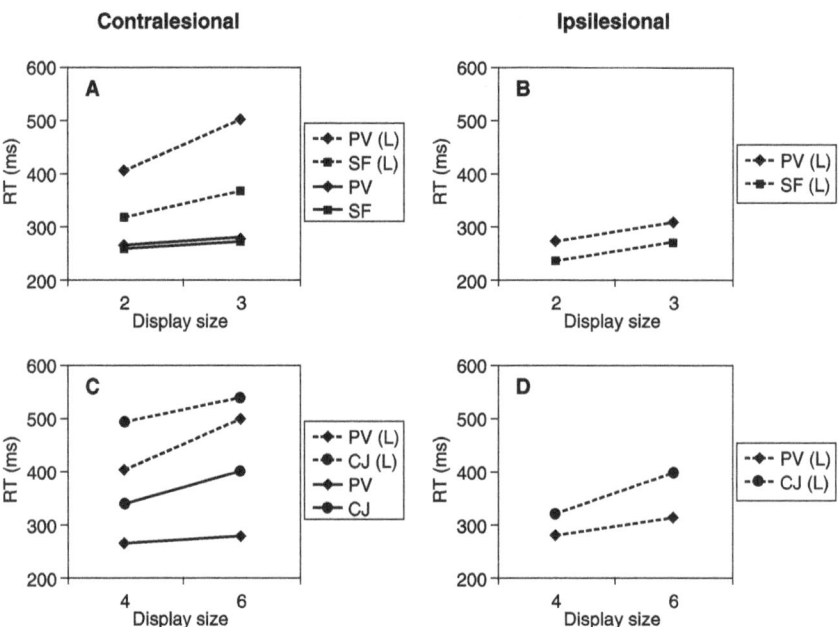

Figure 4.5 The mean correct reaction times (RTs, in ms) for the unlesioned (dotted lines) and lesioned versions (solid lines) of sSoTS. Data for the unlesioned model are plotted for convenience in the "contralesional" panel, but the data are for targets shown on either side of space. (A) RT data for contra-lesional targets (in the lesioned model) for preview and single-feature search, plotted against the display sizes in the search display. (B) RT data for ipsilesional targets for preview compared with single-feature search. (C) Data for contralesional targets (in the lesioned model) plotted against the full display sizes for preview and conjunction search. (D) RT data for ipsilesional targets for the preview condition relative to the conjunction condition.

the location map can lead to the case where old previewed items are not suppressed, partly because their processing is weakened and so the frequency-adaptation process never "kicks in" to dampen-down activation. As a consequence, old distractors can continue to compete for selection, creating more competition than was the case in the unlesioned version of the model. The finding that conjunction search is more disrupted than single-feature search fits with human neuropsychological data (Eglin, Robertson, & Rafal, 1989; Riddoch & Humphreys, 1987). This result has been interpreted in terms of the PPC playing a critical role in binding together visual conjunctions (shape and colour, in this case) (Treisman, 1988). However, in sSoTS the location maps do not explicitly bind information; rather, they modulate competition between feature units to enable a set of winning units to emerge at the selected location. Lesioning the maps disrupts performance by increasing competition. The finding indicating that preview search is also disrupted by lesioning simulates data reported by Humphreys et al. (2006b) and Olivers and Humphreys (2004) with PPC patients.

Figure 4.6 illustrates the activity in the location maps in sSoTS for target and distractor locations for single-feature (a), conjunction (b) and preview

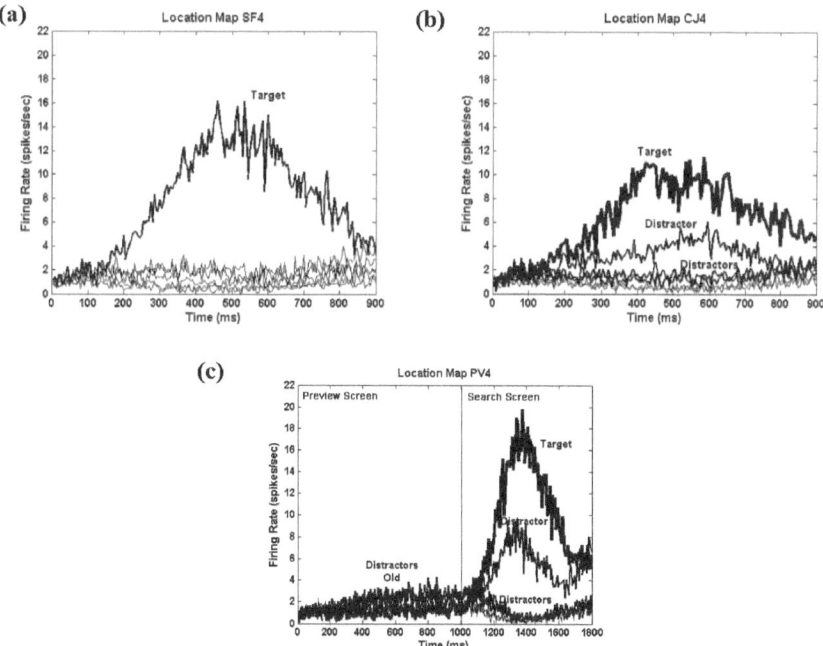

Figure 4.6 Activation in the location map following lesioning, for contralesional targets at display size 4 (a: single-feature search; b: conjunction search; c: preview search).

conditions (c), all at display size 4. Note that the competition for target selection is increased in the conjunction and preview conditions particularly.

Our study of the effects of lesioning sSoTS also generated some counter-intuitive predictions. One concerns the situation outlined in Figure 4.7a, where the old and new items are presented on the same or on opposite sides of space. One influential view of the effects of unilateral PPC lesions on visual selection is that they lead to poor disengagement of attention from distractor stimuli on the unimpaired (ipsilesional) side of space when targets are presented on the impaired (contralesional) side (Posner, Walker, Friedrich, & Rafal, 1984). According to this "attentional disengagement" account, patients should find preview search most difficult when the old items appear on the ipsilesional side and the new items (plus the target) are presented on the contralesional side. The sSoTS model, however, makes a different predic-tion. As illustrated in Figure 4.7b, sSoTS generated a search pattern in which targets were most difficult to detect when both they and the old distractors fell in the contralesional field, and search for new ipsilesional targets was also difficult when the old distractors appeared on the ipsilesional side. That is, search was worse when the old and new stimuli appeared in the same field, relative to when they appeared in different fields. For sSoTS, search efficiency is determined by competition between the distractors and the target. Any competition from the old (previewed) distractor with the target is influenced by whether this distractor is suppressed at the time when the target appears, along with the relative magnitude of activation for this distractor compared with the new target. As we have noted, there is likely to be less suppression (by frequency adaptation), and any suppression will take longer to effect, when the old distractor falls on the contra-relative to the ipsilesional side. Alongside this, the contralesional distractor will also generate reduced activ-ity compared with an ipsilesional target, and thus any accrual of activation in the location units also takes more time. These two effects can combine to generate the observed pattern of results. Search is most difficult for a cont-ralesional target following a contralesional preview (the contralesional, within-field condition) because the target has relatively weak activation and the distractor is not strongly suppressed at the time the target activation is accruing. Search for a contralesional target is better in the across-field condi-tion because the ipsilesional preview is suppressed at the time that activation for the target starts to accumulate. This situation changes, however, with an ipsilesional target. In this case, a target can sometimes accrue activation before the ipsilesional distractor is suppressed, in which case the target suf-fers competition and slowed selection. For sSoTS, there is an interplay of influences that determine search efficiency based on the relative timing of suppression of the preview and of activation accrual and strength for the target. Interestingly, the human data match the predictions of sSoTS. Contrary to the attentional disengagement account, PPC patients are worse at detecting both contra- and ipsilesional stimuli when old distractors appear

a.

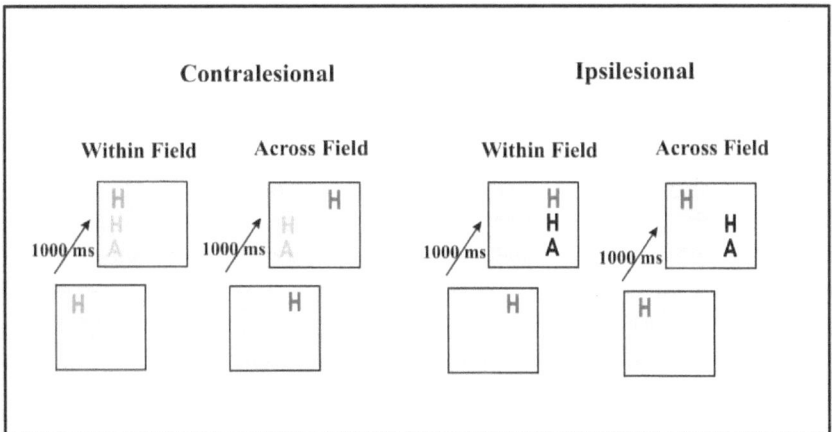

b.

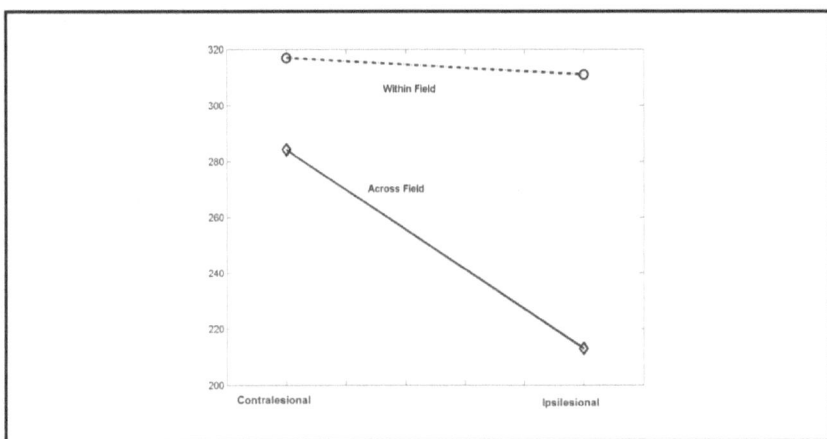

Figure 4.7 (a) Example displays for preview search with the old and new items presented in either the same field (the within-field conditions) or in different
fields (the across-field conditions). (b) Mean correct RTs (ms) (y-axis) for
sSoTS according to whether the target fell in the ipsi- or contralesional
field (x-axis).

in the same field (the within-field conditions shown in Figure 4.7), compared
to when they appear in opposite fields (the between-field conditions in
Figure 4.7) (Olivers & Humphreys, 2004, Expt. 3).

   The confirmation of the prediction also highlights one of the values in
developing explicit models of cognitive processes. The counter-intuitive pattern of results, with the within-field conditions being worse than the
between-field condition with ipsilesional distractors and a contralesional

target, arises due to an interplay of different processes in the model – the slow and weak suppression of contralesional distractors, the fast suppression of ipsilesional distractors, and the differential accrual of activation for contra- and ipsilesional targets. It would be difficult to predict this pattern of performance without the explicit model.

In the Introduction we also discussed evidence that the syndrome of unilateral neglect has also been linked to the presence of non-spatial deficits in patients, including low levels of general arousal (Robertson & Manly, 1999). We provided an existence proof of this proposal in sSoTS by reducing spontaneous activity modulated by the NMDA excitatory current. We assumed that this reduction in neurotransmitter regulation could be caused by a right PPC lesion, in addition to the effect of the lesion on selection mediated by the location map. The resulting patterns of search can be observed in Figure 4.8. When the NMDA current was reduced, there was a decrease in search efficiency in the preview condition even when the target appeared in the ipsilesional visual field, and, in the contralesional field, there was evidence even for single-feature search being disrupted. This last result is of interest because it mimics the emergence of full-blown neglect in the model, where even salient stimuli are missed on the contralesional side of

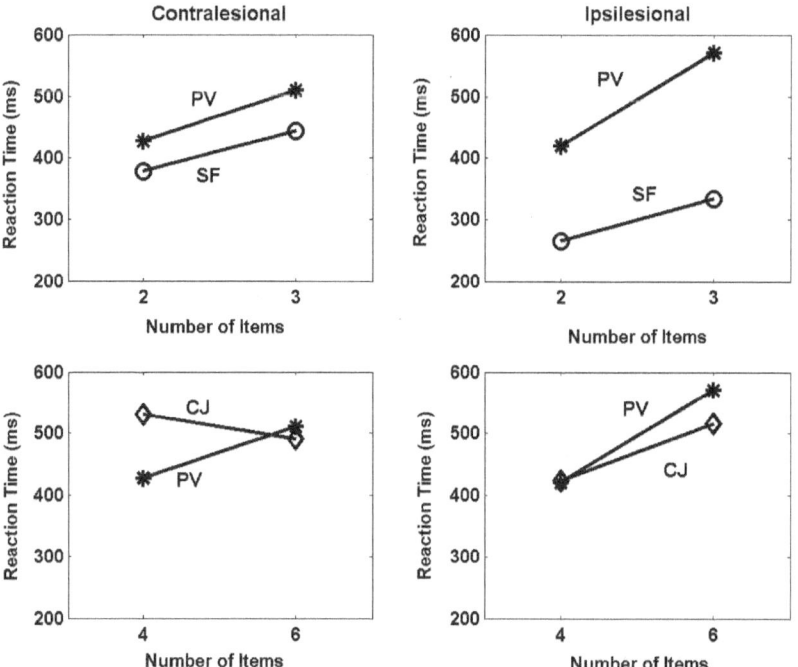

Figure 4.8 The mean correct RTs (ms) when the NMDA parameter was reduced (Humphreys et al., in press). Note the costs on search even in the ipsilesional field (cf. Figure 4.4b).

space. The drop in performance even for ipsilesional targets is exactly what would be expected if there was a non-spatial component in the attentional deficits observed in patients with damage to the right PPC. The simulations of neurotransmitter deficits in the syndrome of unilateral neglect are one of the bonuses produced by modelling using networks that have increased biological plausibility.

4 Simulating fMRI studies of search

As well as simulating deficits, how whole-system behaviour can break down after brain damage, we have also used sSoTS to probe the neural basis of visual selection. Deco, Rolls, and Horwitz (2004) proposed how spiking-level neural networks could be used to simulate fMRI, by convolving outputs from a model with the assumed haemodynamic response function (see Glover, 1999). There have now been numerous studies of both spatial search (see Corbetta & Shulman, 2002, for a review) and visual search over time (Allen & Humphreys, 2008; Humphreys et al., 2004; Olivers, Smith, Matthews, & Humphreys, 2005; Pollmann et al., 2003). These studies have demonstrated the involvement of a fronto-parietal network in search, but the exact roles of the different brain areas in the search process have yet to be defined. Search is a complex process, probably involving grouping and segmentation of the visual field, shifts of spatial attention, working memory for previously attended stimuli and locations, top-down attentional guidance towards the expected features of targets, and suppressive biases of attention away from distractors (see Duncan & Humphreys, 1989; Wolfe, 1994). Preview search enables processes involved in grouping and suppressing the initial set of distractors to be isolated from the process of searching for the target (see Braithwaite, Humphreys, & Hulleman, 2005), but even here the previewed stimuli can be used to set a positive bias for target features as well an inhibitory bias against the old distractors (Braithwaite & Humphreys, 2003). As a consequence, imaging studies that have isolated activation to preview displays (Allen & Humphreys, 2008; Olivers et al., 2005; Pollmann et al., 2003) may still reflect excitatory as well as inhibitory processes in selection. One way forward in such cases is to try to use a model such as sSoTS to provide theory-based decomposition of activation in the brain regions activated in search.

In an attempt to do just this, we (Mavritsaki, Allen, & Humphreys, in press) have used the patterns of top-down excitation and inhibition operating in sSoTS as a means of fractionating BOLD signals associated with excitatory and inhibitory activity in the brain. We analysed data from Allen and Humphreys (2008). In this study, participants searched for a blue house target that appeared among blue face and red house distractors. In the preview version of the task, the blue face distractors appeared before the house stimuli. Allen and Humphreys reported activation in a network of areas in PPC, including the superior and lateral parietal lobe and the precuneus, as well as brain regions responsive to the content of the different displays (e.g., the fusiform face region, when faces were presented) in preview search. Applying the architecture of sSoTS to these experiments, we can propose that the shape feature maps correspond to the face and house areas found in the fusiform and parahippocampal gyri (see Kanwisher & Yovel, 2006). In the model, the positive expectation for the blue house target is associated with top-down excitatory input applied into the blue and house maps, given by the differences in activity in these maps compared with the other (red and face) maps. In contrast, the inhibitory bias against the old distractors is associated with the blue and face maps (relative to the red and house maps). We summed the different activity patterns given across the maps across time (averaged across 20 runs for each of 9 replications), convolved this activation with the assumed haemodynamic response function, and used the resultant vector as a regressor when analysing the fMRI data. We found that the haemodynamic response function in the location map in the preview condition was increased for preview search compared with conjunction search – a result that reflects the greater salience of the target in preview compared with conjunction search, and not the difficulty of the search tasks (conjunction search being the more difficult). This matches data reported by Allen and Humphreys (2008) for the precuneus (see Figure 4.9). In addition, the excitatory regressor predicted activity in the lateral parietal region, while the inhibitory regressor predicted activity in the precuneus (Figure 4.10). This suggests that, although both regions are associated with enhanced activity in preview search compared with other baseline search conditions, the regions may be performing different jobs – lateral parietal cortex being sensitive to excitatory guidance of selection to targets, and the precuneus linked to suppression of old distractors. The model-based analysis highlights different functional roles within the network of areas activated in search.

5 Summary and conclusions

We have reviewed work in which we have used a spiking-level model of visual search, the sSoTS model, to simulate aspects of human performance. We have shown how the model captures patterns of "serial" search across space (although it has a parallel processing architecture) and "parallel" search across time. The temporal parameters of the model arise out of a biologically plausible frequency-adaptation process. As well as capturing aspects of normal, whole-system behaviour, sSoTS shows a pattern of breakdown after lesioning that mimics the performance of human patients after damage to PPC. There are selective deficits for conjunction and preview search, and the spatial relations between distractors and targets in preview search affect performance in counter-intuitive ways that are nevertheless supported by the neuropsychological data (Olivers & Humphreys, 2004). Moreover, sSoTS can simulate non-spatial aspects of neglect when neurotransmitter modulation is reduced, as predicted if lesions reduce general arousal in patients

a. Activation in the precuneus

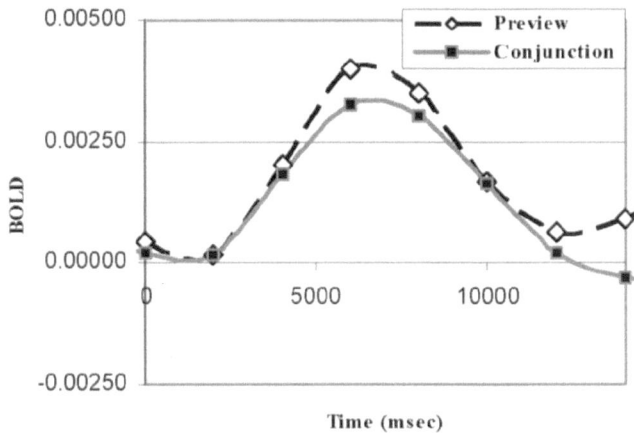

b. Active Inhibition activation (model).

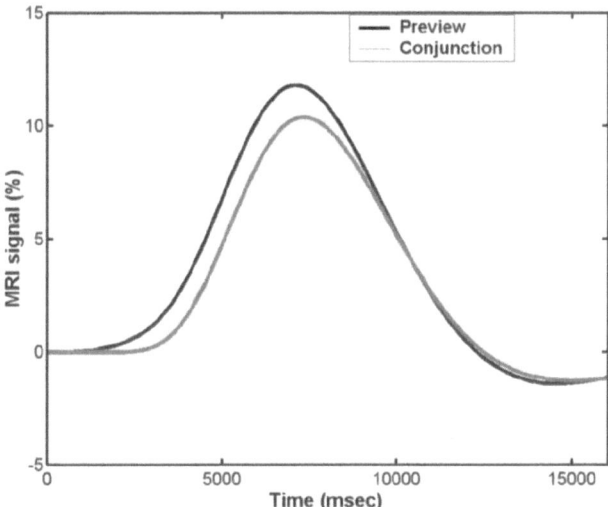

Figure 4.9 (a) The haemodynamic response found in the precuneus in preview and conjunction search, in Allen and Humphreys (2008). (b) The simulated haemodynamic response in the location map in sSoTS.

(Robertson & Manly, 1999). Finally, sSoTS can be used to model fMRI data, and we have shown that it can provide a framework for decomposing networks of activation that are found in brain imaging studies of search.

Like any model, sSoTS has limitations. For example, it employs a very simplified neural architecture. It can be said to receive "pre-processed" data

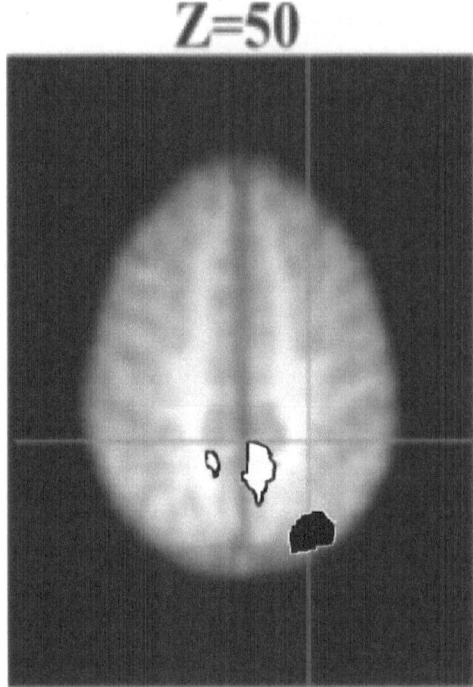

Figure 4.10 Brain regions from the study of Allen and Humphreys (2008) showing reliable correlations with the activity in sSoTS associated with
active inhibition (white patch, in the precuneus) and top-down excitation (black patch, in the lateral parietal cortex).

corresponding to high-level properties of shape and colour, rather than
deriving these properties from a visual image. It is relatively indifferent to
factors such as visual grouping. However, our view is that the worth of a
model should be judged in terms of what questions it attempts to answer.
sSoTS is not a model of early visual processing, and we would argue that
it should not be judged on that basis. Rather sSoTS reflects competitive
interactions for visual selection that operate within and across higher-level
cortical regions. Its aim is to simulate differences between search conditions
that use similar stimuli (and so do not differ in terms of the early visual
information being coded) but differ in how they are presented (e.g., whether
staggered over time or exposed simultaneously). The model attempts to take
seriously factors such as the timing of neural events and the modulatory
effects of different neurotransmitters on neural activity. We believe there is a
pay-off in doing this, since these factors allow the model to capture aspects
of human performance (e.g., the temporal properties of preview search, the
presence of non-spatial as well as spatial deficits in unilateral neglect) that are
not examined in higher-level simulations. By taking such factors into

account, we believe that models such as sSoTS serve a useful purpose in cognitive neuroscience, building a bridge between high-level cognitive theory and lower-level neuronal analyses. When accounting for different neuro-psychological disorders, and for activation within complex neural networks, this bridge allows physiological data to be related to emergent behaviour.

6 Acknowledgements

This work was supported by grants from the BBSRC and MRC (UK).

7 References

Abeles, A. (1991). *Corticonics*. Cambridge, UK: Cambridge University Press.

Allen, H. A., & Humphreys, G. W. (2008). A neural marker for content specific active ignoring. *Journal of Experimental Psychology: Human Perception and Performance*, 34(2), 286–297.

Bisiach, E. (1999). Unilateral neglect and related disorders. In G. Denes & L. Pizzamiglio (Eds.), *Handbook of clinical and experimental neuropsychology*. London: Psychology Press.

Braithwaite, J. J., & Humphreys, G. W. (2003). Inhibition and anticipation in visual search: Evidence from effects of color foreknowledge on preview search. *Perception & Psychophysics*, 65, 213–237.

Braithwaite, J. J., Humphreys, G. W., & Hulleman, J. (2005). Color-based grouping and inhibition in visual search: Evidence from a probe-detection analysis of preview search. *Perception & Psychophysics*, 67, 81–101.

Brunel, N., & Wang, X. (2001). Effects of neuromodulation in a cortical networks model of object working memory dominated by current inhibition. *Journal of Computational Neuroscience*, 11, 63–85.

Corbetta, M., & Shulman, G. L. (2002). Control of goal-directed and stimulus-driven attention in the brain. *Nature Reviews Neuroscience*, 3, 201–215.

Coull, J. T. (1998). Neural correlates of attention and arousal: Insights from electrophysiology, functional neuroimaging and psychopharmacology. *Progress in Neurobiology*, 55, 343–361.

Cowey, A. (1985). Aspects of cortical organization related to selective attention and selective impairments of visual perception. In M. I. Posner & O. S. M. Marin (Eds.), *Attention and performance XI*. Hillsdale, NJ: Lawrence Erlbaum Associates.

Deco, G., & Rolls, E. (2005). Neurodynamics of biased competition and cooperation for attention: A model with spiking neuron. *Journal of Neurophysiology*, 94, 295–313.

Deco, G., Rolls, E., & Horwitz, B. (2004). Integrating fMRI and single-cell data if visual working memory. *Neurocomputing*, 58–60, 729–737.

Deco, G., & Zihl, J. (2001). Top-down selective visual attention: A neurodynamical approach. *Visual Cognition*, 8(1), 119–140.

Desimone, R., & Duncan, J. (1995). Neural mechanisms of selective visual attention. *Annual Review of Neuroscience*, 18, 193–222.

Duncan, J., & Humphreys, G. W. (1989). Visual search and stimulus similarity. *Psychological Review*, 96, 433–458.

Eglin, M., Robertson, L. C., & Rafal, R. D. (1989). Visual search performance in the neglect syndrome. *Journal of Cognitive Neuroscience*, 1, 372–385.

Glover, G. H. (1999). Deconvolution of impulse response in event-related BOLD fMRI. *NeuroImage, 9*, 419–429.

Gottlieb, J. P., Kusunoki, M., & Goldberg, M. E. (1998). The representation of visual salience in monkey parietal cortex. *Nature, 391*, 481–484.

Heinke, D., & Humphreys, G. (2003). Attention, spatial representation and visual neglect: Simulating emergent attention and spatial memory in the selective attention for identification model (SAIM). *Psychological Review, 110*, 29–87.

Heinke, D., & Humphreys, G. W. (2005). Computational models of visual selective attention: A review. In G. Houghton (Ed.), *Connectionist models in cognitive psychology*. Hove, UK: Psychology Press.

Heinke, D., Humphreys, G. W., & C. L. Tweed (2006). Top-down guidance of visual search: A computational account. *Visual Cognition, 14*, 985–1005.

Humphreys, G. W., Kyllinsbæk, S., Watson, D. G., Olivers, C. N. L., Law, I., & Paulson, O. (2004). Parieto-occipital areas involved in efficient filtering in search: A time course analysis of visual marking using behavioural and functional imaging procedures. *Quarterly Journal of Experimental Psychology, 57A*, 610–635.

Humphreys, G. W., Mavritsaki, E., Heinke, D. G., & Deco, G. (in press). Simulating posterior parietal damage in a spiking neural network: A neuropsychological analysis of the sSoTS model. *Cognitive Neuropsychology.*

Humphreys, G. W., Olivers, C. N. L., & Braithwaite, J. J. (2006a). The time course of preview search with color-defined, not luminance-defined stimuli. *Perception & Psychophysics, 68(8)*, 1351–1358.

Humphreys, G. W., Olivers, C. N. L., & Yoon, E. Y. (2006b). An onset advantage without a preview benefit: Neuropsychological evidence separating onset and preview effects in search. *Journal of Cognitive Neuroscience, 18(1)*, 110–120.

Humphreys, G. W., Watson, D. G., & Jolicoeur, P. (2002). Fractionating visual marking: Dual task decomposition of the marking state by timing and modality. *Journal of Experimental Psychology: Human Perception and Performance, 28*, 640–660.

Husain, M., & Rorden, C. (2003). Non-spatially lateralized mechanisms in hemispatial neglect. *Nature Reviews Neuroscience, 4*, 26–36.

Itti, L., & Koch, C. (2000). A saliency-based search mechanism for overt and covert shifts of visual attention. *Vision Research, 40*, 1489–1506.

Kanwisher, N., & Wocjiulik, E. (2000). Visual attention: Insights from brain imaging. *Nature Reviews Neuroscience, 1*, 91–100.

Kanwisher, N., & Yovel, G. (2006). The fusiform face area: A cortical region specialized for the perception of faces. *Philosophical Transactions of the Royal Society, B361*, 2109–2128.

Madison, D., & Nicoll, R. (1984). Control of the repetitive discharge of rate CA1 pyramidal neurons *in vitro. Journal of Physiology, 345*, 319–331.

Mavritsaki, E., Allen, H., & Humphreys, G.W. (in press). Model based analysis of fMRI-data: Applying the sSoTS framework to the neural basis of preview search. In L. Paletta (Ed.), *Lecture notes in computer science*. London: Sage.

Mavritsaki, E., Heinke, D., Humphreys, G. W., & Deco, G. (2006). A computational model of visual marking using an interconnected network of spiking neurons: The spiking Search over Time & Space model (sSoTS). *Journal of Physiology, Paris, 100*, 110–124.

Mavritsaki, E., Heinke, D., Humphreys, G. W., & Deco, G. (2007). Suppressive effects in visual search: A neurocomputational analysis of preview search. *Neurocomputing, 70*, 1925–1931.

Mozer, M. C., & Sitton, M. (1998). Computational modelling of spatial attention. In H. Pashler (Ed.), *Attention*. London: Psychology Press.

Olivers, C. N. L., & Humphreys, G. W. (2004). Spatiotemporal segregation in visual search: Evidence from parietal lesions. *Journal of Experimental Psychology, 30*(4), 667–688.

Olivers, C. N. L., Smith, S., Matthews, P., & Humphreys, G. W. (2005). Prioritizing new over old: An fMRI study of the preview search task. *Human Brain Mapping, 24*, 69–78.

Paus, T., Jech, R., Thompson, C. J., Comeau, R., Peters, T., & Evans, A. C. (1997). Transcranial magnetic stimulation during positron emission tomography: A new method for studying connectivity of the human cerebral cortex. *Journal of Neuroscience, 17*, 3178–3184.

Pollmann, S., Weidner, R., Humphreys, G. W., Olivers, C. N. L., Müller, K., Lohmann, G., et al. (2003). Separating segmentation and target detection in posterior parietal cortex: An event-related fMRI study of visual marking. *NeuroImage, 18*, 310–323.

Posner, M. I., & Petersen, S. E. (1990). The attention system of the human brain. *Annual Review of Neuroscience, 13*, 25–42.

Posner, M. I., Walker, J. A., Friedrich, F. J., & Rafal, R. D. (1984). Effects of parietal injury on covert orienting of attention. *Journal of Neuroscience, 4*, 1863–1874.

Riddoch, M. J., & Humphreys, G. W. (1987). Perceptual and action systems in unilateral neglect. In M. Jeannerod (Ed.), *Neurophysiological and neuropsychological aspects of spatial neglect*. Amsterdam: Elsevier Science.

Robertson, I. H. (1994). The rehabilitation of attentional and hemi-attentional disorders. In M. J. Riddoch & G. W. Humphreys (Eds.), *Cognitive neuropsychology and cognitive rehabilitation* (pp. 173–186). London: Psychology Press.

Robertson, I. H., & Manly, T. (1999). Sustained attention deficits in time and space. In G. W. Humphreys, J. Duncan, & A. Treisman (Eds.), *Attention, space and action: Studies in cognitive neuroscience* (pp. 279–310). Oxford: Oxford University Press.

Robertson, I. H., Mattingley, J. B., Rorden, C., & Driver, J. (1998). Phasic alerting of neglect patients overcomes their spatial deficit in visual awareness. *Nature, 395*, 169–172.

Servan-Schreiber, D., Printz, H., & Cohen, J. D. (1990). A network model of catecholamine effects: Gain, signal-to-noise ratio, and behavior. *Science, 249*, 892–895.

Treisman, A. (1998). Feature binding, attention and object perception. *Philosophical Transactions of the Royal Society, B353*, 1295–1306.

Treisman, A., & Gelade, G. (1980). A feature-integration theory of attention. *Cognitive Psychology, 12*, 97–136.

Watson, D. G., & Humphreys, G. W. (1997). Visual marking: Prioritising selection for new objects by top-down attentional inhibition. *Psychological Review, 104*, 90–122.

Watson, D. G., Humphreys, G. W., & Olivers, C. N. L. (2003). Visual marking: Using time in visual selection. *Trends in Cognitive Sciences, 7*, 180–186.

Wolfe, J. A. (1994). Guided Search 2.0: A revised model of visual search. *Psychonomic Bulletin & Review, 1*, 202–238.

5 The selective attention for identification model (SAIM)

A framework for closing the gap between the behavioural and neurological levels

Dietmar Heinke, Eirini Mavritsaki, Andreas Backhaus, and Martin Kreyling

The selective attention for identification model (SAIM) implements translation-invariant object recognition in multiple-object scenes by employing competitive and cooperative interactions between rate code neurons. With these mechanisms, SAIM can model a wide range of experimental evidence on attention and its disorders (e.g., Heinke & Humphreys, 2003). This chapter presents two new versions of SAIM that address issues at the two sides of the gap between the behavioural and neurological levels. For the behavioural side, we demonstrate that an extension of SAIM can process ecologically valid inputs (natural colour images). For the neurological side, we present a method for replacing rate code neurons with spiking neurons while maintaining SAIM's successes on the behavioural level. Hence, this method allows SAIM to link attentional behaviour with processes on the neurological level, and vice versa.

1 Introduction

Recently, we presented a computational model, termed SAIM (selective attention for identification model; Backhaus, Heinke, & Humphreys, 2005; Heinke, Backhaus, Sun, & Humphreys, 2008; Heinke & Humphreys, 1997, 2003, 2005b; Heinke, Humphreys, & diVirgilo, 2002b; Heinke, Humphreys, & Tweed, 2006; Heinke, Sun, & Humphreys, 2005). SAIM was developed to model normal attention and attentional disorders by implementing translation-invariant object recognition in multiple-object scenes. In order to do this, a translation-invariant representation of an object is formed in a "focus of attention" (FOA) through a selection process. The contents of the FOA are then processed with a simple template-matching process that implements object recognition. These processing stages are realized by non-linear differential equations often characterized as competitive and cooperative interactions between neurons (e.g., Amari, 1977). With these mechanisms SAIM can model a wide range of experimental evidence on attention and its disorders (for a discussion, see Heinke & Humphreys, 2005a).

This chapter aims to demonstrate that SAIM provides a very good framework for closing the gap between behaviour and neurophysiology. In order to do so, we present two new extensions of SAIM tackling important issues when dealing with the two sides of the gap. On one hand, a model should be able to simulate human behaviour with ecologically valid stimuli – for example, natural colour images. On the other hand, a model has to be based on neurophysiologically plausible assumptions – for example, a spiking neuron model. Here, we will demonstrate that SAIM represents a framework for dealing with theses two issues. Before that, we will give an introduction into SAIM, highlighting the crucial mechanisms that form the basis for the two extensions. The first extension can process natural input images tackling the behavioural side of the gap. The second extension of SAIM performs selection and identification with spiking neurons instead of rate code neurons, moving SAIM closer to the neurological end of the gap.

2 SAIM with rate code neurons

SAIM aims at translation-invariant object identification by using a focus of attention. SAIM consists of several modules, including a contents network, a selection network, a knowledge network and a location map (see Figure 5.1).

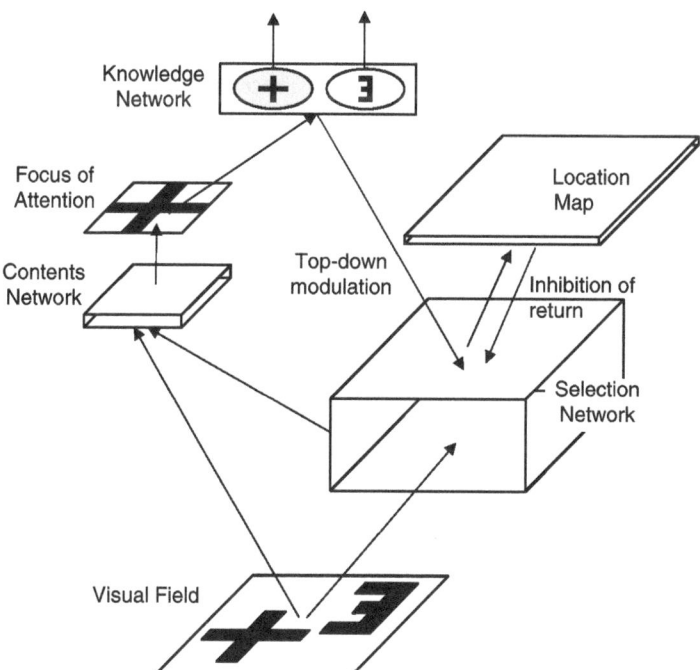

Figure 5.1 The architecture of SAIM.

The contents network enables a translation-invariant mapping from the visual field into the FOA. The selection network directs the FOA to locations, and the knowledge network identifies the contents of the FOA with a simple template matching procedure. The location map implements a mechanism that enables attention to be shifted from an object, once recognized, to other objects in the field. To design SAIM, an energy minimization approach was used (Hopfield & Tank, 1985). In this approach, first, constraints for the successful completion of a task – for example, object recognition – are defined. In turn, these constraints define the activity patterns in SAIM that are permissible and others that are not. Following the procedure suggested by Hopfield and Tank (1985), we define an energy function whose minimal values are generated by just these permissible activity values. To find these minima, a gradient descent procedure was applied, resulting in a differential equation system. The differential equation system defines the topology of the network (i.e, the strengths and signs of the connections between units).

   In the following section we will describe SAIM's operations in a qualitative way and omit the mathematics behind SAIM. The mathematical description will be presented together with the spiking version, as both versions utilize an energy minimization approach.

2.1 Contents network

The contents network enables a translation-invariant mapping from the visual field into the FOA by utilizing a layer of sigma-pi units (for a one-dimensional illustration, see Figure 5.2). Sigma-pi units perform an addition as well as a multiplication of input activations (Rumelhart & McClelland, 1988). In the contents network they receive inputs from the selection network and the visual field. They then multiply the activations from corresponding locations in the selection network and visual field. For instance, in Figure 5.2, Multiplication Point i multiplies activation from Position 1 in the visual field and activation from Unit g. The results of the multiplication are summed up (e.g., the results from Multiplication Points i, iv and vi) and produce the output of the content network, the FOA. Hence, one layer in the selection network controls one pixel in the FOA. With this input–output mapping, the contents network introduces a connection matrix specifying all possible relationships between locations on the retina and locations in the FOA. Hence, the contents network can implement all possible mappings from the visual field into FOA, depending on the distribution of activation in the selection network.

2.2 Selection network

The selection network has a multilayered structure in which each layer determines the activation of one FOA pixel via the contents network (see

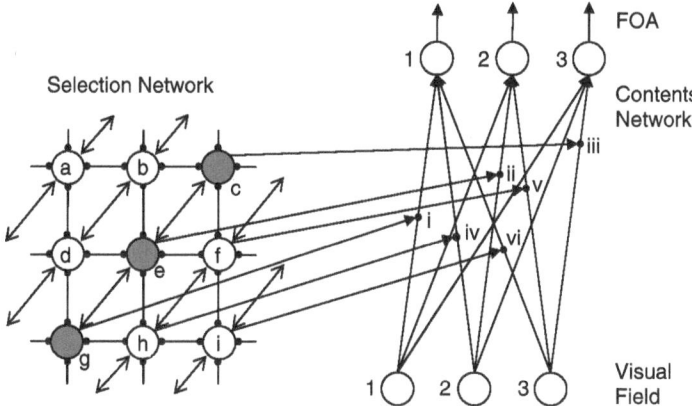

Figure 5.2 An illustration of how connections between units in the selection network impose constraints on the mapping from the retina to the FOA in SAIM. "Vertical" and "horizontal" connections in the selection network are mutually inhibitory, while "diagonal" connections are mutually excitatory (see the text for details). Within the selection network, columns contain units activated by one given position on the retina; rows (layers) contain units that activate one position in the FOA.

Figure 5.2). As explained in the previous section, this structure can set up any arbitrary mapping from the visual field onto the FOA. However, such mappings should not be arbitrary. For example, in order for subsequent recognition to be successful, neighbouring units on the visual field should be mapped into neighbouring units in the FOA (e.g., Units 1 and 2 on the visual field into Units 1 and 2 in the FOA; see Figure 5.2). Hence, the activation patterns in the selection network need to be constrained to preserve such neighbourhood relations. Overall, the following constraints were incorporated into the selection network: (1) neighbourhood relations should be preserved in mapping through to the FOA; (2) one retinal unit should not be mapped more than once into the FOA; and (3) one unit in the FOA should not receive input from more than one retinal unit (the converse of Constraint 2).

After applying the energy minimization approach, the constraints are captured in the following way. First, there are excitatory connections between units in the selection network that would preserve neighbourhood relations in mapping from the retina to the FOA; these are indicated by the arrow connections between units along the diagonals of the selection network shown in Figure 5.2. The strength of the excitatory connections has a Gaussian function with respect to the distance between units: more proximal units have stronger connections than more distant units. This produces the sensitivity of the model to the spatial proximity of pixels in an image. Second, there are inhibitory connections between units in the selection network that would map the same retinal location into different units in the FOA. In

Figure 5.2 these are illustrated by the —• connections along the vertical axes of the selection network. (Although for illustrative purposes only connections between neighbouring units are shown, there are inhibitory connections (with equal weights) between all units along the vertical and horizontal axes of the selection network.) For example, in Figure 5.2, Units e and h would map Retinal Location 2 into different units in the FOA (via Units ii and iv in the contents network); hence Units e and h are mutually inhibitory. Third, there are inhibitory connections between units within one layer that would map from different units in the retina into the same unit in the FOA. These are illustrated by the —• connections along the horizontal axes of the selection network shown in Figure 5.2. For example, Units e and f "compete" to gate activity into Unit 2 in the FOA (via Units ii and v in the contents network). Finally, note that, in essence, the two inhibitory connections – vertical and horizontal – implement a winner-take-all (WTA) mechanism.

2.3 The knowledge network

SAIM also contains a higher level "knowledge network", the activation of which achieves object recognition. The knowledge network contains "template" units, each of which represents a known object. Template units are activated on the basis of learned weights that connect with pixels in the FOA. The introduction of the knowledge network into SAIM adds a top-down constraint on selection in the model. The input will activate most the best-matching template, which, given the WTA properties of the network, leads to inhibition of templates that match less well. To integrate this constraint into SAIM, an energy function was defined for the knowledge network and added to the energy function in the selection network. Here the gradient descent procedure results in a topology that connects the template units in a WTA fashion. The templates in SAIM generate top-down modulation of the selection network. The top-down modulation can be thought of in the following way. A linear combination of the templates is spatially correlated with the visual field. The weighting of the templates in the linear combination is determined by the activation of the template units in the knowledge network (for details, see Heinke & Humphreys, 2003). The result of the correlation is additively fed into the selection network and therefore directs the selection process towards locations where the templates match well.

2.4 Shifting attention and the location map

One further property of SAIM is its ability to shift attention between items, once stimuli are identified. The competitive nature of the model means that, by and large, identification is limited to one item at a time. When multiple objects are present, there needs to be some mechanism to enable each to be selected in turn. In the model, switching of attention is accomplished by top-down inhibition, which takes place once a template unit reaches a given threshold: there is temporary suppression both of the selected template and of the selected location. This enables other items in the field to then win the competition for selection, so that they in turn become attended. Suppression of previous winners provides a form of "inhibition of return" (IOR; see Posner & Cohen, 1984). To enable the location of selected items to be inhibited, activity in units corresponding to each retinal location in the selection network is summed and passed on to a "location map", which essentially represents which locations currently hold items of attentional interest. Early in the processing of a display with multiple objects, several locations in the map will be occupied. However, as the dynamics of competition within the selection network evolve, so active units in the map will decrease until only one remains; this represents the location of the attended object. Once the target is selected, the active unit in the location map inhibits input from relevant location in the visual field, so that it is no longer represented in the selection network (see Figure 5.1, where the inhibitory link from the location map to the selection network is specified). This leads to a discontinuity in the energy function from which the settling process must start again, but in this case there is a decreased likelihood that spatial regions previously attended are subsequently selected. Inhibited positions in the location map also provide SAIM with a form of spatial memory for previously attended regions of field. Finally, note that these mechanisms are not part of the new versions of SAIM introduced in this chapter. However, it would be straightforward to integrate this serial scanning mechanism into the new versions.

2.5 Results

The two results presented here highlight two important aspects of SAIM: the competitive interactions and the multilayered architecture of the selection network (Heinke & Humphreys, 2003). These mechanisms are important for the success of the two extensions presented in this chapter. Also, later we will compare these simulation results with the results from the spiking version.

2.5.1 Basic behaviour

The stimuli presented to the model were a cross (+) and a 2 to test whether the model can deal with two objects in the input. Template units were given an identification threshold (.9), after which identification was considered to have taken place. Once this threshold was passed, inhibition was applied (see above). Figure 5.3a illustrates the time course of activation in the template units when the + stimuli were presented alone. The number of iterations required for the + to reach threshold at the template level was 610 iterations. Figure 5.3b demonstrates the consequences of presenting both the + and the 2 together in the field. The + is selected and identified first (after 670

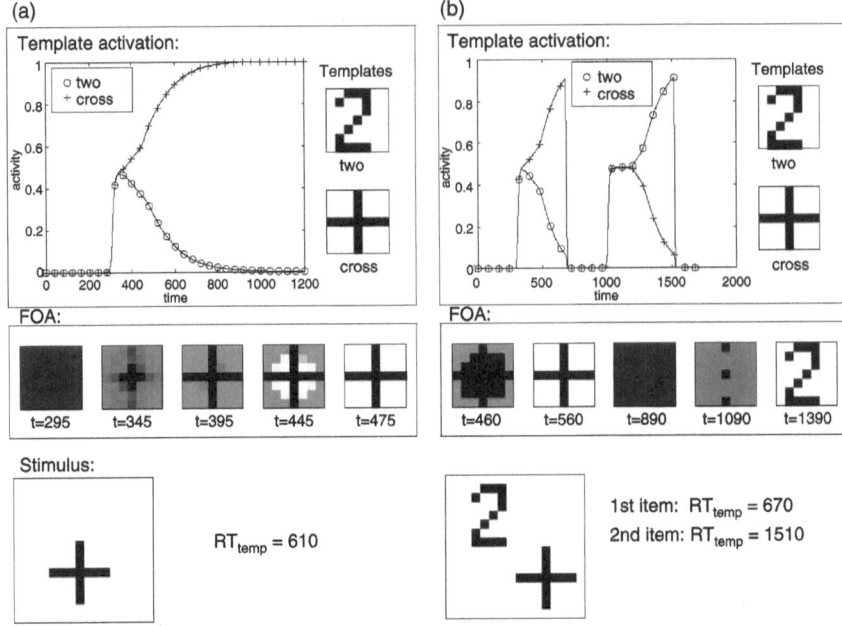

Figure 5.3 Results illustrating SAIM's basic behaviour using rate code neurons. The simulation results show that this version of SAIM can deal with multiple-object, black-and-white scenes. It also show that SAIM's reaction time to identify the cross increases when a second object is placed in the input image. This mimics the two-object cost often found in experiments.

iterations), but following this, selection and identification proceed for the 2. SAIM switches attention from the first to the second item. The study demonstrates that SAIM can deal with multiple-object displays. The selection bias found in this simulation is a result of the Gaussian excitatory connections in the selection network and is proportional to the weighted pixel mass of an object whereby the weight for each pixel is given by its distance from the object's centre of gravity (for details, see Heinke & Humphreys, 2003). More importantly, SAIM's reaction time for identifying the + increases from the one-object display to the two-object display. This two-object cost is an emergent property of the competitive processes in the selection network and mimics experimental findings (e.g., Duncan, 1980).

2.5.2 Simulation of neuropsychology

In order to simulate a broad variety of attentional deficits, Heinke and Humphreys (2003) lesioned SAIM in various ways. Here we focus on what they termed "horizontal" lesion, since it illustrates the usefulness of the multilayered architecture of the selection network. In the horizontal lesion

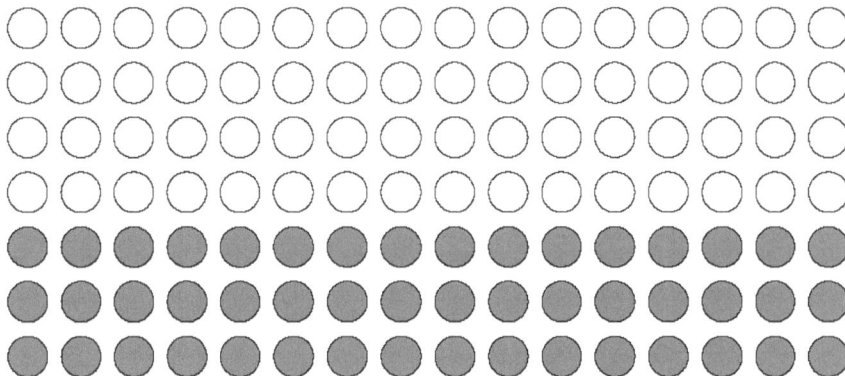

Figure 5.4 Illustration of "horizontal" lesioning in SAIM.

there was damage to layers in the selection network that control, via the contents network, the left side of the FOA (along the y-axis of the matrix in Figure 5.4). The 7 × 7 pixel cross was used as stimulus. The cross could appear at any of six positions in the visual field (moving from left to right). The patterns of convergence achieved in the FOA, along with the times to attend to and identify the cross, are illustrated in Figure 5.5.

Figure 5.5 shows that, across all the field positions, there was neglect of the left-most pixel in the cross, and the cross was in all cases mapped too far to the left of centre of the FOA. Here SAIM exhibits a form of "object-based" neglect, in which the left features of the object are neglected even when the stimulus falls in the right visual field. In a second assessment of horizontal lesion, we examined performance when multiple stimuli were presented, when SAIM should switch attention from one object to another. Studies of neglect patients using simple drawings tasks (Gainotti, d'Erme, Monteleone, & Silveri, 1986; Marshall & Halligan, 1994), and tasks requiring the discrimination of features within shapes (Humphreys & Heinke, 1998), have demonstrated that forms of "sequential neglect" can occur – there is neglect of each object selected in a sequence. For instance, in drawing, a patient may fail to include some details on the left of one item, move onto the next item

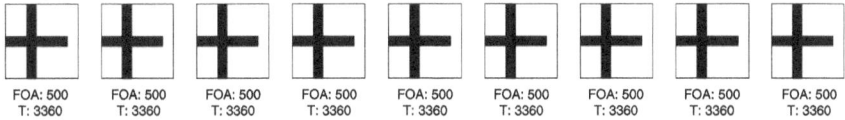

Figure 5.5 Simulation results of object-based neglect within the FOA as the stimulus was positioned at different locations across the visual field (left to right). The numbers below the FOAs state the reaction times for the FOA and the knowledge network.

and then fail to include its left-side details too. This can produce neglect of features on the left of objects that fall further into the right (non-neglected) field than do non-neglected features on the right side of an object. To test whether SAIM can mimic these results, Heinke and Humphreys (2003) used two stimuli: a cross (+) and a bracket (]). When the model was unlesioned, it first selected the + and then the]; also, each stimulus was fully mapped into the FOA. After the horizontal lesion had been imposed, SAIM attended sequentially to the + and then the] as before, but for each stimulus there was a shift in its position within the FOA, with the left-most pixel for each being neglected. This is shown in Figure 5.6. SAIM shows sequential left-neglect of each object selected. Left-neglect occurred for each object because the model has difficulty in mapping input into the left side of the FOA, so that the right-side features play a more dominant role in the mapping process. The object is consequently displaced in the FOA, and neglect occurs. This result matches data from human patients (Driver, Baylis, & Rafal, 1992; Gainotti et al., 1986; Humphreys & Heinke, 1998).

Importantly, this success is due to the multilayered architecture in the selection network, as it allows SAIM to treat neglect of parts of objects

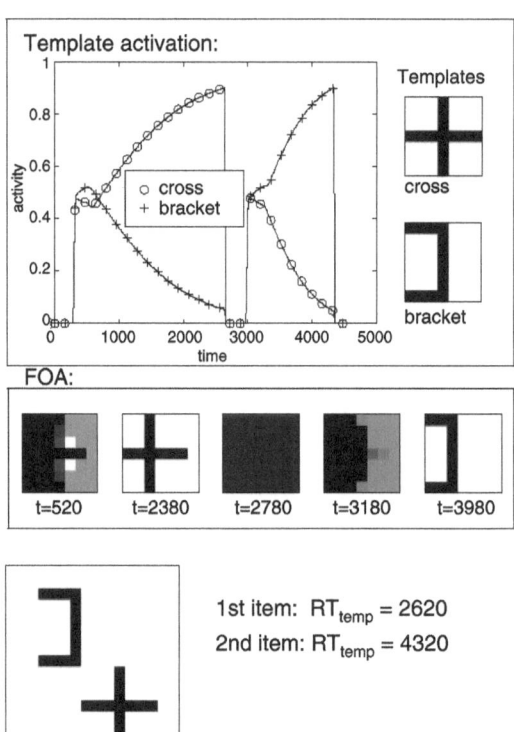

Figure 5.6 Selection and switching attention with a horizontal lesion. Note that there is now left, object-based neglect for each stimulus.

separately from neglect of parts of the field. It is not clear how object-based neglect can be explained with single-layered selection networks (e.g. Heinke, Deco, Zihl, & Humphreys, 2002a; Mozer, Halligan, & Marshall, 1997; Mozer & Sitton, 1998). In these approaches, neglect can be modelled in terms of a graded lesion, affecting attentional activation differentially from left to right across the map (e.g., with the lesion being more severe on the left). Now, within such a model, effects of within-object neglect can be mimicked if there are graded lesions across the attentional network, so that there is always a bias for one side of an object over another irrespective of the object's position on the retina (e.g., Mozer et al., 1997). Competition between units in the attentional network will then ensure that one part of the object gets privileged access to stored knowledge over the other. This can simulate forms of object-based neglect in which there is neglect of left-side features in objects presented within the right field. However, the same graded lesion that produces this object-based effect would also lead to reduced activation of an object placed in the left field, and, indeed, the activation of the right-side features of such an object may be less than the activation of the left-side features of an object in the right field. Perception of the left-field object should be poor. This is because neglect of parts of an object is intrinsically intertwined with neglect of parts of the visual field.

3 SAIM with rate code neurons for natural images

So far, SAIM has only processed black-and-white input images. This limitation questions the validity of SAIM as a general model for the human visual system. The aim of this section is to demonstrate that SAIM is capable of processing natural images by adding an appropriate feature extraction while maintaining the central concepts of the model, such as filtering images through a focus of attention and using competitive interactions between stimuli to generate selection and recognition. As these elements are essential for SAIM's earlier successes, it is likely that the new version presented here will still model the effects captured by previous versions.

3.1 Overview of the major extensions

First, SAIM was extended with a biologically plausible feature extraction, a Gabor filtering and a HSV (hue-saturation-value) colour space. The HSV colour space mimics human colour perception (Gonzalez & Woods, 2002), and Gabor filters are generally seen as a good approximation of receptive fields in V1 (Daugman, 1988). Consequently, FOA pixels and template pixels are represented by feature vectors instead of 0s and 1s in the black-and-white version. To implement a fitting matching metric in the knowledge network, the scalar product was replaced by the Euclidian distance. Furthermore, instead of connecting adjacent units in separate layers (see Figure 5.2), the excitatory connections in the selection network feed into every fifth unit (see also Equation 19). As a consequence, the FOA represents only every fifth pixel of the selected area in the input image. This subsampling was implemented for simple practical reasons, since the objects used here spanned around 30 pixels and a FOA of this size would have led to impractical computer time. However, this change also highlights the flexibility of SAIM's architecture. Possibly, in future, a similar mechanism will be used to implement size-invariance.

The last major change concerned the encoding of object information in SAIM's templates. Previous versions of SAIM used square object templates, which obviously are not suitable for storing information about natural objects. Therefore, we introduced a code for "background" into SAIM's templates, marking pixels that are not part of an object (for examples, see Figure 5.8). Background pixels are encoded by the zero vector. The zero vector can be distinguished from object pixels, because feature values are scaled to values larger than zero. In the top-down projection this difference between background pixels and feature values leads to a suppression of activation in the corresponding layers of the selection network, thereby allowing the selection process to separate object pixels from background pixels in the input image. Again, this new feature highlights the flexibility of SAIM's approach to selection and object identification, especially that of the multi-layered structure of the selection network.

3.2 Results

3.2.1 Images and templates

Figure 5.7 shows examples of the pictures used in the simulations. The pictures were used by one of the authors in earlier work (e.g., Heinke & Gross, 1993). Two objects of similar size were placed by hand onto different backgrounds at different locations. Even though an effort was made to keep the orientation, lighting, colour and size of the objects constant, as can be seen from the examples, variations occurred and images exhibited natural noise.

Figure 5.7 Examples of the natural colour images used in this chapter. [A colour version of this figure can be viewed at www.cognitiveneurosciencearena. com/brain-scans]

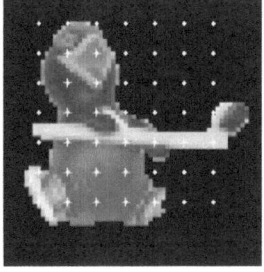 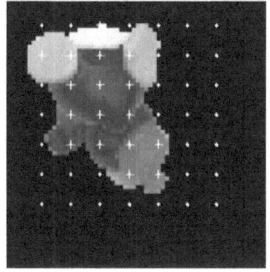

Figure 5.8 Templates: Object 1 (left) and Object 2 (right). The crosses mark locations in the object that are used to extract feature values for the templates. The dots are "background pixels". [A colour version of this figure can be viewed at www.cognitiveneurosciencearena.com/brain-scans]

The aim of this chapter is to show that, in principle, SAIM is capable of simulating results of visual search tasks with natural images, so this limitation is not crucial to testing the proof of principle.

Figure 5.8 shows the two templates (Object 1 and Object 2). Figure 5.9 shows two examples of simulation results. The examples demonstrate that, in principle, the new version of SAIM is capable of processing natural

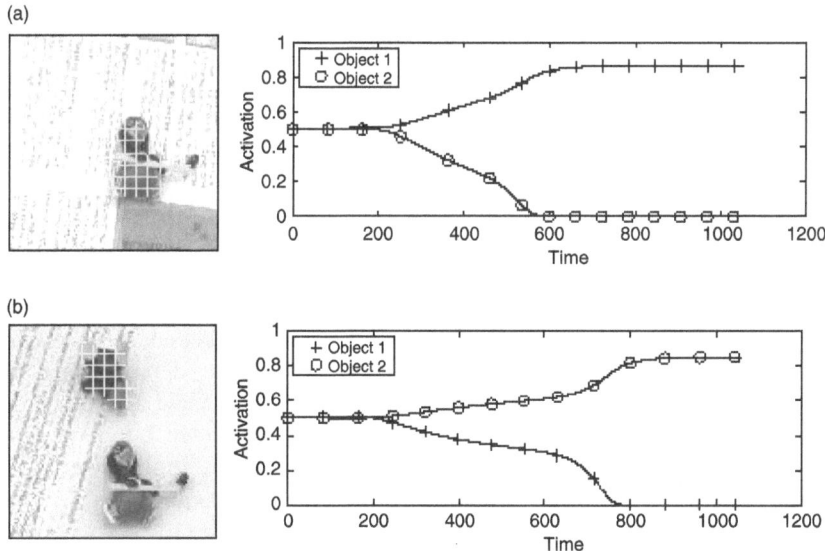

Figure 5.9 Two examples of simulation results. The crosses in the left images indicate the image locations SAIM selected. The plots on the right show the time course of the activation of the templates in the knowledge network. The time scale is arbitrary but can be interpreted as milliseconds. [A colour version of this figure can be viewed at www.cognitiveneurosciencearena. com/brain-scans]

images. In Figure 5.9a, SAIM successfully selected Object 1 from a textured background. Figure 5.9b shows that for a scene with two known objects in front of a textured background, SAIM successfully selected one of the two objects, Object 2. SAIM selected Object 2, because it matched better the corresponding template than the template of Object 1. In both simulations, SAIM's knowledge network correctly identified the selected object. Figure 5.9a also illustrates that SAIM appears to be robust against variations in these scenes, as Object 1 is slightly tilted to the right and SAIM still successfully identifies the object. This is due to the subsampling introduced in the selection network. For instance, the template grid points are still positioned on the (yellow) bat matching its colour, even though the exact positions in the original template were slightly different.

4   SAIM with spiking neurons (SP-SAIM)

So far SAIM has been based on a rate code neuron model. However, in order to move towards closing the gap between macroscopic behaviour and neurophysiology, it is necessary to employ more neurophysiological plausible neuron models (e.g., integrate-and-fire neurons). Moreover, theoretical considerations suggest that the rate code cannot be utilized in human information processing. For instance, in order to process information that a neuron receives from another neuron, single spikes need to be integrated to determine the rate code. However, such temporal integration would make information processing incredibly slow, not matching human speed of processing. Therefore, it seems more plausible that a single spike is more important than just contributing to an average neural activity (for a detailed discussion, see Thorpe, chapter 2, this volume).

In SAIM's top-down approach to modelling, a problem with spiking neurons is that it is difficult to design a predefined network behaviour in a direct fashion. Therefore, Deco and Rolls (2005) developed a mathematical framework that allows one to transform a network of rate-coded neurons (formulated in a mean-field approach) to a network with spiking neurons. Crucially, this transformation preserves the macroscopic behaviour defined by attractor states of the network. Several papers have demonstrated the success of this approach (e.g., Deco & Rolls, 2005; Mavritsaki, Heinke, Deco, & Humphreys, 2006) (see also Deco & Rolls, chapter 3, this volume). However, there are problems with this approach when applied to SAIM. First, to be mathematically valid, the new spiking level requires a large number of integrate-and-fire neurons. This is a consequence of the mean-field approach, in which one unit represents a large pool of spiking neurons. Hence, in the case of large models, such as SAIM, the computer time becomes infeasibly large. Second, so far, their approach is limited to additive spiking neurons, and units like the sigma-pi unit cannot be transformed. Therefore, this chapter suggests an alternative approach. The first step in this approach is to replace each rate code neuron with a single spiking neuron.

Hence, it compromises between the practical issues with using spiking neurons and the modelling necessity of taking into account spikes as an important part of neural information processing. In the second step, elementary functions in SAIM, such as WTA and sigma-pi units, were designed to resemble their rate code behaviour with spiking inputs and outputs. The third step connected these modules to implement a spiking version of SAIM. Details of these three steps will be given in the following sections.

4.1 Spiking neuron

The spiking version of SAIM is based on a simple integrate-and-fire model, the SRM_0 model (e.g., Gerstner & Kistler, 2002). In SRM_0 the cell potential of a neuron i is described as:

$$s_i(t) = s_{syn} s_i^{syn}(t) + s_{ref} s_i^{ref}(t). \tag{1}$$

s_i^{syn} is the input from other neurons, and s_i^{ref} takes into account the refractory process from previously produced spikes:

$$s_i^{syn}(t) = \sum_{j=1}^{n} w_{ij} \sum_{f=1}^{F} e(t - t_j^f), \tag{2}$$

where n is the number of neurons that are connected to the neuron i, and w_{ij} weights the influence of each neuron j on the neuron i. The function $\varepsilon(t)$ describes the synaptical transmission between the neurons:

$$\varepsilon(t) = \frac{1}{\tau} \cdot e^{-t/\tau} \tag{3}$$

For the refractory process, the following mathematical description was chosen:

$$s_i^{ref}(t) = \sum_{f=1}^{F} h(t - t_i^f), \tag{4}$$

with

$$h(t) = -h_0 \times e^{-t/t_{ref}}. \tag{5}$$

The relationship between the probability of a neuron producing a spike and the cell potential is modelled by the following equation:

$$p_i^{fire} = \frac{1}{1 + e^{-\sigma_{fire} \cdot (s_i(t) - \theta)}}. \tag{6}$$

The temporal convolutions describing the synaptic transmission and the refractory process were implemented with an Euler approximation.

4.2 Winner-take-all

The rate code SAIM utilized a WTA network in the knowledge network and the selection network that was based on an energy function suggested by Mjolsness and Garrett (1990):

$$E_{WTA}(y_i) = a \times \left(\sum_i y_i - 1 \right)^2 - \sum_i y_i \times I_i. \tag{7}$$

Here I_i are the inputs and y_i the outputs of the units. This energy function is minimal when all y_is are zero except one y_i, and the corresponding I_i has the maximal value of all I_is. To find the minimum, a gradient descent procedure can be used (Hopfield, 1984):

$$\tau \times \dot{x}_i = -\frac{\partial E(y_i)}{\partial y_i}. \tag{8}$$

The factor τ is antiproportional to the speed of descent.

In the Hopfield approach, x_i and y_i are linked together by the sigmoid function:

$$y_i = \frac{1}{1 + e^{-m \times (x_i - s)}},$$

and the energy function includes a leaky integrator, so that the descent turns into:

$$\tau \times \dot{x}_i = -x_i - \frac{\partial E(y_i)}{\partial y_i}. \tag{9}$$

Hence the WTA network is described by the following set of differential equations:

$$\dot{x}_i = -x_m - a \times \left(\sum_i^K y_i - 1 \right) + I_m. \tag{10}$$

Here K is the number of neurons in the WTA layer. The corresponding topology is shown in Figure 5.10. It is important to note that the term $\sum_i^K y_i - 1$ is a fast inhibitory neuron feeding back the sum of output activation to

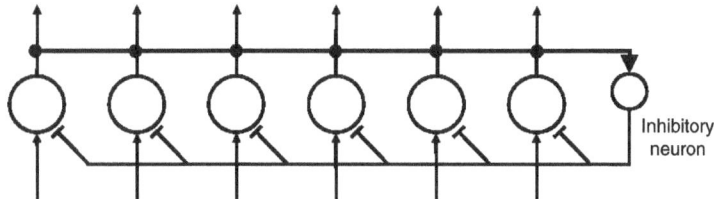

Figure 5.10 Topology of the WTA network. The large circles are SRM_0 neurons. The inhibitory neuron is a "fast" neuron that sums up the output from the spiking neurons.

each neuron. The −1 ensures that only one winner is produced by the network.

To employ spiking neurons in the WTA, the leaky integrator neurons were replaced by the SRM_0 neurons. The fast inhibitory neuron in the spiking version summed the spikes at each time step and returned this sum as input to the SRM_0 neurons. The −1 term was implemented with a neuron spiking at the maximal possible frequency, as defined by the refractory time. Since this maximal frequency is equivalent to the value 1 in the rate code domain, this implementation is analogous in the spiking domain.

Figure 5.11 shows a successful simulation with this spiking WTA network, demonstrating that the transformation from a rate code network to a spiking network was successful.

4.3 Sigma-pi unit

The rate code version of the contents network used sigma-pi units (Rumelhart & McClelland, 1988) to map a section from the visual field into the FOA. Each sigma-pi unit is a pairwise multiplication of inputs:

$$y^{out} = \sum_{k=1}^{N} y_k^{in1} \times y_k^{in2} \tag{11}$$

To implement the multiplication in the spiking domain, every input spike train (t_i^f) is multiplied after a "synaptic transmission":

$$In_i(t) = \sum_{f=1}^{F} \frac{1}{t} \times e^{(t-t_i^f)/t}, \tag{12}$$

$$Out(t) = \prod_i In_i^{spu}(t), \tag{13}$$

whereby $Out(t)$ forms another additive term in Equation 1.

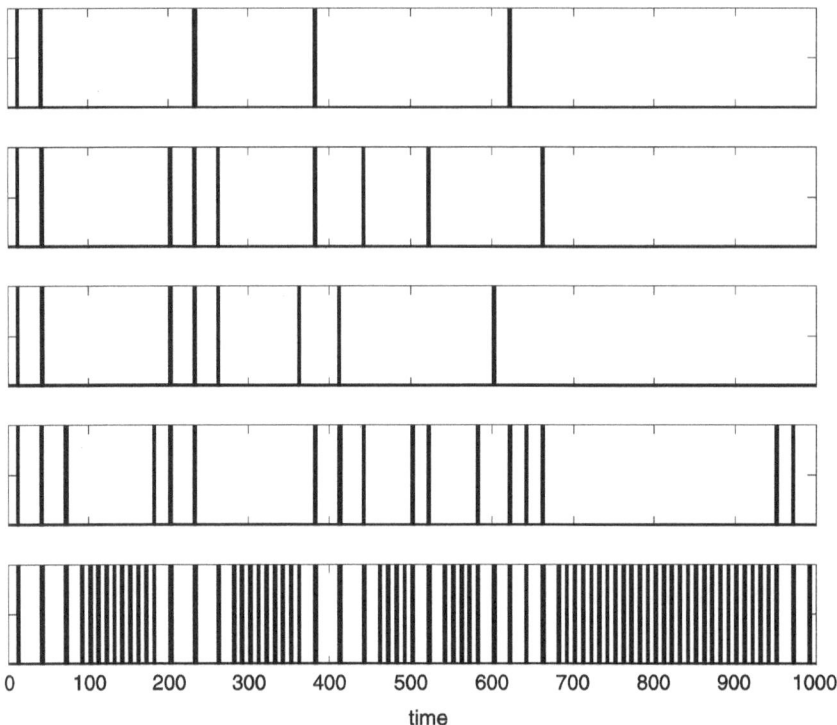

Figure 5.11 Simulation results of a spiking WTA network with 5 neurons. The bottom neuron received the highest input of all neurons. This is detected by the WTA network by settling in a state where the bottom neuron produces a very high spike frequency and all other neurons show little output activity.

To test whether this neuron approximates a sigma-pi unit, we implemented a unit with two inputs and ran simulations with all possible pairs of input frequencies. Figure 5.12 shows the resulting average output frequency for each pairing. The result shows that it is not a perfect implementation of a sigma-pi unit, but at crucial values the unit behaves like a sigma-pi: if one of the two inputs is 0, the output frequency is 0. If one of the two inputs is 1, the output frequency roughly reflects the activation of the second input.

Note that this implementation mainly resulted from practical considerations. Because a direct multiplication of two spike trains does not lead to enough resulting spikes, we decided to filter the spike trains through a "synaptic transmission" and then multiply the resulting activation. Obviously, the multiplication would now produce enough activation to be added by the cell body. However, interestingly, this implementation is compatible with the two-compartment model of neurons (for a recent summary, see Häusser

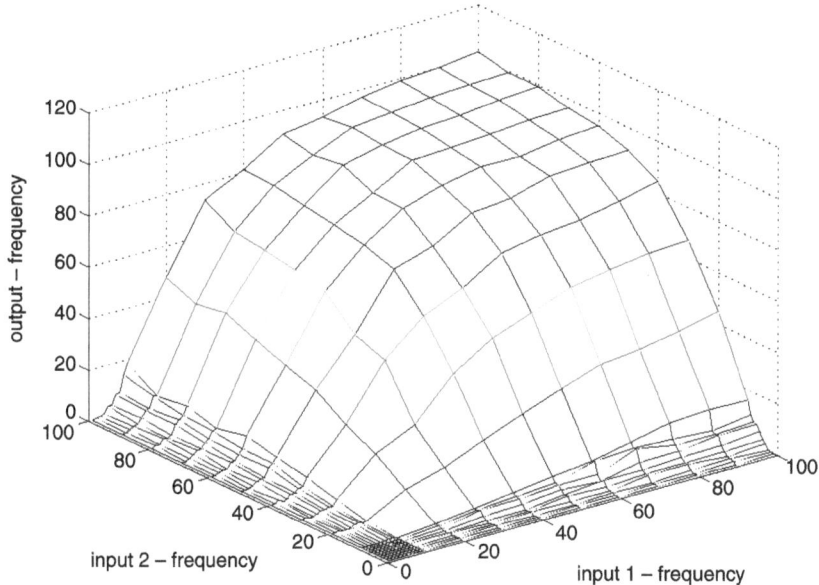

Figure 5.12 Input–output relationship of the spiking sigma-pi unit.

& Mel, 2003).[1] According to this model, a neuron can be separated into a proximal compartment consisting of soma, basal dendrites and axons, and a distal compartment accounting for distal apical dendrites. Inputs into these two compartments interact with each other before they lead to spike generations at the output of the neuron. Even though the real processes in dendritic trees are far more complex, multiplicative interactions can be considered as a good approximation of these processes (Häusser & Mel, 2003). Hence, the spiking sigma-pi unit suggested here can be seen as a good model of neural processes in dendritic trees.

4.4 Network

In the following, the equations for the rate code version of SAIM are stated, as they take up less space than the spiking version. However, the equations can be simply transformed into the spiking version by applying the rules explained for the WTA network and the sigma-pi unit.

4.4.1 Contents network

As explained earlier, SAIM uses one-to-one mapping from the retina to all locations in the FOA, mediated by the connections between the contents

---

1 We would like to thank Kevin Gurney for pointing this out.

and selection networks. This allows any arbitrary mapping to be achieved from the retina to the FOA to generate translation invariance. The following equation describes the mapping produced through the sigma-pi units in the contents network:

$$
y_{ij}^{FOA} = \sum_{k=1}^{N} \sum_{l=1}^{N} y_{kl}^{VF} \times y_{ikjl}^{SN}. \tag{14}
$$

Here, y_{ij}^{FOA} is the activation of units in the FOA, y_{kl}^{VF} the activation of units in the visual field and y_{ikjl}^{SN} the activation of units in the selection network. N is the size of the visual field. The indices k and l refer to retinal locations, and the indices i and j refer to locations in the FOA. Note that this notation is kept through all of the equations.

4.4.2 Selection network

The selection network implements three constraints to ensure a veridical mapping from input image to the FOA. To incorporate the first constraint – that units in the FOA should receive the activity of only one visual field unit – the WTA equation turns into:

$$
E_{WTA}^{(1)} (y_{ikjl}^{SN}) = \sum_{ij} \left(\sum_{k,l} y_{ikjl}^{SN} - 1 \right)^{2}, \tag{15}
$$

where the term $\left(\sum_{k,l} y_{ikjl}^{SN} - 1 \right)^{2}$ ensures that the activity of only one retinal unit is put through to FOA units (compare with Equation 7).

To incorporate the second constraint, the WTA equation turns into:

$$
E_{WTA}^{(2)} (y_{ikjl}^{SN}) = \sum_{kl} \left(\sum_{i,j} y_{ikjl}^{SN} - 1 \right)^{2}, \tag{16}
$$

where the term $\left(\sum_{i,j} y_{ikjl}^{SN} - 1 \right)^{2}$ makes sure that the activity of visual field units is mapped only once into the FOA (again, compare with Equation 7).

So far, the last term of Equation 7 ("input term") has been taken out of Equations 15 and 16. It is now reintroduced by a common input from the visual field:

$$
E_{input}(y_{ikjl}^{SN}) = -\sum_{kl} y_{ikjl}^{SN} \times y_{kl}^{VF}. \tag{17}
$$

For the neighbourhood constraint, the energy function was based on the Hopfield associative memory approach:

$$E(y_i) = -\sum_{iji'j} T_{ij} \times y_i \times y_j \tag{18}$$

The minimum of the function is determined by the matrix T_{ij}. For T_{ij}s greater than zero the corresponding y_is and y_js should either stay zero or become active in order to minimize the energy function. In the associative memory approach, T_{ij} is determined by a learning rule. Here, we chose the T_{ij} so that the selection network fulfils the neighbourhood constraint. The neighbourhood constraint is fulfilled when units in the selection network that receive input from the adjacent units in the visual field, and control adjacent units in the FOA, are active at the same time. Hence, the T_{ij} for these units in Equation 18 should be greater than zero and for all other units T_{ij} should be less than or equal to zero. This leads to the following equation:

$$E_{neighbour}\left(y_{ikjl}^{SN}\right) = -\sum_{i,j,k,l} \sum_{s=-Ls'0}^{L} \sum_{r=-Lr'0}^{L} g_{sr} \times y_{ikjl}^{SN} \times y_{i+r,k+s,j+r,l+s}^{SN}, \tag{19}$$

with g_{sr} being defined by a Gaussian function:

$$g_{sr} = \frac{1}{A} \cdot e^{-\frac{s^2+r^2}{\sigma^2}}. \tag{20}$$

where A was set, so that the sum over all g_{sr} is 1.

The effect of g_{sr} is equivalent to the effect of T_{ij} in Equation 18. However, due to the properties of g_{sr}, energy is minimized not only by the activation of immediate neighbours ($s = -1$, $r = -1$, $s = 1$, $r = -1$) but also by the activation of units in the selection network controlling units in FOA that receive input from more widely space pixels in the visual field (all other values of r and s). Note that in SAIM for natural images, the term $y_{i+r,k+s,j+r,l+s}^{SN}$ is replaced by $y_{i+r,k+s,j+5\cdot r,l+5\cdot s}^{SN}$ to implement the mapping of every fifth pixel.

4.4.3 Knowledge network

In order to introduce effects of stored knowledge into SAIM, a simple template matching approach was used. Here, the match between templates and contents of the FOA was determined using a scalar product as a similarity measure:

$$I_m^{temp} = \sum_{l=1}^{M} \sum_{j=1}^{M} y_{ij}^{FOA} \times w_{ij}^m,$$ (21)

where the w_{ij}^ms are the templates and M is the size of the FOA. A WTA is used to detect the best-matching template. The same energy function as in Equation 7 was used with I_m^{temp} as input:

$$E_{knowledge}(y_m^{KN}) = \frac{a^{KN}}{2} \times \left(\sum_m y_m^{KN} - 1\right)^2 - \sum_m^K y_m^{KN} \times I_m^{temp}$$ (22)

Here, K is the number of template units.

4.4.4 The complete model

The complete energy function of SAIM, which satisfies all constraints, is simply the sum of the different energy functions:

$$E_{total}(y_m^{KN}, y_{ikjl}^{SN}) = a_1 \cdot E_{WTA}^{(1)}(y_{ikjl}^{SN}) + a_2 \cdot E_{WTA}^{(2)}(y_{ikjl}^{SN}) + b_2 \cdot E_{input}(y_{ikjl}^{SN})$$

$$+ b_1 \cdot E_{neighbour}(y_{ikjl}^{SN}) + b_3 \cdot E_{knowledge}(y_m^{KN}, y_{ikjl}^{SN})$$ (23)

The coefficients of the different energy functions weight the different constraints against each other.

4.5 Simulation results and discussion

The aim of the simulations was to test whether the new spiking version of SAIM behaved similarly to the old version. So we ran two simulations with the same displays: a single cross, and a cross together with a 2 (see Figure 5.13). The results showed that, in principle, the spiking SAIM produces the same results. In the one-object condition, SAIM successfully identified the object. In the two-objects condition, it selected and identified one of the two objects. (Note that we did not implement the location map in the spiking version, as it would not add anything to the overall aim of this work.) Interestingly, SAIM's reaction time also increased from the one-object condition to the two-objects condition, indicating that the new version can also simulate attentional effects.

However, there is an interesting difference between the two models. SAIM, as expressed by the activation in the FOA, reached the stable state (energy minimum) on a different trajectory. In the rate code version, the selected object slowly emerged in the FOA in a coherent fashion. In contrast, the spiking version only showed the complete object when it reached the stable state; during the selection period, parts of objects appeared and disappeared.

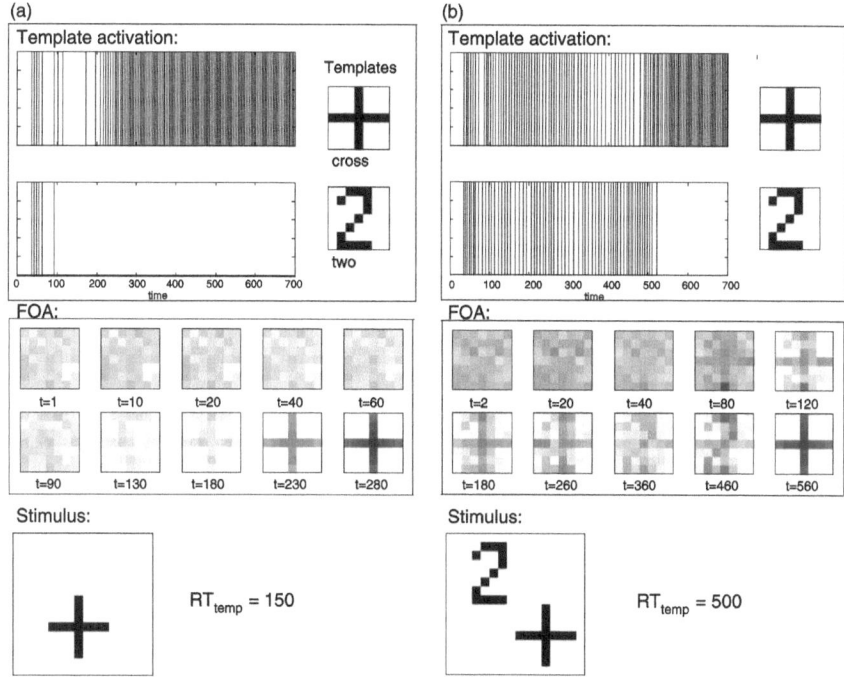

Figure 5.13 Simulation results with spiking SAIM. This version of SAIM replicates qualitatively the behaviour of the rate code version (see Figure 5.3). However, interestingly, the trajectories of the two systems differ. Whereas the rate code version shows a gradual appearance of objects in the FOA, the spiking SAIM exhibits a "chaotic" behaviour in which parts of objects occasionally appear and disappear (the time course of the FOA activation was computed by counting the number of spikes in a sighting window).

Hence, the behaviour is not simply more noisy as might have been expected from SRM_0-neuron's behaviour, but it exhibits a different qualitative behaviour, which may be loosely characterized as "chaotic" (not in the mathematical sense).

This result also reveals an interesting parallel to Deco and Rolls's (2005) approach. In their approach, a one-to-one relationship between rate code level and spiking level exists only for the attractors of the system. In order to determine the trajectories into these attractors for the spiking level, the spiking level has to be simulated. Deco and Rolls (2005) see this difference as an important reason for making the transition from the rate-coded level to the spiking level, as it constitutes a more veridical model behaviour. The same relationship between rate code SAIM and spiking SAIM also became apparent in the simulations. The energy functions from the rate code version defined the attractors in the spiking SAIM. However, the trajectories of the two systems are different. Hence, our pragmatic approach shows similar properties to the more mathematical approach favoured by Deco and Rolls (2005) (see also Deco & Rolls, chapter 3, this volume).

5 General discussion

This chapter presented two extensions of the selective attention for identification model (SAIM) to demonstrate that SAIM represents a good framework for closing the gap between macroscopic behaviour and neurophysiology. With the first extension, we showed that SAIM could identify objects in natural colour images. This is important since a good approach for closing the gap needs to be able to deal with ecologically valid stimuli such as natural images, thus tackling one side of the gap, behaviour. With the second extension of SAIM, we moved SAIM closer towards simulating data on the other side of the gap, neurophysiology. This was done by replacing the rate code neurons with spiking neurons, at the same time maintaining SAIM's ability to select and identify objects in multiple-object scenes.

So what are the pillars for SAIM's successes? To begin with, the competitive and cooperative processes in the selection and identification stage allow SAIM to deal successfully with multiple-object scenes. These types of processes are also responsible for SAIM's ability to mimic a broad variety of attentional behaviour and attentional disorders (e.g., Heinke & Humphreys, 2003). This has also been shown by other models (e.g., Deco & Zihl, 2001; Heinke et al., 2002a; Mozer & Sitton, 1998). Especially for attentional disorders such as visual neglect, all models show that an unbalanced competitive process due to lesions can lead to behaviour found in stroke patients (for a discussion, see Heinke & Humphreys, 2005b). However, interestingly and in contrast to other models, SAIM's results suggest that, in order to achieve a fuller account of attentional deficits, the selection processes should be implemented in a multilayered architecture. In such a multilayered architecture different parts of the FOA are controlled by different layers in a selection network. In this chapter we demonstrated this with simulation results of object-centred neglect. Heinke and Humphreys (2003) presented further results mimicking extinction, field-dependent neglect, within- and between-objects neglect, etc. In addition, we showed in the current chapter that the multilayered approach can implement selection of arbitrarily shaped objects in natural images. Here, specific layers can be selectively suppressed by the top-down modulation from the knowledge network. Moreover, we demonstrated that the multilayered selection network can realize subsampling of selected objects, thereby dealing with the high amount of data occurring in natural images. At present the neurological basis for the selection network is unclear. Heinke and Humphreys (2003) argued that a possible corresponding structure could be the pulvinar (for a supporting view, see Shipp, 2004). However, further research needs to be conducted to link the selection network with the neurophysiological level. In this chapter we aimed to advance SAIM's validity with respect to the neurophysiological level by replacing rate code neurons with spiking neurons.

Another crucial pillar for SAIM's success is the soft constraint satisfaction using energy minimization. In this approach, first, constraints are motivated by the intended target behaviour – for example, recognizing objects in multiple-object scenes. These constraints define permissible attractor states of the network. In turn, an energy function is designed whose minima are these permissible activations. Finally, in order to find the minima, a gradient descent procedure is applied that defines connections between rate-coded neurons (Hopfield & Tank, 1985). Now when SAIM was extended from processing simple black-and-white-pixel images to colour images, the energy minimization approach provided an ideal framework for this advancement. On a mathematical level, moving from black-and-white-pixel images to colour images is expressed by using vectors instead of scalar values as inputs. It turned out that simply replacing the scalar values with vectors in the energy function was sufficient to achieve this transition. However, the template matching needed to be modified, since the Euclidian metric performs a better comparison for vectorial templates than does the scalar product originally used. This also underlines the strength of the energy function approach, as the new network topology for the new template matching can be simply derived from the gradient descent.

The energy function approach also provided a crucial framework for the spiking version of SAIM. Here, the topology defined by the energy function was maintained, but instead of employing rate code neurons, fire-and-integrate neurons were used. The simulations demonstrated that this simple pragmatic approach was successful. From a mathematical viewpoint it is unclear at the moment why this approach worked, since, strictly speaking, the energy function approach only applies to rate code neurons and not to spiking neurons. However, the synaptic transmission, with its temporal integration, may lead to an approximation of a rate code behaviour in the cell body. Hence, it is possible that, under the parameter setting chosen here, cell potentials define the energy of the model and that spike signals between the units are not crucial for the energy minimization in the model. On the other hand, the simulation results show that the trajectories (gradient descent) into the attractors are qualitatively different between rate-coded SAIM and spiking SAIM, suggesting that spiking SAIM does not simply approximate rate code behaviour. To explore this issue further, future work needs to establish a mathematical framework for the type of transition between rate code level to spiking level suggested here, along similar lines developed by Deco and Rolls (2005; see also Deco & Rolls, chapter 3, this volume) for the mean-field approach. Apart from embedding our method in a mathematical framework, the simulation results also need to be verified through electrophysiological data in future work.

However, at the current stage, it is clear that this is a promising approach as, on the one hand, it allows us to set up complex network models that are able to deal with behaviourally valid stimuli, and, on the other, spiking neurons can be integrated to construct a more neurologically plausible model. Hence, our approach represents a good framework for linking human behaviour to the neurophysiological level, and vice versa.

6 Acknowledgement

This work was supported by grants from the European Union, the BBSRC and the EPSRC (UK) to Dietmar Heinke. The authors would like to thank Glyn Humphreys and Gustavo Deco for invaluable discussions.

7 References

Amari, S. I. (1977). Dynamics of pattern formation in lateral-inhibition type neural fields. *Biological Cybernetics, 27*, 77–87.

Backhaus, A., Heinke, D., & Humphreys, G. W. (2005). Contextual learning in the selective attention for identification model (CL-SAIM): Modeling contextual cueing in visual search tasks. In: *Proceedings of the 2005 IEEE Computer Society conference on computer vision and pattern recognitione (CVPR'05): 3rd international workshop on attention and performance in computer vision (WAPCV'05), San Diego, CA*. Washington, DC: IEEE Computer Society.

Daugman, J. G. (1988). Complete discrete 2-D Gabor transforms by neural networks for image analysis and compression. *IEEE Transaction on Acoustics, Speech, and Signal Processing, 36*(7), 1169–1179.

Deco, G., & Rolls, E. T. (2005). Neurodynamics of biased competition and cooperation for attention: A model with spiking neurons. *Journal of Neurophysiology, 94*, 295–313.

Deco, G., & Zihl, J. (2001). Top-down selective visual attention: A neurodynamical approach. *Visual Cognition, 8*(1), 119–140.

Driver, J., Baylis, G. C., & Rafal, R. D. (1992). Preserved figure–ground segmentation in visual matching. *Nature, 360*, 73–75.

Duncan, J. (1980). The locus of interference in the perception of simultaneous stimuli. *Psychological Review, 87*, 272–300.

Gainotti, G., d'Erme, P., Monteleone, D., & Silveri, M. C. (1986). Mechanisms of unilateral spatial neglect in relation to laterality of cerebral lesions. *Brain, 109*, 599–612.

Gerstner, W., & Kistler, W. M. (2002). *Spiking neuron models: Single neurons, populations, plasticity*. New York: Cambridge University Press.

Gonzalez, R., & Woods, R. E. (2002). *Digital image processing*. Upper Saddle River, NJ: Prentice Hall.

Häusser, M., & Mel, B. (2003). Dendrites: Bug or feature? *Current Opinion in Neurobiology, 13*(3), 372–383.

Heinke, D., Backhaus, A., Sun, Y., & Humphreys, G. W. (2008). The selective attention for identification model (SAIM): Simulating visual search in natural colour images. In L. Paletta & E. Rome (Eds.), *Attention in cognitive systems* (pp. 143–172). Berlin: Springer-Verlag, LNAI 4840.

Heinke, D., Deco, G., Zihl, J., & Humphreys, G. W. (2002a). A computational neuroscience account of visual neglect. *Neurocomputing, 44*, 811–816.

Heinke, D., & Gross, H. M. (1993). A simple selforganizing neural network architecture for selective visual attention. In *Proceedings of the international conference on artificial neural network – icann'93* (pp. 63–66). Amsterdam: Springer Verlag.

Heinke, D., & Humphreys, G. W. (1997). SAIM: A model of visual attention and neglect. In *Proceedings of the 7th international conference on artificial neural networks – icann'97* (pp. 913–918). Lausanne, Switzerland: Springer Verlag.

Heinke, D., & Humphreys, G. W. (2003). Attention, spatial representation and visual neglect: Simulating emergent attention and spatial memory in the selective attention for identification model (SAIM). *Psychological Review, 110*(1), 29–87.

Heinke, D., & Humphreys, G. W. (2005a). Computational models of visual selective attention: A review. In G. Houghton (Ed.), *Connectionist models in psychology* (pp. 273–312). Hove, UK: Psychology Press.

Heinke, D., & Humphreys, G. W. (2005b). Selective attention for identification model: Simulating visual neglect. *Computer Vision and Image Understanding, 100* (1–2), 172–197.

Heinke, D., Humphreys, G. W., & diVirgilo, G. (2002b). Modeling visual search experiments: Selective attention for identification model (SAIM). *Neurocomputing, 44*, 817–822.

Heinke, D., Humphreys, G. W., & Tweed, C. L. (2006). Top-down guidance of visual search: A computational account. *Visual Cognition, 14*(4/5/6/7/8), 985–1005.

Heinke, D., Sun, Y. R., & Humphreys, G. W. (2005). Modeling grouping through interactions between top-down and bottom-up processes: The grouping and selective attention for identification model (G-SAIM). In L. Paletta, J. K. Tsotsos, E. Rome, & G. W. Humphreys (Eds.), *Attention and performance in computational vision, 2nd international workshop (WAPCV'04), Prague, Czech Republic, May 15, 2004* (pp. 148–158). Berlin: Springer-Verlag.

Hopfield, J. J. (1984). Neurons with graded response have collective computational properties like those of two-state neurons. *Proceedings of the National Academy of Sciences, 81*, 3088–3092.

Hopfield, J. J., & Tank, D. (1985). "Neural" computation of decisions in optimization problems. *Biological Cybernetics, 52*, 141–152.

Humphreys, G. W., & Heinke, D. (1998). Spatial representation and selection in the brain: Neuropsychological and computational constraints. *Visual Cognition, 5*(1/2), 9–7.

Marshall, J. C., & Halligan, P. W. (1994). The yin and yang of visuo-spatial neglect: A case study. *Neuropsychologia, 32*, 1037–1057.

Mavritsaki, E., Heinke, D., Deco, G., & Humphreys, G. W. (2006). A computational model of visual marking using an interconnected network of spiking neurons: The spiking search over time & space model (sSoTS). *Journal of Physiology, Paris, 100-* (1/2/3), 110–124.

Mjolsness, E., & Garrett, C. (1990). Algebraic transformations of objective functions. *Neural Networks, 3*, 651–669.

Mozer, M. C., Halligan, P. W., & Marshall, J. C. (1997). The end of the line for a brain-damaged model of unilateral neglect. *Journal of Cognitive Neuroscience, 9*(2), 171–190.

Mozer, M. C., & Sitton, M. (1998). Computational modeling of spatial attention. In H. Pashler (Ed.), *Attention* (pp. 341–393). London: Psychology Press.

Posner, M. I., & Cohen, Y. (1984). Components of visual orienting. In H. Bouma & D. G. Bouwhuis (Eds.), *Attention and performance X: Control of language processes* (pp. 531–556). Hove, UK: Lawrence Erlbaum Associates.

Rumelhart, D. E., & McClelland, J. L. (1988). *Parallel distributed processing. Explorations in the microstructure of cognition: Vol. 1. Foundations.* Cambridge, MA: MIT Press.

Shipp, S. (2004). The brain circuitry of attention. *Trends in Cognitive Sciences, 8*(5), 223–229.

6 Computational models in neuroscience

From membranes to robots

Kevin N. Gurney

The basal ganglia are a set of subcortical nuclei that are believed to play a critical role in solving the problem of action selection – the competition between brain systems requesting behavioural expression through limited motor resources. Thus, the basal ganglia are hypothesised to play the role of a central "switch" that examines the salience, or urgency, of action requests and allows some actions to proceed while others are prohibited. We present a series of models of the basal ganglia (and associated circuits) that demonstrate how they might help solve the computational problem of action selection. The models range from conductance-based models of individual neurons, through spiking neuronal networks, to systems-level models with rate codes and embodied (robotic) models exhibiting behaviour. At each level of description, we seek to test the computational hypothesis that the basal ganglia implement action selection. This may be done by "mining" the known biology of these circuits for mechanisms that could perform selection. This bottom-up approach to modelling is contrasted with a complementary top-down approach that seeks to "map" abstract neuronal mechanisms for a given computation onto the biology.

1 Introduction

This chapter describes a principled approach in computational neuroscience for building models at multiple levels of description – from those constrained by the physiology of membranes, to those that exhibit behaviour. An important guiding principle in our approach is to ensure that, no matter what level one is working at, there is a computational function ascribed to the system under study (Marr, 1982). This then needs to be augmented with other kinds of analyses such as algorithms and mechanisms, and there are different (but complementary) ways of discovering what these might be. One approach works from the "bottom up" and seeks to discover mechanisms that perform the target computation directly from the biological data; here, algorithm discovery is a secondary goal. In contrast, we may work "top-down", as it were, and impose an algorithm *ab initio*. We must then seek to discover whether this maps onto the biological substrate via biologically plausible supporting mechanisms.

This entire programme is illustrated using examples from work by the author's group on models of the basal ganglia and related brain circuits. In the context of our methodological prescription, we hypothesise that the basal ganglia play a critical role in solving the computational problem of action selection. However, the reader does not have to share our interest in the basal ganglia to appreciate the methodological principles we propose, for these have quite general applicability.

1.1 Levels of description and computational analysis

The title of the workshop giving rise to this volume alludes to a "closing of the gap" between physiology and behaviour. The need for this process is a direct consequence of the multiple levels of structural description that may be used to describe the brain (Churchland & Sejnowski, 1992) (see Figure 6.1). Each level is each characterised by a spatial scale and functional description and lends itself naturally to certain well-known modelling tools.

At Level 1 are intracellular processes that describe, for example, the second messenger cascades (of chemical reactions) initiated by neuromodulators like dopamine (DA). Modelling at this level is the domain of what is now known as computational systems biology (Kitano, 2002).

Levels 2 and 3 deal with individual neurons. At Level 2, patches of neural membrane or single neurons are modelled using, typically, the *Hodgkin–Huxley formalism* (see, e.g., Koch, 1999). This describes the dynamics of the membrane potential V_m, as determined by a multiplicity of ionic currents $I(V_m, t)$ whose properties depend, in turn, on V_m. This reciprocal causal relationship leads to a rich repertoire of neural dynamics. At the next level, we deal only with whole neurons and are more interested in neural firing patterns. Models are couched in a simplified or *reduced* form that capture, in a phenomenological way, essential neurodynamics without recourse to a

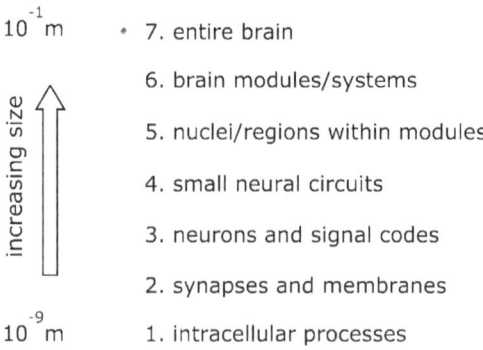

10^{-1} m • 7. entire brain

6. brain modules/systems

increasing size 5. nuclei/regions within modules

4. small neural circuits

3. neurons and signal codes

2. synapses and membranes

10^{-9} m 1. intracellular processes

Figure 6.1 Levels of structural analysis in the brain.

detailed biophysical description of the membrane currents $I(V_m, t)$ (Izhikevich, 2007). Typically, these models have a pair of coupled ordinary differential equations (ODEs) describing V_m and one other variable.

At Level 4, we consider microcircuits within a single brain nucleus. The example *par excellence* here is the cortical microcircuit that extends over six neuronal layers. Models may use the reduced forms of Level 3 or use even simpler leaky integrate-and-fire (LIF) neuron models that make use of only a single ODE for V_m.

At Level 5, microcircuits are agglomerated into brain nuclei and, beyond that, into entire functional modules. Examples at Levels 5 and 6 include individual hippocampal regions (CA1–CA3) and the hippocampal formation, respectively; an individual basal ganglia nucleus (described later in this chapter); and the basal ganglia *in toto*. Models at these levels typically use rate-coded neurons (leaky integrators) but may also use LIF spiking neurons.

Level 7 is the entire brain (or CNS), and while not usually the subject of detailed models *per se*, we may wish to establish sufficient functionality to exhibit behaviour in an embodied (robotic) model. This will require implementing complete sensorimotor competences, thereby endowing the robot with whole-brain-like capabilities. The rationale for this approach and a principled methodology therein, are described in section 5

In practice, models may draw on features from adjacent levels, and there is nothing special about the division into seven levels described here. However, irrespective of the particular stratification used, how should one go about developing a model of a brain system? Our answer is based on the proposal by David Marr (1982) that brain functions address the solution of computational problems. Furthermore, these problems decompose (at least in the first instance) into three levels of analysis. At the top level is a description of "what" is being computed and "why" the computational task. The next level describes "how" the computation is carried out in terms of an algorithm and any associated representations. The final level specifies "where" the computation is carried out – which brain system implements the computation.

Examples of this scheme in terms of neuronal and brain systems will be given in the context of our basal ganglia models described later. However, Marr's original example in his 1982 book *Vision* provides a very clear illustration. Consider the computation of a bill in a supermarket with a cash register. In answer to the top-level question of "what" is being computed, it is the arithmetical operation of addition. As to "why" this is being done, it is simply that the laws of addition reflect or model the way we should accumulate prices together when shopping; it is incorrect, for example, to multiply them together. Algorithmically, we use the normal procedure (add digits representing the same order of magnitude and "carry" any overflow if needed). Furthermore, in cash registers, this will be done in the decimal representation rather than binary (normally encountered in machine arithmetic) because rounding errors are incurred when converting between the everyday (decimal) representation of currency and binary. As for the implementation, this occurs using logic gates made out of silicon, silicon-oxide and metal. Note that choices at different levels are, in principle, independent of each other. For example, we could have chosen to use a binary representation, and alternative implementations might make use of mechanical machines or pencil and paper. The importance of discovering good representations for solving the problem is crucial: the ancient Romans failed to develop a positional number system and so struggled to fully develop arithmetic.

In the context of computational neuroscience, we will see that a more appropriate framework makes use of an additional level of analysis that deals with "mechanism" as distinct from "algorithm"; this idea is fleshed out in section 6. In any case, whether we use three of four levels of analysis, how does an *analytic* framework of this kind mesh with the multilevel *structural* hierarchy in Figure 6.1? One possibility is to suppose that computations are defined at a high (or systems) level dealing with modules, say, and that implementation occurs at the neural and molecular levels. Algorithms (and any other intermediate levels of analysis) then sit somewhere in between. In contrast we, and others (Churchland & Sejnowski, 1992), believe it makes more sense to think of each level of structural description as a potential seat of computational function. It may then be possible to detail other analytic levels at the same structural level or, at most, one or two structural levels below that defining the computational function.

A final issue concerns the methods by which one navigates the analytic computational framework in practice. As described in section 6, it is not necessary to work top down from computation to implementation, and bottom-up approaches are perfectly valid. Nevertheless, the common anchor point of any approach is the notion that we are testing computational hypotheses about brain function. We now look at how this might work in one particular brain system – the basal ganglia – with respect to a functional hypothesis concerning action selection.

2 Action selection and the basal ganglia

2.1 *Action selection*

All animals in natural environments are continually bombarded with a plethora of external (sensory) and internal (cognitive and homeostatic) information. The animal must use this information to guide its behaviour in an integrated and goal-directed way, but it has only limited motor resources (one set of limbs, head–eye system, etc.) with which to enact this behaviour. Often, different information streams are demanding different behaviours with the same resources (e.g., to flee from, or fight with, a preda-tor). This situation, encountered moment-to-moment by the animal, defines the problem of action selection: the resolution of conflicts between independent brain systems that are in competition for behavioural expression via limited motor resources. Stated informally, it is the problem of how we decide "what to do next".

The problem of action selection appears in many forms across various disciplines. Within ethological studies of animal behaviour, it is often referred to as the problem of "behavioural switching" or "decision-making" (McFarland, 1989). The latter terminology is also used by psychologists in perceptual tasks (Bogacz, Brown, Moehlis, Holmes, & Cohen, 2006) – a theme taken up elsewhere in this volume. The problem extends beyond biology and is recognised as a central part of the research programme in autonomous robotics (Brooks, 1994). In the wider class of embedded computer systems, similar problems arise and their solution is sought in the engineering of real-time operating systems. Control of industrial plant also requires a continuous sensing of the plant (the "environment") and computation of appropriate control operations ("actions"). We conjecture that the analogies with real-time operating systems and control engineering may offer powerful tools for the future study of action selection in animals.

We now analyse potential solutions to the problem of action selection from a computational perspective. As illustrated in Figure 6.2, there would appear to be two broad approaches. Figure 6.2a shows a distributed scheme in which brain systems or modules, competing for control of behavioural expression, communicate directly with each other. Such a solution might work using, for example, reciprocal inhibitory connections. The disadvantages of this scheme are twofold. First, given *n* existing modules, every new module competing for behavioural expression needs an additional 2*n* long-range connection. Second, the intrinsic functionality of the module may be compromised by its having to deal also with selection.

A better solution to action selection appears in Figure 6.2b. Here each system sends a "bid" for its behavioural expression into a central "switch" or selection module. Now, when a new module is added, only two extra

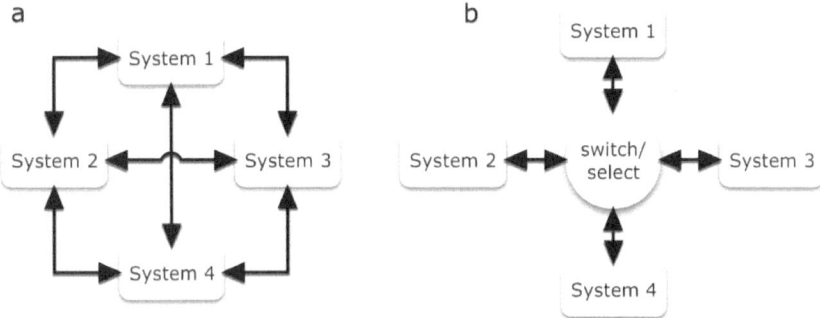

Figure 6.2 Two possible solutions to the action selection problem. The arrows show long-range connections.

long-range connections are required. In addition, the function of selection has been passed to a dedicated system so that the behavioural subsystems no longer need to mediate selection themselves. Thus, we suppose that there is a "common currency" of action *salience* or urgency, which is passed to the central selection module, and that those actions with high salience are selected for expression. In addition, projections from the selection mechanism back to the brain systems making action requests control the ability of these systems to enact behaviour.

2.2 The basal ganglia and biological selection

Given the advantages of using a central switch to solve the problem of action selection, we might expect to find that biology has adopted a solution of this form. The basal ganglia are a set of subcortical nuclei that are evolutionarily highly conserved throughout the vertebrate lineage, and in Redgrave, Prescott, and Gurney (1999) we outlined the proposal that the basal ganglia are well suited to play the role of the central selection circuit in the vertebrate brain. The link between the basal ganglia and action selection is a recurring theme in the literature (see, e.g., Cools, 1980; Mink, 1996; Kropotov & Etlinger, 1999). However, our recent work (Prescott, Redgrave, & Gurney, 1999; Redgrave et al., 1999) has developed this idea as a major unifying hypothesis of basal ganglia function, showing how it relates to the known anatomy and physiology of this system. We now outline some of these ideas here.

The internal anatomy of the basal ganglia is shown in Figure 6.3 (for a recent review, see Redgrave, 2007). The main input nuclei of the basal ganglia are the striatum and the subthalamic nucleus (STN). These receive excitatory input from many areas of the brain, including extensive parts of cerebral cortex and intralaminar areas of thalamus. This widespread input is consistent with the idea that the basal ganglia are a central selection device.

In the primate, the major output nuclei are the internal segment of the globus pallidus (GPi) and substantia nigra pars reticulata (SNr). These nuclei

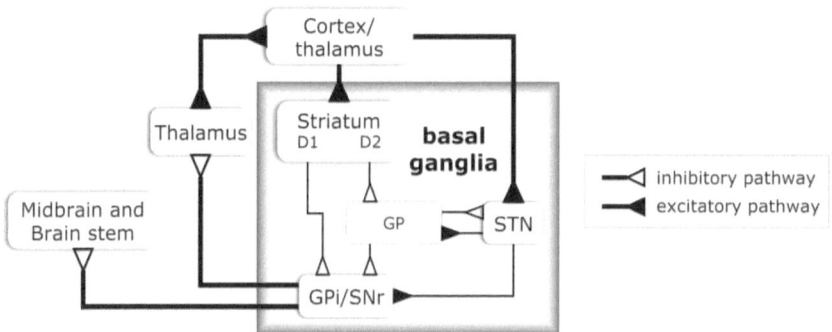

Figure 6.3 Basal ganglia anatomy and its connectivity with the rest of the brain.

provide extensively branched GABAergic efferents to functionally related zones of the ventral thalamus (which in turn projects back to the cerebral cortex), the midbrain and brainstem areas critical for movement (including the superior colliculus). Once again, this widespread projection is consistent with control of brain systems requesting behavioural control.

Internally, basal ganglia connectivity is dominated by the existence of two separate striatal populations: (1) a population that contains the neuropeptides substance P and dynorphin, and that preferentially expresses the D1 subtype of dopamine receptors; and (2) a population that contains enkephalin, and preferentially expresses the D2 subtype of dopamine receptors. In most accounts of basal ganglia anatomy, the D1-preferential population is usually associated with projections to SNr and GPi alone, while its D2 counterpart is associated with projections to a source of inhibition wholly within the basal ganglia – the globus pallidus (GP).

In order to enact selection, the basal ganglia must be able to selectively control, or gate, its targets. Basal ganglia output is inhibitory and tonically active, so that prevention of behavioural expression is the default. Selection is then supposed to occur via selective disinhibition of target structures when basal ganglia output to those targets is reduced (Deniau & Chevalier, 1985). Thus, in the basal ganglia, the default condition is for an inhibitory "brake" to be applied to all subsystems, and for this brake to be taken off only if the subsystem has been selected. This mechanism should be contrasted with those models of selection that rely on selective excitation of subsystems singled out for behavioural expression.

Another requirement for supporting selection is that the basal ganglia must be able to process signals for different actions in a relatively discrete way. That is, we suppose that the brain is processing, in parallel, a large number of sensory, cognitive and motivational streams or *channels*, each of which may be requesting a different action be taken. This channel-based scheme is consonant with the view that there are a series of coarsely defined, parallel processing streams or loops through the basal ganglia (Hoover & Strick, 1993). Here, cortical subsystems innervate basal ganglia, which, in turn, control these subsystems via thalamic feedback (Figure 6.3). Basal ganglia loops are associated with somatic motor control, oculomotor control and control of limbic structures (others are associated with control of prefrontal cortex). It is also possible to identify several loops through subcortical structures (McHaffie, Stanford, Stein, Coizet, & Redgrave, 2005). Furthermore, within each coarsely defined basal ganglia loop, we suppose there may be a finer granularity of action representation. This is consistent with the existence of motor somatotopy maintained throughout the basal ganglia within the main motor loop (Romanelli, Esposito, Schaal, & Heit, 2005).

A quite general requirement for the hypothesis that a brain system supports a given computation is that it should be endowed with appropriate neuronal mechanisms. In the case of action selection and the basal ganglia, we require that these nuclei contain mechanisms capable of a selection function, and Figure 6.4 shows that there is a range of such mechanisms.

The projection neurons in striatum are highly nonlinear and tend to exist in one of two states (Wilson & Kawaguchi, 1996): a relatively inactive down-state and an excitable up-state, initiated by massive correlated synaptic input, in which action potentials may occur. Computationally, this mechanism could act to filter out action requests (striatal inputs) with small salience. Striatum also contains a complex microcircuit, with interconnectivity between its projection neurons and at least three types of interneuron (Bolam & Bennett, 1995). Much of this interconnectivity is GABAergic and so may be computationally equivalent to local recurrent inhibition, which can act to contrast, enhance or select its inputs. The output nuclei (GPi, SNr) contain local recurrent collaterals that could also act in a similar way.

However, aside from these microcircuits and neuronal mechanisms, there is a network-wide mechanism that could support selection – namely an off-centre, on-surround feed-forward network constituted by striatum, STN and the output nuclei (Mink & Thach, 1993). The "off-centre" here is provided by focused striatal projections to GPi/SNr and the "on-surround" by diffuse projections to these nuclei from STN. A channel i with high salience input appearing at striatum and STN will therefore tend to inhibit its corresponding neural targets in GPi/SNr and excite its neighbours therein. In this way, basal ganglia targets in channel i are relieved of their inhibition (and thereby selected) while other channels are subject to increased inhibition.

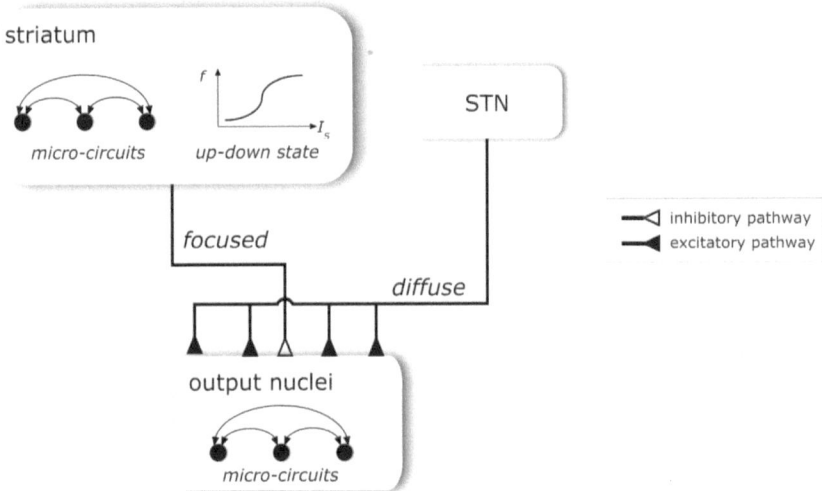

Figure 6.4 Selection mechanisms in the basal ganglia.

3 A systems-level model

How should we first attempt to test a computational hypothesis about a brain circuit quantitatively in a model? We propose that a first model should address the circuit *principally* at its intrinsic level of description and, in general, leave features at lower levels of description for subsequent studies. The exceptions will include mechanisms that occur naturally in the description of the model components (illustrated below in the context of our model of striatal neuronal populations). A complete brain module, which consists of several distinct nuclei, is defined by its inter-nucleus connectivity. The primary mechanistic features in such a model will therefore be derived from its overall architecture. We refer to modelling brain modules in this way as *systems-level modelling*. We now go on to describe a systems-level model of the basal ganglia that was based on the computational hypothesis of action selection (Gurney, Prescott, & Redgrave, 2001a, 2001b).

If we are interested in inter-nucleus connectivity, the selection mechanism of primary concern in the basal ganglia will be the feed-forward (off-centre, on-surround) network described in the previous section. In order to realise this we need to represent neuronal populations by rate-coded model neurons. The natural candidates for such populations are the sets of neurons responsible for a single channel within each of the basal ganglia nuclei (details of their implementation are given below). The feed-forward network shown in Figure 6.4 is now articulated using this scheme in Figure 6.5a. This network implements five channels, but for simplicity only one is shown in the input nuclei. Furthermore, while the inputs to this network are shown as originating in the cortex, cortex is not modelled *per se*, and we assume that channel salience has already been extracted from phasic excitatory input by neuronal processes in the input nuclei. The input to the model is, therefore, simply the scalar-valued salience of each channel.

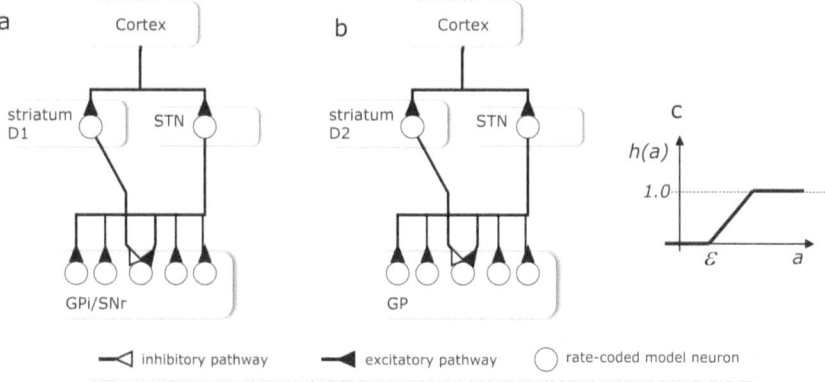

Figure 6.5 Development of the systems-level model of the basal ganglia in Gurney et al. (2001a).

Note this network finesses the scheme shown in Figure 6.4 because it identifies a subpopulation of striatal projection neurons involved – namely, that which preferentially expresses D1-type dopamine receptors. However, if this circuit can perform selection, what is the function of the rest of the basal ganglia? The remaining anatomy is shown in Figure 6.5b (STN has been replicated to show its relation to GP – no anatomical division is implied).

This question is resolved by observing that the relative levels of excitation and inhibition required for selection in the network in Figure 6.5a *alone* depend critically on the weights. Thus, if there are n channels, then unless the weights from STN to the output nuclei are roughly $1/n$ times the weights from striatum to the outputs, subthalamic excitation overwhelms striatal inhibition and no sensible selection can take place. A careful crafting of the weights would appear biologically unrealistic and non-robust. An alternative, however, is to automatically scale the excitation in STN, according to the number of channels, using a dynamically determined degree of inhibition. It may be shown analytically (and in simulation) that the inhibition supplied by GP, when embedded in the circuit of Figure 6.5b, exactly fulfils this criterion, and we know that GP innervates STN (see Figure 6.3). The unification of the two subsystems shown in Figure 6.5a and 6.5b, is shown in the full basal ganglia architecture of Figure 6.6a

Note that GP also innervates GPi/SNr and that it has also been established that GP sends projections back to striatum (not shown in Figure 6.6 – but see Bevan, Booth, Eaton, & Bolam, 1998). Given these projections, and the specific functional role ascribed to the GP–STN pathway of "automatic gain

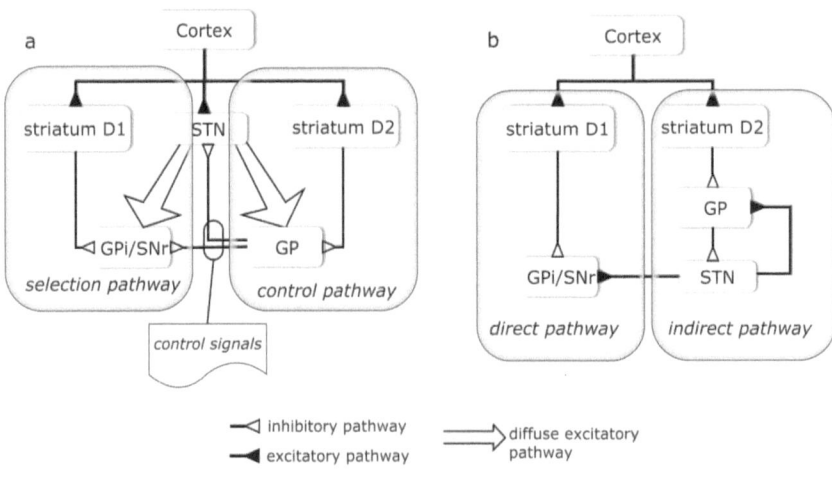

Figure 6.6 Basal ganglia functional architectures. (a) The new architecture in Gurney et al. (2001a), based on the computational hypothesis of selection. (b) The "direct–indirect" pathway model of Albin et al. (1989).

control" for STN excitation, we hypothesise that the circuit comprising striatum D2, STN, and GP constitutes a *control pathway* and that GP acts as a source of control signals to ensure the correct functioning of the *selection pathway* – striatum D1, STN, and GPi/SNr.

The new functional architecture in Figure 6.6a is quite different from the pre-existing "direct/indirect" pathway model (Albin, Young, & Penney, 1989), shown in Figure 6.6b. The latter is not motivated by a computational hypothesis and simply emphasises an inhibitory ("direct") and excitatory ("indirect") influence on the output nuclei. Further, it excludes the role of the STN as a major input nucleus to the basal ganglia.

We now turn to the details of the neuron population model. Each such population is represented as a "leaky integrator" with dynamics given by $da/dt = -ka + u$, where a is an activation variable, k determines the characteristic time constant and u is the input (typically a weighted sum of afferents from other neurons) The output is given by $y = h(a)$, where y is interpreted as the average rate of firing of the neural population. In our case we made $h()$ a piecewise linear function of the form shown in Figure 6.5c. Thus $h(a)$ is zero if a is less than some threshold ε, and it has a constant positive slope, m, until it saturates at $h(a) = 1$. This form is not the more normal sigmoid and was chosen to facilitate analysis of equilibrium states. The analytic tractability of a model can be a powerful feature and should not be eschewed if it requires minor embellishments that do not sacrifice essential aspects of functionality (note that the sigmoidal form is often chosen for convenience rather than on grounds of principle).

The threshold ε is assigned to ensure that relative tonic firing rates of each neuron type are approximately preserved. For example, in striatum, we put $\varepsilon > 0$, since these neurons require substantial input before they fire (a feature of their up–down state behaviour). This is an example where a mechanism that is normally associated with a lower level of description is more naturally incorporated rather than omitted at the systems level. Thus, the up–down state behaviour in striatal projection neurons is most appropriately described at the membrane level in the Hodgkin–Huxley formalism. However, this mechanism has a phenomenological description at whatever level of complexity is used in the model neurons (including rate-coded leaky integrators), and it would be perverse to wilfully ignore it.

The neurotransmitter dopamine plays a key role in basal ganglia function, as evidenced by the high density of dopamine receptors, especially in striatum (Beckstead, Wooten, & Trugman, 1988). Dopamine is released in striatum via axon terminals associated with cell bodies in the substantia nigra pars compacta (SNc). Cell death in SNc is a primary cause of Parkinson's disease, symptoms of which include akinesia (failure to initiate movement) and bradykinesia (slowness of movement). Dopamine imbalance has also been associated with schizophrenia (Laruelle, Abi-Dargham, Gil, Kegeles, & Innis, 1999). We therefore sought to include the effects of striatal dopamine transmission in our model and used a phenomenological description of this process as one of facilitation and attenuation of cortico-striatal drive at D1 and D2 receptors, respectively.

Results for a simple selection scenario in a six channel model are shown in Figure 6.7a. Two channels (labelled 1 and 2) are stimulated and so are in competition with one another. Channel 3 is representative of all other channels. At time $t = 1$, a moderate salience input (0.4) is applied to Channel 1, resulting in a decrease in GPi/SNr output y_1 on that channel and an increase in output of other channels. In general, we suppose that if y_i is reduced below some selection threshold θ_s, then channel i is selected. In the instance shown, if $\theta_s > 0.05$, then Channel 1 is selected for $1 < t < 2$. At $t = 2$, Channel 2 receives a salience input of 0.6. This results in clear selection of Channel 2 ($y_2 = 0$) and clear deselection of Channel 1. There has therefore been a switching event between Channels 1 and 2. At $t = 3$, Channel 1 increases its salience to 0.6 – equal to that of channel 2 – and so $y_1 = y_2$. Note that $y_1, y_2 > 0$ as a result of competition or "interference" between the two channels, and both may fail to be selected as result.

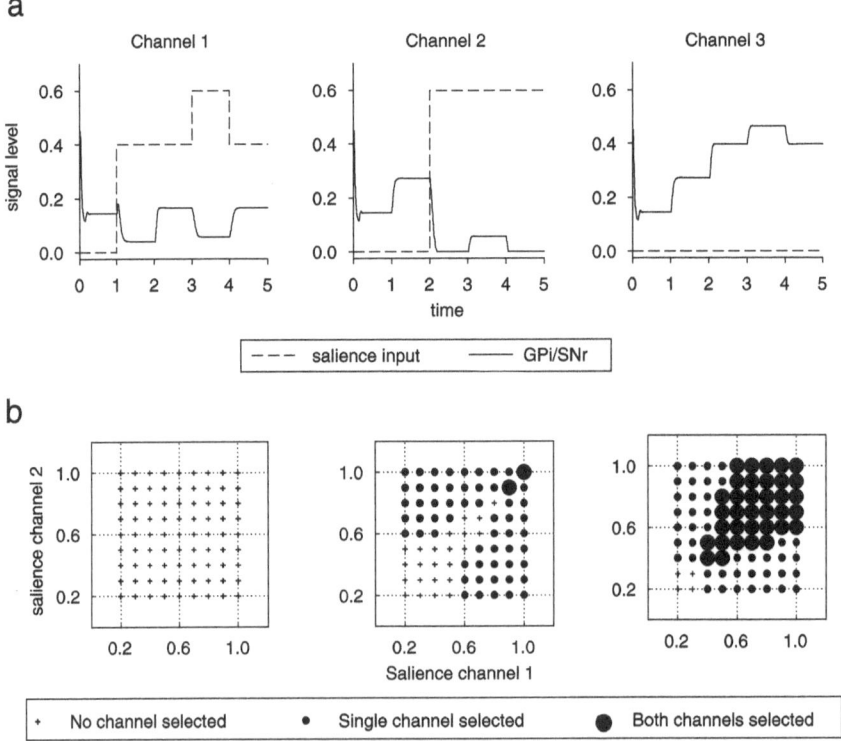

Figure 6.7 Results from systems-level model in Gurney et al. (2001a). (a) Single selection experiment with normal levels of dopamine. (b) Summary of multiple experiments with different levels of dopamine.

Now consider an experiment of the form just described but with no salience transient on Channel 1 for time $3 < t < 4$. At the end of this experiment, we classify the equilibrium outcome into one of three possibilities: (1) neither channel is selected, (2) a single channel is selected, (3) both channels are selected. The centre panel of Figure 6.7b shows the summary outcome of many experiments of this form, with varying salience inputs on Channels 1 and 2, and a constant level of simulated dopamine identical to that used to generate the results of Figure 6.7a. Very small levels of input result in no selection, and most other inputs give single-channel selection. Only two instances of very high and equal salience give simultaneous selection of both channels. In sum, the model displays desirable selection properties. In contrast, the left-hand panel in Figure 6.7b shows the outcome with low levels of simulated dopamine. No selection ever takes place; the model is completely "akinetic" and mimics the Parkinsonian state. In the right-most panel, a high level of simulated dopamine has been used, and there are now many instances of multiple selection. In a dynamic environment with continually changing saliences (resulting, in part, from the animal's interaction with the environment), simultaneous selection may become rapid switching between two easily selected alternatives. This is consistent with some of the positive symptoms of schizophrenia, such as excited motor activity, and incoherent thoughts and speech.

In summary, a biologically constrained systems-level model of the basal ganglia, constructed under the hypothesis that these structures have a selection function, forces us to conceive of a new functional architecture. Furthermore, the model displays behaviour that provides confirmatory evidence for the selection hypothesis. In particular, high-quality selection and switching are available with moderate levels of simulated dopamine, while extremes of dopaminergic control display behaviours consistent with pathologies implicating dopamine imbalance in the basal ganglia.

4 Testing the hypotheses further: increased biological realism

Having established confirmation of a hypothesis in a systems-level model, what should be the next step in a programme to refine and test the hypothesis further? If the biological solution to implementing the target computation is in some sense optimal, we would expect that increasing the realism of the model by incorporating further neurobiological constraints should improve its performance. There are two dimensions in which this may be accomplished. First, additional anatomically identified pathways may be included that were not present in the original model, or which place the model in a wider anatomical context with other brain areas. Second, we may drop down a level and include extra physiological detail. We now explore the first of these options and highlight some of the methodological issues arising therein.

4.1 Incorporating additional constraints at the systems level

Figure 6.8 shows an anatomically extended model that includes thalamic nuclei and their connections with cortical areas (Humphries & Gurney, 2002). This circuit finesses the anatomy of the return loops through thalamus, shown in Figure 6.3, by articulating a division of the thalamic nuclei into the thalamic reticular nucleus (TRN) and ventrolateral thalamus (VL).

Notice that the TRN projects diffusely to VL (Pinault & Deschenes, 1998), thereby giving rise to another selection mechanism. The extended model shows enhanced selection properties in several respects. First, inputs with lower level salience are able to force selection (although very small inputs are still limited in their effect by the up–down state mechanism in striatal neurons). Second, the range of salience values over which actions can be switched between is increased. Third, the contrast between the selected and non-selected actions is enhanced via improved differentiation of outputs from the basal ganglia. Fourth, transient increases in the salience of a non-selected action are prevented from interrupting the ongoing action (unless the transient increase is of sufficient magnitude). Finally, the selection of the ongoing action persists when a new closely matched salience action becomes active. All these desirable selection properties (except the first) are contingent on invoking the TRN.

In line with the discussion above, we take the selection advantage accruing from biologically realistic additions to the original basal ganglia model as confirmatory evidence of the selection hypothesis.

In another model, we included a variety of additional connections internal to the basal ganglia (Gurney, Humphries, Wood, Prescott, & Redgrave,

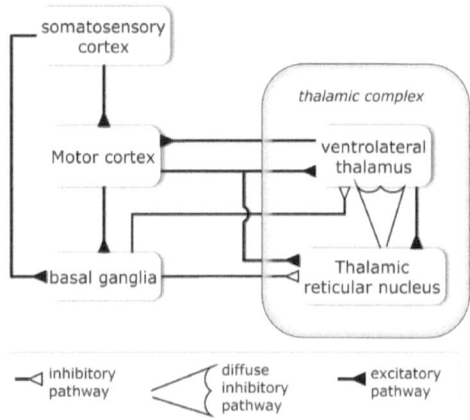

Figure 6.8 Extended model that includes a thalamic complex and connections with cortical areas (Humphries & Gurney, 2002). The box labelled "basal ganglia" is the entire architecture in Figure 6.6.

2004a). The details of the connectivity are less important here than the methodological issues that were raised in that study. First, it became apparent that, in order to quantitatively compare several model variants, it was necessary to define a metric for "selection". Such a metric serves to sharpen what was loosely referred to as "optimal selection" in the discussion at the start of this section. The details of the metric definition are not important, but, in brief, it was based on proposing idealised outcomes in terms of bubble plots like those in Figure 6.7b and determining the extent to which the model outcome matched these ideals. Then, if a model based on anatomically realistic additions shows superior performance (according to the selection metric), we take this as confirmatory evidence for the selection hypothesis. However, this claim is made stronger if, in a model with additional pathways that are not realistic, selection (metrically defined) deteriorates. These artificially contrived models therefore act as controls, and in Gurney et al. (2004a) the control models were indeed found to display poorer performance.

4.2 Incorporating additional physiological constraints

In working at the systems level, many details of neuronal operation are ignored or poorly approximated. The most obvious example is that the basic neuronal signalling mechanism – namely discrete action potentials or "spikes" – is replaced by a population or ensemble firing rate. In addition, nothwithstanding the general nonlinearity in the output function $h(a)$ of the rate-coded models, idiosyncratic nonlinear membrane properties cannot be captured. Nor can any nonlinear dendritic integration, because the total input is a linear function of the individual afferents. These physiological and biophysical properties of the neuron present a raft of potential additional constraints.

Several possible relationships exist between a systems-level model and one that incorporates this lower level of detail. First, the nonlinear mechanisms of the biological neuron may work synergistically to produce an overall semilinear function of the kind found in our systems-level "neurons" (Koch, 1999). A weaker variant of this supposes that the neuron is roughly linear over some part of its operating range. In either case, this emergent linearity means that our systems-level model is "right for the right reasons" – our computational hypothesis is supported and is mechanistically correct at the systems level. Second, it is possible that while the individual neural populations are highly nonlinear, their behaviour when embedded in the entire system is comparable to that of the systems-level model. This global equivalence means we are "right for the wrong reasons" – our computational hypothesis is still supported, but its mechanistic explanation needs revising. Finally, it is possible that the neural nonlinearities lead to entirely different system behaviours and that our systems-level model is simply wrong – our computational hypothesis is refuted. Only by building the model at the lower level of description can we be sure which possibility applies. A possible bonus here is that the additional mechanisms incorporated may further support the putative computation implemented by the system as a whole.

Working at a level of description that respects the spiking behaviour of the neuron also enables a model to make contact with physiological data describing spike-train patterning. This kind of data offers further constraints. Thus, confirmatory evidence of the computational hypothesis is obtained if the model can simultaneously implement the target function and display spike patterning observed *in vivo*. Conversely, if this is not possible, the hypothesis is brought into question.

We have recently (Humphries, Stewart, & Gurney, 2006) implemented a model of the basal ganglia using spiking neurons which incorporates much of the known physiology of individual neuronal species, including synaptic currents with time constants for AMPA, NMDA and GABAergic synapses; spontaneous currents; dopamine in STN and GP as well as striatum; and inter-neuronal delays. In addition, while the neurons were based on simple LIF dynamics, they incorporated phenomenological descriptions of shunting (multiplicative) inhibition at proximal dendrites and the soma (Koch, 1999) and rebound bursting in STN neurons.

The model had the same overall architecture as the systems-level model in Figure 6.6a and used three channels. Each channel in each nucleus had 64 neurons, and, within a channel, connectivity was partial and stochastic (rather than all-to-all). The model's selection performance under differing levels of simulated dopamine is shown in Figure 6.9. The patterning of mean firing rates of neurons in SNr is the same as that of the outputs of the systems-level model. Insofar as the latter captures the kind of selection

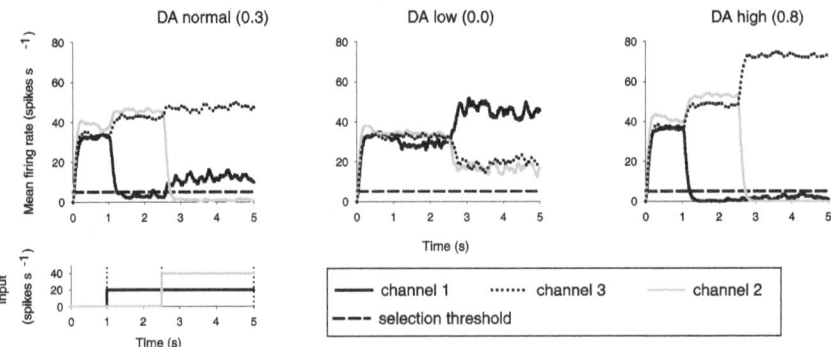

Figure 6.9 Selection results from the model in Humphries et al. (2006) under differing levels of simulated dopamine (DA parameter in parentheses). The stimulus protocol is illustrated in the bottom left (it is the same as that used to obtain the results in Figure 6.7b).

behaviour expected in each condition, the new model provides confirmatory evidence for the selection hypothesis.

The model was also pitted against two sets of data describing oscillations in the basal ganglia. Here, we focus on just one of these data sets, generated by Magill, Bolam, and Bevan (2001), which relates to low-frequency oscillations (LFOs) in rat basal ganglia under urethane anaesthesia. Figure 6.10 shows typical single-unit recordings in STN and GP in control animals. The cortex shows a slow-wave EEG with a strong peak in the power spectrum at about 1 Hz. The STN spikes in bursts that appear to follow this slow-wave activity. This is not surprising, since cortex drives STN. What is surprising, however, is that GP (which is nominally tightly coupled to STN) does not mirror STN activity. Magill et al. (2001) also gathered LFO data under three other conditions: after cortex had been removed by ablation; and after 6-OHDA dopamine lesion, with both lesion and ablation.

Figure 6.11b presents a summary of the percentage of cells showing LFO behaviour in each experimental condition (dark bars are experimental data, and light bars model data). Cortical ablation removes all LFO activity, while dopamine depletion induces strong LFO activity in GP (the intuitive result based on tight GP–STN coupling). Subsequent cortical ablation does not, however, completely obliterate all LFO activity. Note that there are no error bars here because the data points are based on total cell counts. Figure 6.11a shows the mean firing rates of cells in each nucleus over all conditions; there are error bars here because the data are averaged over cells. Model results (light bars), for both LFO counts and firing rates, are shown alongside their

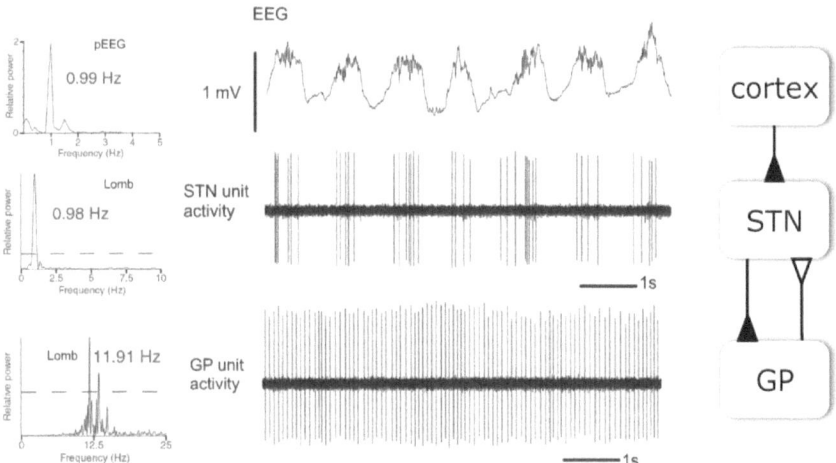

Figure 6.10 Low-frequency oscillations in basal ganglia – typical spike trains in STN and GP (adapted from Magill et al., 2001). The diagram on the right shows the relevant internucleus connectivity

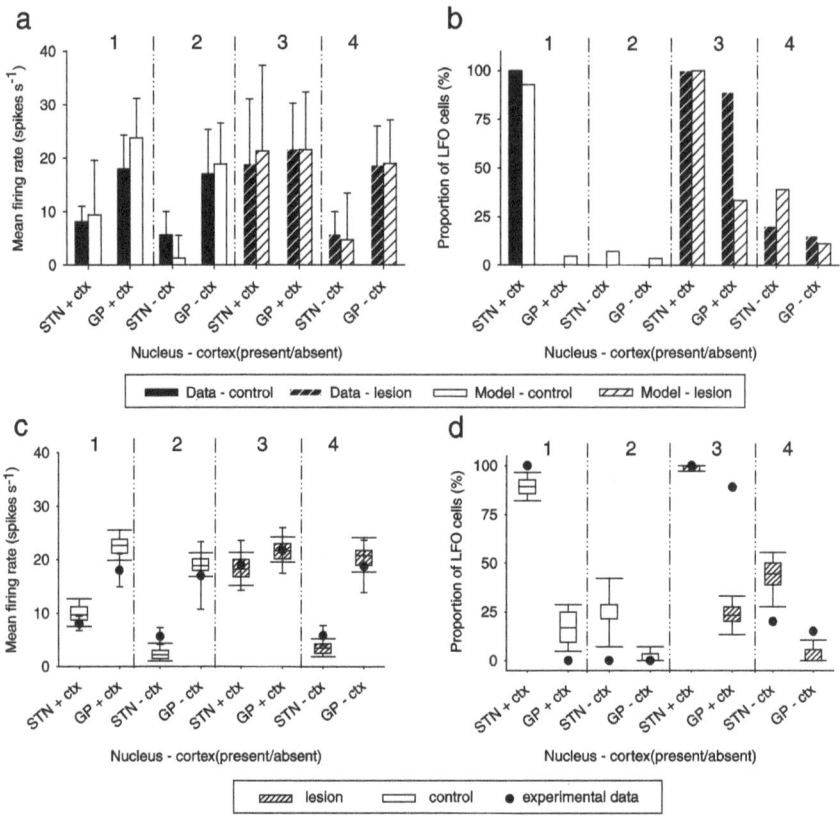

Figure 6.11 LFOs in basal ganglia – summary of data, and modelling results from Humphries et al. (2006). *Conditions 1–4*: 1, control; 2, cortical ablation; 3, dopamine lesion; 4, ablation and lesion.

experimental counterparts. There is good general agreement with the experimental data and the model parameters used for the LFO simulations were the same as those for the selection experiments. The model is therefore able to *simultaneously* satisfy the constraints of the LFO data set and display appropriate selection.

The modelling of the LFO data raises an important issue. In producing model results, should we use all the available model neurons to generate each mean for comparison with the experimental data? Recently, we have argued this should not be the approach adopted (Humphries & Gurney, 2006). Rather, we should sample neurons from the model in a way that emulates the experimental protocol as closely as possible. Thus, because each model is stochastically connected, we identify each separate instantiation of the model with an experimental animal. Furthermore, we sample as many neurons in each model instantiation as were sampled in each animal *in vivo*. Each collection of "models as animals" and neural sampling constitutes a *virtual experiment*. One reason for proceeding in this way is that any para-metric statistical comparison of model and experimental data will rely on there being equal numbers of data points in each case. The model results in Figure 6.11a and 6.11b are for a single virtual experiment. The results in Figure 6.11c and 6.11d show the results of 50 virtual experiments. This gives a measure of the spread of our model results and may also be indicative of the spread we might expect *in vivo*.

A further issue concerns the mechanistic complexity of the model. A model should not gratuitously include an arbitrary set of mechanisms but principally those directed at testing hypotheses. Redundant mechanistic complexity will become apparent in the failure of certain features to impact on the fit of the model with the data. We had hypothesised at the outset that the failure of GP to reflect STN LFO activity was due to a decoupling of these nuclei under dopaminergic modulation in these nuclei. The full model also included recurrent collaterals in GP and GPi/SNr – because their functional significance is in dispute (Bar-Gad, Heimer, Ritov, & Bergman, 2003a) – and a model of the effects of urethane – because of the anaesthetic contingency of the LFO date (Magill, Bolam, & Bevan, 2000). Figure 6.12 shows the percentage of significant correlations with the data over a batch of 50 virtual experiments for each tested variant of the

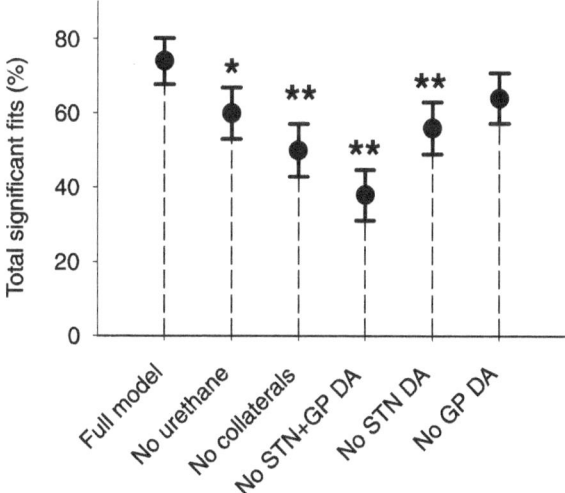

Figure 6.12 Overall correlation between virtual experiments in Humphries et al. (2006) and animal data from Magill et al. (2001) across variations of the model. A single asterisk denotes a $p = .05$ significant difference between the expected value for the model variant and that of the full model; a double asterisk denotes a significant difference at $p = .01$.

model (for details of the correlation procedure, see Humphries et al., 2006).

All mechanisms contributed to the fit of the model with the data, and only one mechanism produced results that were not significantly different from the full model (removing dopamine from GP alone). We also hypothesised that the residual LFO activity seen under cortical ablation (and DA lesion) was due to an intrinsic pacemaking mechanism supported by rebound bursting in STN (Humphries & Gurney, 2001; Plenz & Kitai, 1999). Removing this mechanism did indeed result in loss of all LFO activity from the model under these conditions.

In summary, a large-scale spiking neuron network of basal ganglia that incorporates many physiological constraints is able to show selection properties similar to those of its system-level counterpart. Using the same parameter set, the model is also able to fit data sets associated with oscillatory (LFO and γ-band) phenomena in the basal ganglia and give a mechanistic account of their origin. This, in turn, leads to several testable predictions (details of which may be found in Humphries et al., 2006).

4.3 *Constraints at the microcircuit and neuronal level*

The large-scale model of the basal ganglia described above is a significant advance on its systems-level predecessor in terms of biological realism. In particular it included some of the constraints at the microcircuit and neuronal level (Levels 3 and 4 in Figure 6.1). However, much of the mechanistic apparatus at these levels still remains excluded. Similar arguments to those used at the start of the previous section (in regards to the relation between systems-level and spiking models) may be made again here to motivate the study of models at the level of microcircuits and neurons. Thus we might ask whether our spiking neuron model is "right for the wrong reasons" by omitting microcircuit and neuronal detail.

The complex microcircuit of the striatum (Bolam & Bennett, 1995) is a particularly attractive candidate for attention here. Models of striatum typically invoke a selection function (Gurney & Overton, 2004; Wickens, Kotter, & Alexander, 1995) but have been hampered by a lack of data on functional connectivity – a situation that is starting to be remedied.

At the neuronal level there are physiologically detailed, conductance-based models of several of the major neuron types in basal ganglia (Gillies & Willshaw, 2006; Hanson, Smith, & Jaeger, 2004; Kotaleski, Plenz, & Blackwell, 2006; Wolf et al., 2005; Wood, Gurney, & Wilson, 2004). These models include accurate representations of the many ionic currents across the neural membrane in the conductance-based formalism of Hodgkin–Huxley. In addition, the models may also include detailed representations of neuronal morphology by partitioning the cell into a series of iso-potential compartments (Koch, 1999). In this way, these models aim to capture as many mechanisms at the neuronal level as possible. In many instances, it remains to interpret these mechanisms from a computational viewpoint, but, by way of example, Figure 6.13 shows a phenomenon observed in the medium spiny striatal projection neuron with a putative computational function.

Here, a current injection just sufficient to elicit a spike produces the spike only after the membrane has depolarised gradually over several hundred milliseconds. This "slow-ramp depolarisation" indicates a low-pass filter mechanism at work on inputs that are barely superthreshold, and it may act to prevent small, transient inputs to these neurons from being interpreted as an action request.

5 Testing models in embodied form

Any model of an isolated component of a complete sensorimotor pathway suffers from the drawback that either the model inputs are artificially crafted (not derived from realistic environmental stimuli) and/or the interpretation of its outputs is left unresolved as far as enacting behaviour is concerned. Most models in computational neuroscience (including those described above) are of this kind. To remedy this, we need to model, at least in some guise, a complete sensorimotor system. However, even this does not guarantee a full validation of the model because the sensory input not only drives behaviour, but is also contingent on that behaviour. Ensuring adequate causality therefore demands that the agent–environment–agent loop is closed. These imperatives are all satisfied by a model that is *situated* in a complex environment and is also *embodied* in autonomous robotic form (Brooks, 1991; Webb, 2001).

At first glance, this approach would seem to require the modelling of large parts of the entire brain (or at least a complete sensorimotor competence). This programme is clearly, at best, challenging and, at worst, intractable. However, there is a strategy for testing a model in embodied form that circumvents this difficulty. Thus, suppose we are interested in testing a

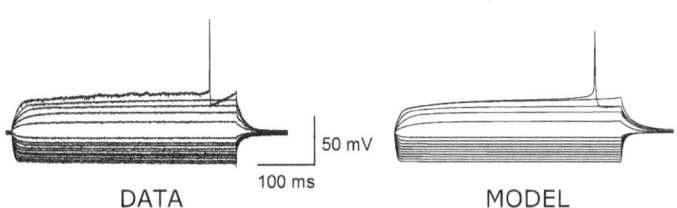

Figure 6.13 Slow-ramp depolarisation in medium spiny projection neurons in striatum. The plots show membrane voltage traces under current injection from experimental data (C. Wilson, personal communication) and a multi-compartment, conductance-based model with 9 currents and 106 compartments (Wood et al., 2004)

model of a particular brain module. This model will contain all the biological realism we have targeted, and we refer to it as the *biomimetic core* (Prescott, Gonzales, Gurney, Humphries, & Redgrave, 2005). In order to ensure sensory input to, and behavioural output derived from, this core, it must be augmented with relevant sensory and motor competencies. However, we argue that these do not have to be biologically realistic. Rather, the only constraint on their construction is that they must interface with the biomimetic core using signal representations that are shared by the core (compare the practice in software engineering where the contents of a software module may be changed arbitrarily, so long as its interface remains consistent with its environment). We refer to the model beyond the biomimetic core as the *embedding architecture*

As an example, consider the basal ganglia model of Figure 6.6. In order to embody this, we require "command" modules associated with actions (and driven by sensory subsystems) that present signals at the basal ganglia model inputs representing the salience or urgency for performing the actions. We also require motor systems that can be gated by disinhibition (from the basal ganglia outputs) and can enact the selected behaviour. The command and motor systems do not have to be biologically plausible, but the interface constraint demands that they provide salience and respond to disinhibition, respectively, as these are the signal representations used by the basal ganglia.

Using this approach, we have embodied the extended model architecture of Figure 6.8 in a small autonomous robot that mimics the foraging behaviour of food-deprived rats immediately after exposure to food in an open arena (Prescott et al., 2005). The model was endowed with motivational states determined by simulated "fear" and "hunger", and food was represented by small reflective cylinders (Figure 6.14). Behavioural command units request the actions of "wall follow", "wall seek", "cylinder pickup", "cylinder deposit" and "cylinder seek". Input to the model encoded the salience of the actions based on their perceptual affordances and motivational states, while model output removed inhibition from the selected motor command.

The time series of salience inputs was complex and noisy, unlike the stylised rectangular inputs in Figure 6.7, and provided a strong test of the model. The basal ganglia outputs were found to be most effective when supporting gating by shunting (divisive) rather than subtractive inhibition, a feature that leads to a prediction about the innervation of basal ganglia targets in the brain. Note that consideration of this issue only becomes imperative and testable in an embodied model with overt behaviour. Furthermore, while the embedding architecture is not particularly biologically plausible, general system requirements of any such architecture (realistic or otherwise) may emerge in the overall system behaviour. For example, we discovered that motor efference copy signals could be a crucial component of salience input.

Figure 6.14 Action selection in a robot controlled by the biologically constrained model of basal ganglia thalamocortical loops in Figure 6.8.

6 Model development: bottom up versus top down

At the start of this chapter, we advocated use of Marr's (1982) framework with three levels: computational, algorithmic and implementational. Subsequently we developed models of the basal ganglia and associated circuits under the hypothesis that the computational purpose of these systems was one of selection. In terms of Marr's scheme, the brain systems constitute the implementation, but no mention has been made yet of any algorithms. Rather, there has been much talk of *mechanisms* – but where do these figure in the computational framework? First, we distinguish what we mean by "algorithm" and neural "mechanism" in the current context. An algorithm is a procedure that may be captured in a high-level computer programming language or mathematical formalism with no reference to any implementation whatsoever. In contrast, a neural mechanism is exactly that – it is tied to a particular neuronal network or constellation of single neuronal features. This is illustrated below in the example of the basal ganglia model by Bar-Gad, Morris, and Bergman (2003b).

Figure 6.15 shows how the tri-level scheme may be augmented to include another level dealing with mechanisms (Gurney, Prescott, Wickens, &

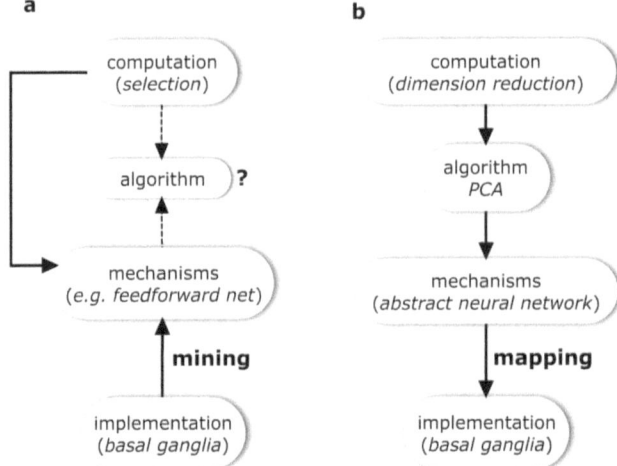

Figure 6.15 Four-level framework and alternative strategies for model construction therein. (a) Mechanism mining (mainly bottom-up) illustrated via the systems-level model in Figure 6.6. (b) Mechanism mapping illustrated via the dimension reduction model of Bar-Gad et al. (2003b).

Redgrave, 2004b).[1] There are now two ways of discovering the contents of each layer in this scheme. In the models described in this chapter, we have started with a computational hypothesis and proceeded to "mine" directly the biological implementation for mechanisms supporting the computation. In this approach, no algorithm *per se* emerges, and its existence at this stage remains in question (however, an algorithmic description of the basal ganglia is described by Bogacz and Gurney (2007)).

This principally bottom-up approach of *mechanism mining* is to be contrasted with a top-down approach that may be illustrated with reference to a model of the basal ganglia proposed by Bar-Gad et al. (2003b). Here, it is proposed that the basal ganglia perform some kind of dimension reduction on their input, as evidenced by the high level of convergence of signals from cortex onto basal ganglia. This hypothesis is not, we propose, at odds with the notion of action selection but, rather, deals with the representation of actions used within basal ganglia as they perform selection. In any case, Bar-Gad et al. go on to propose a specific algorithm for dimension reduction: principal components analysis (PCA). PCA makes no reference to neurons – abstract or biological – and it therefore fits the description of "algorithm"

1 In Marr's original formulation of the computational framework, which appeared in an MIT technical report (Marr & Poggio, 1976), a fourth level was described. However, this was dropped in the more popular account (Marr, 1982). Independently, Gurney proposed a four-level account (Gurney, 1997, pp. 149–150), which was subsequently developed in Gurney et al. (2004b).

given above. However, the algorithm may be realised in a mechanism consisting of a neural network with very specific connectivity and a training procedure to discover the correct weights. Furthermore, several such neural mechanisms have been proposed (see, e.g., Fiori, 2000), all of which implement the same algorithm.

There is now one further step to go before we make contact with the basal ganglia. Thus, the neural network mechanism is an abstract one and does not, *a priori* have anything to do with the basal ganglia. It therefore remains to perform a *mechanism mapping* from the neural network onto the biological substrate of these circuits. Bar-Gad et al. (2003b) argue that this is plausible in the cortico-striatal connectivity and striatal microcircuits.

Note that while mechanism mining and mapping are two complementary approaches to navigating the four-level framework, of prime importance in both is the putative computational function of the brain system in question.

7 From physiology to behaviour

Physiology has made its way into the models described here at several levels of description. In the systems-level and large-scale spiking network model, several physiological features were invoked (e.g., dopamine modulation and rebound bursting in STN), but these were described using simplifying phenomenological mechanisms. Nevertheless, insofar as these simplifications are valid models of the original physiological features, we should expect their impact on behaviour to be reasonably well described.

In general, however, the above approach is at best a partial solution, since a full and reliable account of the effects of a physiological feature will rely on incorporating it at its native level of description. For example, STN bursting relies on a current cascade best described at the neuronal level using conductance-based modelling. On the other hand, contact with behavioural expression will best be realised in high-level models that extend over complete brain systems rather than individual microcircuits or neurons. How can we resolve the tension between these two requirements?

We propose a solution that is an extension of the methodology used in developing embodied models. Thus, at any level of description, we consider a model as constituting a biomimetic core at that level. This model may then be embedded into another, higher level model that, unlike the embodied methodological prototype, may be biologically constrained and will (like the embodied case) honour the signal interface between the two levels.

Recently (Stafford & Gurney, 2007), we used a "hybrid of levels" model of this kind to examine the hypothesis that our basal ganglia model could account for human reaction time (RT) data in cognitive tasks. Here, a neuroscientifically constrained model of the basal ganglia and thalamocortical loop (Figure 6.8) was used as the decision mechanism on the output side of a psychological, connectionist-level model of Stroop processing (Cohen, Dunbar, & McClelland, 1990). While the latter model was not biologically grounded, it made use of rate-coded units and therefore satisfied the interface requirements of the basal ganglia model. The hybrid model was indeed able to account for RT data in the Stroop task, and it showed greater robustness over a wider range of experimental protocols than did the original Stroop model (which used a simple evidence accumulator as its decision mechanism).

The process of model embedding could be iterated at multiple levels of the structural description hierarchy. Furthermore, in going up such a hierarchy, it is reasonable to suppose that the anatomical extent of each level will increase. Thus, detailed conductance-based neuronal models may constitute only a few elements in a single microcircuit domain. This may then be embedded into a larger scale microcircuit using reduced neuronal models, and these are embedded, in turn, into a large-scale spiking network using LIF neurons. In this process of *nested embedding*, physiological processes at a lower level of description are able to express themselves at higher levels – all the way to behaviour – via the support offered by their embedding environments. Reliable expression of this type may, however, require judicious choice of the location and extent of a core model. For example, in the basal ganglia, discovering the implications of a particular ionic current using a conductance-based neuronal model would probably require that model to be implemented in a microcircuit whose extent covered the encoding of a complete action channel in the nucleus of interest.

8 Summary

Throughout this chapter, we have advocated the general applicability of Marr's dictum that, before attempting to build a model, we should strive to define the computational purpose or function of the brain system we are interested in. Having done that, we may then choose to work in a more bottom-up fashion and mine for mechanisms supporting the computation, or proceed top down and propose an algorithm and an abstract network, before attempting to map that network onto the neural substrate. Computation is not the preserve of the higher levels of structural description we ascribe to the brain, and so our proposed methodology is of universal applicability. In general, however, the first model of a brain system is best completed at a relatively high or systems level of description.

If our hypothesis appears to be confirmed by this first foray, subsequent modelling exercises should strive to incorporate more biological constraints, to see if the hypothesis stands up to this further scrutiny. Constraints can take one of two forms. First, we can increase the anatomical richness or extent of the model. In this case, it is useful to define a quantitative metric to define precisely the computation being performed. Then, the realistic anatomical model extensions can be pitted against non-realistic control variants. A second form of biological constraint is provided by "tunnelling down" to incorporate more physiological detail. In this case, we are able to make contact with a variety of physiological data, and our hypothesis is validated by the model being able to simultaneously fit this data and perform the putative function. In the process of fitting data here, we advocate a "models as animals" approach that matches the use of modelling and experimental data. In building models at the lower levels, we must guard against gratuitous use of too many features and must seek to test hypotheses about specific mechanisms.

A strong test of any model is to implement it in embodied (robotic) form. This closes the agent–environment–agent loop, thereby providing realistic input time series and forcing an interpretation of model outputs in behavioural terms. This programme is made possible in a conceptually satisfactory way by making a clean demarcation between the target brain system of interest (the biomimetic core) and an embedding architecture. A generalisation of this idea (nested embedding) promises to facilitate a principled approach to multiple-level models in which the effects of a physiological feature at any level of description may be expressed all the way to behaviour.

9 Acknowledgements

This work was supported in part by EPSRC grant EP/C516303/1. I would like to acknowledge the members of the Adaptive Behaviour Research Group (past and present) who contributed to the work presented here: Peter Redgrave, Tony Prescott, Paul Overton, Ric Wood, Rob Stewart, Fernando Montes-Gonzales, Tom Stafford, and Mark Humphries.

10 References

Albin, R. L., Young, A. B., & Penney, J. B. (1989). The functional anatomy of basal ganglia disorders. *Trends in Neurosciences, 12*, 366–375.

Bar-Gad, I., Heimer, G., Ritov, Y., & Bergman, H. (2003a). Functional correlations between neighboring neurons in the primate globus pallidus are weak or nonexistent. *Journal of Neuroscience, 23*(10), 4012–4016.

Bar-Gad, I., Morris, G., & Bergman, H. (2003b). Information processing, dimensionality reduction and reinforcement learning in the basal ganglia. *Progress in Neurobiology, 71*(6), 439–473.

Beckstead, R. M., Wooten, G. F., & Trugman, J. M. (1988). Distribution of d1 and d2 dopamine receptors in the basal ganglia of the cat determined by quantitative autoradiography. *Journal of Comparative Neurology, 268*(1), 131–145.

Bevan, M. D., Booth, P. A., Eaton, S. A., & Bolam, J. P. (1998). Selective innervation of neostriatal interneurons by a subclass of neuron in the globus pallidus of the rat. *Journal of Neuroscience, 18*, 9438–9452.

Bogacz, R., Brown, E., Moehlis, J., Holmes, P., & Cohen, J. D. (2006). The physics of optimal decision making: A formal analysis of models of performance in two-alternative forced-choice tasks. *Psychological Review*, *113*(4), 700–765.

Bogacz, R., & Gurney, K. N. (2007). The basal ganglia and cortex implement optimal decision making between alternative actions. *Neural Computation*, *19*, 442–477.

Bolam, J. P., & Bennett, B. D. (1995). Microcircuitry of the neostriatum. In M. A. Ariano & D. J. Surmeier (Eds.), *Molecular and cellular mechanisms of neostriatal function* (pp. 1–19). Austin, TX: R. G. Landes.

Brooks, R. A. (1991). Intelligence without reason. In *Proceedings of 12th international joint conference on artificial intelligence* (pp. 569–595). Sydney: Morgan Kauffmann.

Brooks, R. A. (1994). Coherent behaviour from many adaptive processes. In *From animals to animats 3: Proceedings of the 3rd international conference on the simulation of adaptive behaviour* (pp. 22–29). Cambridge, MA: MIT Press.

Churchland, P. S., & Sejnowski, T. J. (1992). *The computational brain*. Cambridge, MA: MIT Press.

Cohen, J. D., Dunbar, K., & McClelland, J. L. (1990). On the control of automatic processes: A parallel distributed-processing account of the Stroop effect. *Psychological Review*, *97*(3), 332–361.

Cools, A. R. (1980). Role of the neostriatal dopaminergic activity in sequencing and selecting behavioural strategies: Facilitation of processes involved in selecting the best strategy in a stressful situation. *Behavioral Brain Research*, *1*, 361–378.

Deniau, J. M., & Chevalier, G. (1985). Disinhibition as a basic process in the expression of striatal functions II. The striato-nigral influence on thalamocortical cells of the ventromedial thalamic nucleus. *Brain Research*, *334*, 227–233.

Fiori, S. (2000). An experimental comparison of three PCA neural networks. *Neural Processing Letters*, *11*, 209–218.

Gillies, A., & Willshaw, D. (2006). Membrane channel interactions underlying rat subthalamic projection neuron rhythmic and bursting activity. *Journal of Neurophysiology*, *95*(4), 2352–2365.

Gurney, K. N. (1997). *An introduction to neural networks*. London: UCL Press.

Gurney, K. N., Humphries, M., Wood, R., Prescott, T. J., & Redgrave, P. (2004a). Testing computational hypotheses of brain systems function: A case study with the basal ganglia. *Network*, *15*(4), 263–290.

Gurney, K. N., & Overton, P. G. (2004). A model of short and long range selective processing in neostriatum. *Neurocomputing*, *58–60C*, 555–562.

Gurney, K. N., Prescott, T. J., & Redgrave, P. (2001a). A computational model of action selection in the basal ganglia I: A new functional anatomy. *Biological Cybernetics*, *84*, 401–410.

Gurney, K. N., Prescott, T. J., & Redgrave, P. (2001b). A computational model of action selection in the basal ganglia II: Analysis and simulation of behaviour. *Biological Cybernetics*, *84*, 411–423.

Gurney, K. N., Prescott, T. J., Wickens, J. R., & Redgrave, P. (2004b). Computational models of the basal ganglia: From robots to membranes. *Trends in Neuroscience*, *27*(8), 453–459.

Hanson, J. E., Smith, Y., & Jaeger, D. (2004). Sodium channels and dendritic spike initiation at excitatory synapses in globus pallidus neurons. *Journal of Neuroscience*, *24*(2), 329–340.

Hoover, J. E., & Strick, P. L. (1993). Multiple output channels in the basal ganglia. *Science*, *259*, 819–821.

Humphries, M. D., & Gurney, K. N. (2001). A pulsed neural network model of bursting in the basal ganglia. *Neural Networks, 14*, 845–863.

Humphries, M. D., & Gurney, K. N. (2002). The role of intra-thalamic and thalamo-cortical circuits in action selection. *Network: Computation in Neural Systems, 13*, 131–156.

Humphries, M. D., & Gurney, K. N. (2006). A means to an end: Validating models by fitting experimental data. *Neurocomputing, 70*(10–11), 1892–1896.

Humphries, M. D., Stewart, R. D., & Gurney, K. N. (2006). A physiologically plausible model of action selection and oscillatory activity in the basal ganglia. *Journal of Neuroscience, 26*(50), 12921–12942.

Izhikevich, E. M. (2007). *Dynamical systems in neuroscience: The geometry of excitability and bursting.* Cambridge, MA: MIT Press.

Kitano, H. (2002). Computational systems biology. *Nature, 420*(6912), 206–210.

Koch, C. (1999). *The biophysics of computation: Information processing in single neurons.* New York: Oxford University Press.

Kotaleski, J. H., Plenz, D., & Blackwell, K. T. (2006). Using potassium currents to solve signal-to-noise problems in inhibitory feedforward networks of the striatum. *Journal of Neurophysiology, 95*(1), 331–341.

Kropotov, J. D., & Etlinger, S. C. (1999). Selection of actions in the basal ganglia – thalamocortical circuits: Review and model. *International Journal of Psychophysiology, 31*, 197–217.

Laruelle, M., Abi-Dargham, A., Gil, R., Kegeles, L., & Innis, R. (1999). Increased dopamine transmission in schizophrenia: Relationship to illness phases. *Biological Psychiatry, 46*(1), 56–72.

Magill, P. J., Bolam, J. P., & Bevan, M. D. (2000). Relationship of activity in the subthalamic nucleus–globus pallidus network to cortical electroencephalogram. *Journal of Neuroscience, 20*, 820–833.

Magill, P. J., Bolam, J. P., & Bevan, M. D. (2001). Dopamine regulates the impact of the cerebral cortex on the subthalamic nucleus–globus pallidus network. *Neuroscience, 106*, 313–330.

Marr, D. (1982). *Vision: A computational investigation into human representation and processing of visual information.* New York: W. H. Freeeman.

Marr, D., & Poggio, T. (1976). *From understanding computation to understanding neural circuitry* (Tech. Rep. No. AIM-357). Cambridge, MA: MIT Press.

McFarland, D. (1989). *Problems of animal behaviour.* New York: Longman Scientific and Technical.

McHaffie, J. G., Stanford, T. R., Stein, B. E., Coizet, V., & Redgrave, P. (2005). Subcortical loops through the basal ganglia. *Trends in Neuroscience, 28*(8), 401–407.

Mink, J. W. (1996). The basal ganglia: Focused selection and inhibition of competing motor programs. *Progress in Neurobiology, 50*, 381–425.

Mink, J. W., & Thach, W. T. (1993). Basal ganglia intrinsic circuits and their role in behaviour. *Current Opinion in Neurobiology, 3*, 950–957.

Pinault, D., & Deschenes, M. (1998). Anatomical evidence for a mechanism of lateral inhibition in the rat thalamus. *European Journal of Neuroscience, 10*(11), 3462–3469.

Plenz, D., & Kitai, S. T. (1999). A basal ganglia pacemaker formed by the subthalamic nucleus and external globus pallidus. *Nature, 400*, 677–682.

Prescott, A. J., Gonzales, F. M., Gurney, K. N., Humphries, M., & Redgrave, P. (2005). A robot model of the basal ganglia: Behavior and intrinsic processing. *Neural Networks, 19*(1), 31–61.

Prescott, T. J., Redgrave, P., & Gurney, K. N. (1999). Layered control architectures in robots and vertebrates. *Adaptive Behaviour, 7*, 99–127.

Redgrave, P. (2007). *Basal ganglia.* Available at http://www.scholarpedia.org/article/Basal_Ganglia

Redgrave, P., Prescott, T. J., & Gurney, K. N. (1999). The basal ganglia: A vertebrate solution to the selection problem? *Neuroscience, 89*, 1009–1023.

Romanelli, P., Esposito, V., Schaal, D. W., & Heit, G. (2005). Somatotopy in the basal ganglia: Experimental and clinical evidence for segregated sensorimotor channels. *Brain Research. Brain Research Reviews, 48*(1), 112–128.

Stafford, T., & Gurney, K. N. (2007). Biologically constrained action selection improves cognitive control in a model of the Stroop task, *Philosophical Transactions of the Royal Society B: Biological Sciences, 362*(1485), 1671–1684.

Webb, B. (2001). Can robots make good models of biological behaviour? *Behavioural and Brain Sciences, 24*(6), 1033–1050.

Wickens, J. R., Kotter, R., & Alexander, M. E. (1995). Effects of local connectivity on striatal function: Simulation and analysis of a model. *Synapse, 20*, 281–298.

Wilson, C. J., & Kawaguchi, Y. (1996). The origins of the two-state spontaneous membrane potential fluctuations of neostriatal spiny neurons. *Journal of Neuroscience, 16*(7), 2397–2410.

Wolf, J. A., Moyer, J. T., Lazarewicz, M. T., Contreras, D., Benoit-Marand, M., O'Donnell, P., et al. (2005). NMDA/AMPA ratio impacts state transitions and entrainment to oscillations in a computational model of the nucleus accumbens medium spiny projection neuron. *Journal of Neuroscience, 25*(40), 9080–9095.

Wood, R., Gurney, K. N., & Wilson, C. J. (2004). A novel parameter optimisation technique for compartmental models applied to a model of a striatal medium spiny neuron. *Neurocomputing, 58–60C*, 1109–1116.

7 Some fingerprints of V1 mechanisms in the bottom-up saliency for visual selection

Li Zhaoping, Keith A. May, and Ansgar Koene

A unique vertical bar among horizontal bars is salient and pops out perceptually regardless of the observer's goals. Physiological data have suggested that mechanisms in the primary visual cortex (V1) contribute to the high saliency of such a unique basic feature, but the data fail to indicate whether V1 plays an essential or peripheral role in input-driven or bottom-up saliency. Meanwhile a biologically based V1 model has suggested that V1 mechanisms can also explain bottom-up saliencies beyond the pop out of basic features (Li, 1999a, 2002). For instance, the low saliency of a unique conjunction feature like a red-vertical bar among red-horizontal and green-vertical bars is explained under the hypothesis that the bottom-up saliency at any location is signaled by the activity of the most active cell responding to it, regardless of the cell's preferred features such as color and orientation. While some recent experimental data have provided support for this V1 saliency hypothesis, higher visual areas such as V2 and V4 also contain neurons tuned to similar basic features that can pop out in the bottom-up manner. Furthermore, previous saliency models can capture much of the visual selection behavior using generic rather than V1-specific neural mechanisms. It is therefore important to ascertain V1's role in saliency by identifying visual selection behavior that shows specific identifying characteristics – that is, fingerprints – of V1 or other cortical areas. In this chapter, we present our recent findings on bottom-up saliency-based behavior of visual search and segmentation that directly implicate V1 mechanisms. The three specific fingerprints are: (1) an ocular singleton captures attention despite being elusive to awareness, (2) V1's collinear facilitation is manifested in texture segmentation, and (3) there is a match between the redundancy gains in double-feature singleton search and V1's conjunctive cells.

1 Introduction

Limitations in cognitive resources force us to select only a fraction of the visual input for detailed attentive processing. Naturally, we are more aware of intentional selections such as directing our gaze to text while reading or being attracted to red colors when looking for a red cup. Indeed, many models of visual attentional mechanisms, such as the stimulus similarity framework (Duncan & Humphreys, 1989), the selective tuning theory of attention (Tsotsos, 1990), and the biased competition model (Desimone & Duncan, 1995), focus mainly on goal-directed or top-down attention and treat selection based on bottom-up saliency as something given, without detailed exploration of its mechanisms. Nevertheless, much of the visual selection is carried out in a bottom-up manner, which can be dominant in selections very soon after visual stimulus onset (Jonides, 1981; Nakayama & Mackeben, 1989; Yantis, 1998). For instance, a vertical bar among horizontal ones or a red dot among green ones automatically pops out to attract perceptual attention (Treisman & Gelade, 1980), typically regardless of the task demands. Such pop-out stimuli are said to be highly salient pre-attentively. Indeed, goal-directed or top-down attention has to work with or even against the bottom-up selection (Zhaoping & Dayan, 2006; Zhaoping & Guyader, 2007). In this chapter, we focus on understanding the mechanisms underlying the bottom-up saliency that automatically guides visual selection.

Physiologically, a neuron in the primary visual cortex (V1) gives a higher response to its preferred feature (e.g., a specific orientation, color, or motion direction) within its receptive field (RF) when this feature is unique within the display, rather than when it is surrounded by neighbors identical to itself (Allman, Miezin, & McGuinness, 1985; DeAngelis, Freeman, & Ohzawa, 1994; Jones, Grieve, Wang, & Sillito, 2001; Knierim & Van Essen, 1992; Li & Li, 1994; Nothdurft, Gallant, & Van Essen, 1999, 2000; Sillito, Grieve, Jones, Cudeiro, & Davis, 1995; Wachtler, Sejnowski, & Albright, 2003; Webb, Dhruv, Solomon, Tailby, & Lennie, 2005). This is the case even when the animal is under anesthesia (Nothdurft et al., 1999), suggesting bottom-up mechanisms. The responsible mechanism is iso-feature suppression – in particular, iso-orientation or iso-color suppression – so that nearby neurons tuned to the same feature suppress each other's activities via intracortical connections between nearby V1 neurons (Gilbert & Wiesel, 1983; Hirsch & Gilbert, 1991; Rockland & Lund, 1983). The same mechanisms also make V1 cells respond more vigorously to an oriented bar when it is at the border, rather than the middle, of a homogeneous orientation texture, as physiologically observed (Nothdurft et al., 2000), since the bar has fewer iso-orientation neighbors at the border. These observations have prompted suggestions that V1 mechanisms contribute to bottom-up saliency for pop-out features like the unique-orientation singleton or the bars at an orientation-texture border (e.g., Knierim & Van Essen, 1992; Nothdurft et al., 1999, 2000; Sillito et al., 1995). This is consistent with observations that highly salient inputs can bias responses in extra-striate areas receiving inputs from V1 (Beck & Kastner 2005; Reynolds & Desimone, 2003).

Behavioral studies have extensively examined bottom-up saliencies in visual search and segmentation tasks (Duncan & Humphreys, 1989; Treisman & Gelade, 1980; Wolfe, Cave, & Franzel, 1989), showing more complex, subtle, and general situations beyond basic feature pop-outs. For instance, a unique feature conjunction – for example, a red-vertical bar as a color-orientation conjunction among red-horizontal and green-vertical bars – is typically less salient; ease of searches can change with target-distractor swaps; and target salience decreases with background irregularities. However, few physiological recordings in V1 have used stimuli of comparable complexity, leaving it open as to how generally V1 mechanisms contribute to bottom-up saliency.

Recently, a model of contextual influences in V1 (Li, 1999a, 1999b, 2000, 2002), including physiologically observed iso-feature suppression and collinear facilitation (Kapadia, Ito, Gilbert, & Westheimer, 1995), has demonstrated that V1 mechanisms can feasibly explain the complex behaviors mentioned above, assuming that the highest response among V1 cells to a target, relative to all other responses to the scene, determines its salience and thus the ease of a task. Accordingly, V1 has been proposed to create a bottom-up saliency map, such that the RF location of the most active V1 cell is most likely selected for further detailed processing (Li, 1999a, 2002). We call this proposal the V1 saliency hypothesis. This hypothesis is consistent with the observation that microstimulation of a V1 cell can drive saccades, via superior colliculus, to the corresponding RF location (Tehovnik, Slocum, & Schiller, 2003) and that higher V1 responses are associated with quicker saccades to the corresponding receptive fields (Super, Spekreijse, & Lamme 2003). This can be clearly expressed algebraically. Let $(O_1, O_2, \ldots, O_M)$ denote outputs or responses from V1 output cells indexed by $i = 1, 2, \ldots,$ M, and let each cell cover receptive field location $(x_1, x_2, \ldots, x_M)$ respectively. Then, the highest response among all cells is $\hat{O} \equiv \max_i O_i$. Let this response $\hat{O}$ be from a cell indexed by $\hat{i}$; mathematically, $\hat{i} \equiv \text{argmax}_i\ O_i$. This cell's receptive field is then at $\hat{x} \equiv x_{\hat{i}}$ and is the most salient or most likely selected by the bottom-up visual selection. The receptive field location of the second-most-responsive cell is the second-most-salient or second-most-likely selected by the bottom-up mechanism, and so on. Note that the interpretation of $x_i = x$ is that the receptive field of cell i covers location x and is centered near x. Defining $\hat{O}(x) \equiv \max_{x_i = x} O_i$ as the highest response among neurons whose receptive field covers location x, then $\hat{O} = \max_x \hat{O}(x)$ – that is, the highest response among all cells is the maximum of $\hat{O}(x)$ among all x. Now define SMAP(x) as the saliency of a visual location x, such that the value of SMAP(x) increases with the likelihood of location x to be selected by bottom-up mechanisms. From the definitions above, it is then clear that, given an input scene,

1   SMAP(x) increases with the maximum response

$$\hat{O}(x) \equiv \max_{x_i = x} O_i \tag{1}$$

to x, and, regardless of their feature preferences, the less activated cells responding to x do not contribute;

2   $\hat{O}(x)$ is compared with $\hat{O}(x')$ at all x' to determine SMAP(x), since

$$\hat{O} = \max_x \hat{O}(x); \tag{2}$$

3   the most likely selected location is

$$\hat{x} = \text{argmax}_x \text{SMAP}(x), \tag{3}$$

where SMAP(x) is maximum.

As salience merely serves to order the priority of inputs to be selected for further processing, only the order of the salience is relevant. However, for convenience we could write Equation (1) as SMAP(x) = $\hat{O}(x)/\hat{O}$, with the denominator $\hat{O}$ as the normalization.

Meanwhile, some experimental observations have raised doubts regarding V1's role in determining bottom-up saliency. For instance, Hegde and Felleman (2003) found that, from V1 cells tuned to both orientation and color to some degree, the responses to a uniquely colored or uniquely oriented target bar among a background of homogeneously colored or oriented bars are not necessarily higher than the responses to a target bar defined by a unique conjunction of color and orientation. According to the V1 hypothesis, the saliency of a target is determined by its evoked response relative to that evoked by the background. Hence, the response to the most salient item in one scene is not necessarily higher than the response to the intermediately salient item in a second scene, especially when the second scene, with less homogeneity in the background, evokes higher responses in a population. Hence, Hegde and Felleman's finding does not disprove the V1 hypothesis, although it does not add any confidence to it. Furthermore, neurons in the extra-striate cortical areas, such as V2 and V4, are also tuned to many of the basic features that pop out pre-attentively when uniquely present in a scene, so it is conceivable that much of the bottom-up selection may be performed by these visual areas higher than V1. Previous frameworks on pre-attentive visual selection (Duncan & Humphreys, 1989; Itti & Koch, 2000, 2001; Koch & Ullman, 1985; Treisman & Gelade 1980; Wolfe et al., 1989) have assumed separate feature maps that process individual features separately. These feature maps are considered to be more associated with the extra-striate areas, some of which seem to be more specialized for some features than for others, compared with V1. Since the different basic features in these models, such as orientation and size, act as apparently separable features in visual search, and early (striate and extra-striate) visual areas have cells tuned to conjunctions of features (such as orientation and spatial frequency), the previous frameworks suggest a relatively late locus for the feature maps. Additionally, previous models for bottom-up saliency (e.g., Itti & Koch, 2000; Koch & Ullman, 1985; Wolfe et al., 1989) assume that the activities in the feature maps are summed into a master saliency map that guides attentional selection, implying that a bottom-up saliency map should be in a higher cortical area such as the parietal cortex (Gottlieb et al., 1998).

1.1 Conceptual characteristic of the V1 saliency hypothesis

Based on the previous experimental observations and the previous models, it is not clear whether or not V1 contributes only marginally to visual saliency, so that the computations of saliency are significantly altered in subsequent brain areas after V1. If this were the case, the behavior of bottom-up selection would be devoid of characteristics of V1 other than some generic mechanisms of iso-feature surround suppression, which, necessary for basic feature pop out, could perhaps be implemented just as well in extra-striate or higher cortical areas (Allman et al., 1985). To address this question, we identify the characteristics of V1 and the V1 saliency hypothesis that are different from other visual areas or from other saliency models. First, let us consider the conceptual characteristics. The V1 hypothesis has a specific selection mechanism to select the most salient location from the V1 responses – namely, that the RF location of the most activated cell is the most salient location regardless of the preferred feature of this cell. This means that the activities of the neurons are like universal currency bidding for selection, regardless of the neuron's preferred features (Zhaoping, 2006). As argued above, this character led to the saliency value SMAP(x) $\propto \max_{x_i = x} O_i$ at location x, and we will call this the MAX rule for calculating saliency. It therefore contrasts with many previous bottom-up saliency models (Itti & Koch, 2000; Koch & Ullman, 1985; Wolfe et al., 1989) that sum activities from different feature maps – for example, those tuned to color, orientation, motion directions, etc. – to determine saliency at each location, and we will refer to this as the SUM rule, SMAP(x) $\propto \Sigma_{x_i = x} O_i$. Recently, Zhaoping and May (2007) showed that the MAX rule predicts specific interference by task-irrelevant inputs on visual segmentation and search tasks, and that such predictions are confirmed psychophysically (see Figure 7.1). Note that the MAX rule acts on the responses from V1 rather than imposing mechanisms within V1 for creating the responses to select from. It arises from the unique assumption in the V1 saliency hypothesis that no separate feature maps, nor any combination of them, are needed for bottom-up saliency (Li, 2002). In other words, the MAX rule would not preclude a saliency map in, say, V2, as long as no separate feature maps or any summation of them are employed to create this saliency map. Hence, while the MAX rule supports the V1 hypothesis, this rule by itself cannot be a fingerprint of V1.

1.2 Neural characteristics that can serve as fingerprints of V1

We identify three neural characteristics of V1. The first characteristic is the abundance of monocular cells. These cells carry the eye-of-origin information. Most V1 neurons are monocular (Hubel & Wiesel, 1968), whereas any higher visual area has only a few monocular cells (Burkhalter & Van Essen, 1986). The second characteristic is collinear facilitation – that is, a neuron's

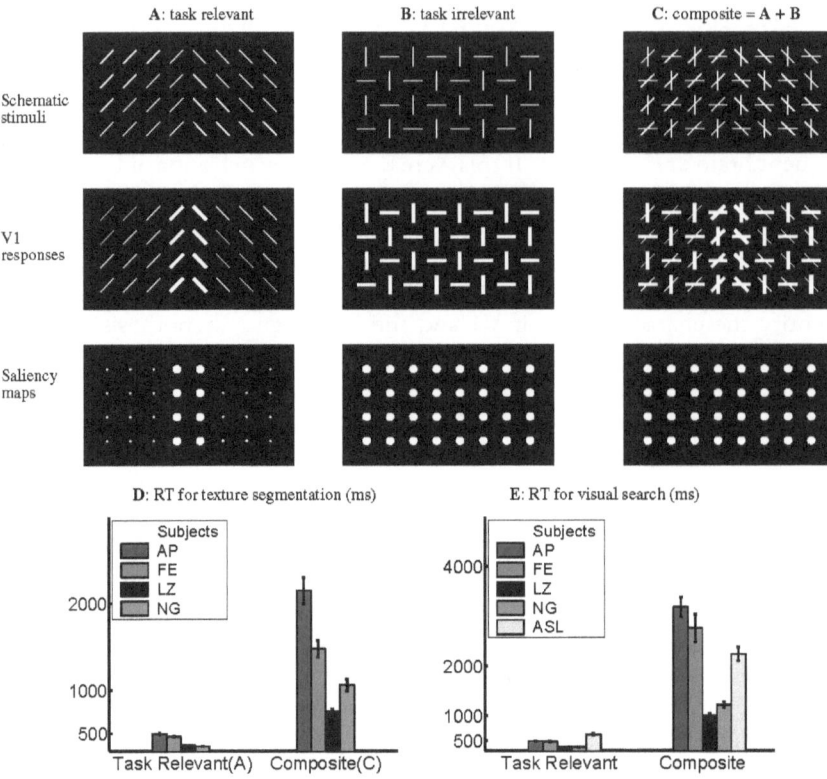

Figure 7.1 Prediction of the MAX rule by the V1 saliency hypothesis – interference by task-irrelevant features, and its psychophysical test (adapted from Zhaoping & May, 2007). Panels A, B, C are schematics of texture stimuli (extending continuously in all directions beyond the portions shown), each followed by schematic illustrations of its V1 responses, in which the orientation and thickness of a bar denote the preferred orientation and response level, respectively, of the most activated neuron by an input bar. Below each V1 response pattern is a saliency map, in which the size of a disk corresponds to the response of the most activated neuron at the texture element location. The orientation contrasts at the texture border in Panel A and everywhere in Panel B lead to less suppressed responses to the stimulus bars since these bars have fewer iso-orientation neighbors to evoke iso-orientation suppression. The composite stimulus C, made by superposing A and B, is predicted to be difficult to segment, since the task-irrelevant features from B interfere with the task-relevant features from A, giving no saliency highlights to the texture border. Panel D gives the reaction times for texture segmentation testing the prediction (differently colored bars denote different subjects). Panel E is like Panel D, but for a task to search for an orientation singleton. The stimuli were made from those in the segmentation task by shrinking one of the two texture regions into a single texture element. RT for the composite condition is significantly higher ($p < .001$). Stimuli for the experiments in Figures 7.1, 7.3, 7.4, and 7.5 consist of 22 rows × 30 columns of items (of single or double bars) on a regular grid with unit distance 1.6° of visual angle. [A color version of this figure can be viewed at www.cognitiveneurosciencearena.com/brain-scans]

response to an optimally oriented bar within its RF is enhanced when a neighboring bar outside the RF is aligned with the bar within the RF such that they could be seen as segments of a smooth contour. It has been observed in V1 since the 1980s (Kapadia et al., 1995; Nelson & Frost, 1985) and is inherited by V2 (Bakin, Nakayama, & Gilbert, 2000; von der Heydt, Peterhans, & Baumgartner 1984), but it could not exist in visual stages before V1 without any cells tuned to orientation. Collinear facilitation is observed psychophysically only when target and flankers are presented to the same eye, suggesting that the phenomenon depends on links between monocular cells (Huang, Hess, & Dakin, 2006), and thus a V1 origin. The third characteristic is the feature-specific conjunctive cells in V1. V1 has cells tuned conjunctively to a specific orientation and motion direction, or conjunctively to specific orientation and color (Hubel & Wiesel, 1959; Livingstone & Hubel, 1984; Ts'o & Gilbert, 1988) but has almost no cells tuned conjunctively to specific color and motion direction (Horwitz & Albright, 2005). This is not the case in V2 or V3, where there are cells tuned conjunctively to all of the three pairwise-possible conjunctions of feature dimensions (Gegenfurtner, Kiper, & Fenstemaker, 1996; Gegenfurtner, Kiper, & Levitt, 1997; Shipp, 2007 private communication; Tamura, Sato, Katsuyama, Hata, & Tsumoto, 1996), and the higher cortical neurons are expectedly selective to more complex input features than is V1.

If V1's responses indeed dictate saliency by evoking responses (e.g., via superior colliculus to drive eye movement and selection) before the involvement of the subsequent visual areas, these V1 specific characteristics should be reflected in the corresponding selection behavior. These fingerprints, corresponding to the three neural characteristics of V1, are specifically as follows. First, given V1's monocular cells, and its mechanism of iso-ocular suppression (DeAngelis et al., 1994; Webb et al., 2005) as an instantiation of iso-feature suppression responsible for a feature singleton to pop out, the V1 saliency hypothesis predicts that an ocular singleton should capture attention automatically. It is known that eye-of-origin information is typically elusive to visual awareness (Kolb & Braun, 1995; Morgan, Mason, & Solomon, 1997; Wolfe & Franzel, 1988). This is consistent with the idea that, unlike higher cortical areas, information available in V1 is usually at most weakly associated with awareness (see reviews by Crick & Koch, 1995; Tong, 2003). Hence, attention capture by an ocular singleton even without awareness would be a hallmark of V1 and, perhaps, the ultimate bottom-up or exogenous visual selection. Second, collinear facilitation suggests that, between oriented bars, contextual influences that determine saliency are not isotropic. Consequently, the selection behavior in stimuli consisting of orientation textures should depend on the spatial configuration in these stimuli in specific non-isotropic ways that are signatures of the collinear facilitation mechanism in V1. Third, consider the saliency of a red-vertical bar among green-vertical bars, and of a red-vertical bar among red-horizontal bars, and of a red-vertical bar among green-horizontal bars. We will refer to the first two as single-feature (color or orientation) saliency and the last as the double-feature (color–orientation) saliency and expect that the double-feature singleton should be somewhat more salient than the single-feature ones. The magnitude of this double-feature advantage, or the feature redundancy gain, should depend on whether the conjunctive cells for the two features concerned exist. Hence, the existence of V1's conjunctive cells in some combinations of feature dimensions and not others should create a corresponding, feature-dimension-specific, pattern of double-feature advantages.

In the next section, we review the behavioral fingerprints of V1 mechanisms in detail and illustrate how they arise as predictions of the V1 hypothesis. The experimental data confirming these predictions are then shown. All the details of the experiments have been published (Koene & Zhaoping, 2007; Zhaoping, 2008; Zhaoping & May, 2007; Zhaoping & Snowden, 2006). The presentation in this chapter not only reviews the published results for the purpose of summarizing and highlighting the fingerprints of V1 in saliency, but also presents some different perspectives and analysis of the published results. We summarize with discussions in section 3.

2   Predicted fingerprints and their experimental tests

To predict behavioral consequences of V1's responses that are used for bottom-up saliency, we need to know the most relevant V1 characteristics, which are summarized as follows: (1) neural tuning to basic features within its RF, such that, for example, a neuron tuned to color responds more to preferred than to non-preferred colors; (2) iso-feature suppression, which suppresses a neuron's response to a preferred feature within its RF when there are inputs of the same feature outside and yet near its RF; (3) general surround suppression: that is, a neuron's response is suppressed by activities in all nearby neurons regardless of their feature preferences (this suppression is weaker than the iso-feature suppression but introduces interactions between neurons tuned to different features); (4) collinear facilitation: enhancement of a neuron's response to an optimally oriented bar within its RF when a contextual bar outside its RF is aligned with the bar within; (5) neural tuning to conjunctions of orientation and motion direction (OM), or color and orientation (CO), but not to color and motion direction (CM); (6) some V1 neurons are monocular and thus are tuned to eye of origin. Mechanisms (1) and (2) are essential for unique feature pop-out – for example, a singleton red pops out of many green items since a red-tuned cell responding to the singleton does not suffer from the iso-color suppression imposed on the green-tuned neurons responding to the background items (Li, 1999a, 1999b). Mechanism (3) will modify the contextual influences to modulate but typically not dictate the saliency outcome, as will be discussed later. One may argue that Mechanisms (1) to (3) – except for the neural tuning to eye of origin and iso-ocular suppression, as specific examples of (1) and (2) – are generic and are also present in higher visual areas (Allman et al.,

1985). V1's fingerprints on saliency behavior will have to arise from Mechanisms (4), (5), and (6), which we will show to manifest in the saliency outcome in a predictable way. (Even though V2 also manifests Mechanism (4), we consider Mechanism (4) as special for V1, given the psychophysical evidence – Huang et al., 2006 – for its V1 origin.)

2.1 The fingerprint of V1's monocular cells

V1 is the only cortical area that has a substantial number of cells tuned to ocular origin – that is, being differentially sensitive to inputs from the different eyes or receiving inputs dominantly from one eye only. Since a V1 neuron's response is suppressed more by contextual inputs presented to the same rather than a different eye (DeAngelis et al., 1994; Webb et al., 2005), a V1 neuron responding to an ocular singleton – that is, an input item with a different eye of origin from that of all other input items – is expected to give a higher response than the V1 neurons responding to any of many identical input items seen through the other eye. In other words, an ocular singleton should be salient to capture attention.

Since inputs that differ only in their eye of origin typically appear identical to human subjects, it is difficult to directly probe whether an ocular singleton pops out by asking subjects to search for it (Kolb & Braun, 1995; Wolfe & Franzel, 1988). This fingerprint can, however, be tested by making the ocular-singleton task-irrelevant and observing its effect on the performance in a task that requires attention to a task-relevant location (Zhaoping, 2007, 2008). In one experiment, observers searched for an orientation singleton among background horizontal bars. The search display was binocularly masked after only 200 ms, and the subjects were asked to report at their leisure whether the tilt singleton was tilted clockwise or anticlockwise from horizontal. The display was too brief for subjects to saccade about the display looking for the target, which was tilted only 20° from the background bars. Hence, this task was difficult unless subjects' attention was somehow covertly guided to the target. Unaware to the subjects (except one, the first author), some trials were dichoptic congruent (DC), when the target was also an ocular singleton; some were dichoptic incongruent (DI), when a distractor on the opposite lateral side of the target from the display center was an ocular singleton; and the other trials were monocular (M), when all bars were seen by the same single eye (see Figure 7.2A). If the ocular singleton can exogenously cue attention to itself, subjects' task performance should be better in the DC condition. This was indeed observed (Figure 7.2B). A control experiment was subsequently carried out to probe whether the same observers could detect the attention-capturing ocular singleton if they were informed of its existence. The stimuli were the same, except that all bars were horizontal. Randomly, in half of the trials, an ocular singleton was at one of the same locations as before, and the observers were asked to report by forced choice whether an ocular singleton existed. Their performance was better than the chance level only

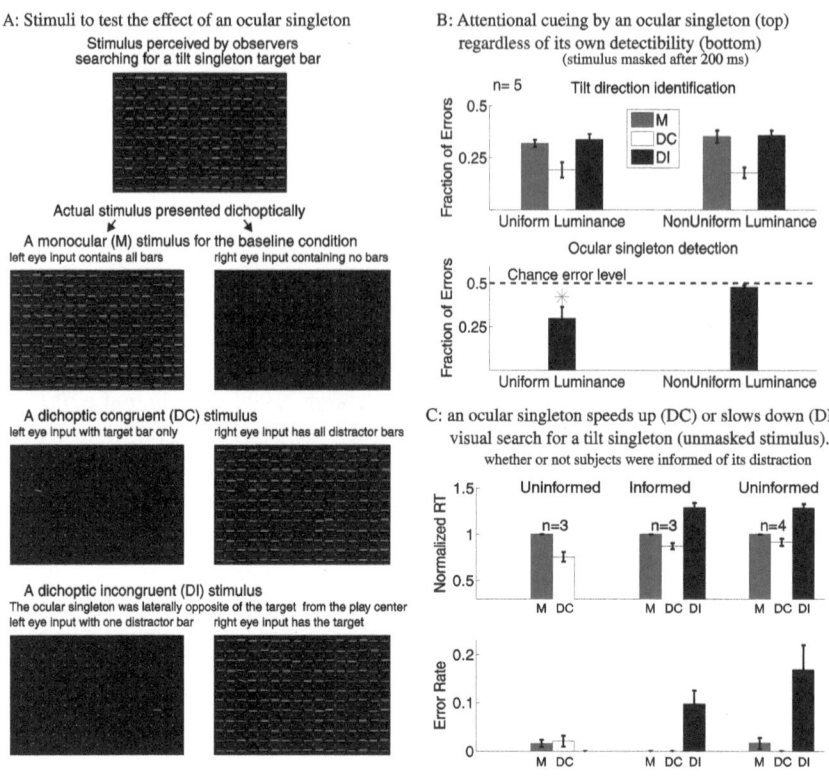

Figure 7.2 An ocular singleton captures attention automatically. (A) Illustrations of the stimulus for visual search for an orientation singleton, and the various dichoptic presentation conditions: dichoptic congruent (DC), dichoptic incongruent (DI), and monocular (M). The actual stimuli used had 22 rows × 30 columns of bars, spanning 34° × 46° in visual angle. From the display center, the search target had an eccentricity of ~15°, and at least 12° horizontal eccentricity. The eye of origin of the task-critical bar(s) was random. (B) Fractions of error trials for reporting the tilt direction of the tilt singleton (top) in a brief (200-ms) display (like in Panel A, except that in half of the trials, all bars had the same, uniform, luminance), and for reporting whether the ocular singleton was present in the same stimuli without the tilt singleton (bottom; "*" denotes significant difference from the chance error level). The left and right halves of the (top and bottom) plots are for when the stimulus bars had uniform or non-uniform (as in Panel A) luminance values, respectively. Tilt identification was best in the DC condition (top), independent of the ability (depending on whether the bars had uniform luminance values) to detect the ocular singleton beyond the chance level (bottom). Data are averages from $n = 5$ subjects, who, before the ocular-singleton detection task, were acquainted with the ocular singleton in an example stimulus displayed for as long as necessary. (C) Reaction times (top, RT_M, RT_{DC}, and RT_{DI} in M, DC, and DI conditions, respectively) and fractions of error trials (bottom) for reporting whether the tilt singleton was in the left or right half of the display, which stayed unmasked before subjects' reports. Each subject's RT was normalized by

when all bars had the same (uniform) luminance value (Figure 7.2B, bottom). Meanwhile, the same ocular singleton, whether it was detectable by forced choice or not, had demonstrated the same degree of cueing effect (Figure 7.2B, top). This suggests that the ocular singleton cued attention completely exogenously to its location, facilitating the identification of the tilt singleton in the DC condition. The M and DI conditions can be seen as the uncued and invalidly cued conditions, respectively. Note that the tilt singleton in the M condition, with a 20° orientation contrast from the background bars, should pop out in a typical visual search task when the search stimulus stays unmasked, at least when the bars had uniform luminance. Our data suggest that the ocular singleton was more salient than the orientation singleton.

In three additional experiments, the search display stayed unmasked until the subjects responded, and the orientation contrast between the target and distractors was 50°. Observers were asked to report as soon as possible whether the tilt singleton was in the left or the right half of the display. Their reaction times (RTs) in reporting were shorter in the DC, and longer in the DI, than in the M condition, regardless of whether the observers were aware or informed of the existence of the different task-irrelevant dichoptic conditions (Figure 7.2C). Note that RT_{DI} (the RT in the DI condition) was about 200 ms longer than RT_{M} (the RT in the M condition). This 200-ms difference is about an average fixation duration in typical visual search tasks (Hooge & Erkelens, 1998). Hence, our findings suggest that, in typical trials, attention was more quickly attracted to the target in the DC condition, and initially distracted from the target in the DI condition. In particular, our data suggest that, in a typical DI trial, subjects saccaded to the ocular singleton distractor first, realizing that it was not the target, before shifting attention to the target. Hence, an ocular singleton, though elusive to awareness, can effectively compete for attention with an orientation singleton of even 50° contrast from the background bars. This is consistent with the finding that subjects also made more errors in the DI condition: presumably, in a hurry to respond, they easily mistook the ocular singleton distractor as the target. Furthermore, the high error rates persisted even when the subjects were informed that

---

his/her RT_{M} (~700 ms). All data are averages among subjects. Stimuli were as in A except that all bars had the same (uniform) luminance, and the target and non-target bars were tilted 25° from the horizontal in opposite directions. The left, middle, and right parts of the (top and bottom) plots are results from three different experiments, respectively, employing $n = 3$, 3, and 4 subjects, respectively. The first experiment (*left*) included M and DC conditions; both the second (*middle*) and third (*right*) experiments included M, DC, and DI conditions. In the first and third experiments, subjects were uninformed (and unaware, except for one subject in the third experiment, who became aware of an attention-capturing distractor) of the existence of the various task-irrelevant dichoptic conditions. In the second experiment, subjects were informed (before data taking) of a possible attention-capturing distractor in some trials. Note that $RT_{DC} < RT_{M} < RT_{DI}$. The DI condition, when included, caused the most errors. [A color version of this figure can be viewed at www.cognitiveneurosciencearena.com/brain-scans]

an attention-capturing distractor could appear in some trials (Figure 7.2C, the middle group of the data). This suggests that it is not easy to suppress the saliency by an ocular singleton by top-down control.

If the eye-of-origin feature was as visible to visual awareness as some of the basic features such as color and orientation, it would be considered a basic feature, defined as one when a singleton in the feature dimension has a negligible set-size effect in visual search (i.e., when RT does not depend on the number of background items). Its elusiveness to awareness means that subjects cannot do a visual search for an ocular singleton effectively, as shown by Wolfe and Franzel (1988). However, an ocular singleton should make a difficult search easier by eliminating the set-size effect. Indeed, we found that the set-size effect in searching for a T among Ls can be eliminated when the target was also an ocular singleton (Zhaoping, 2008).

2.2 Fingerprints of V1's collinear facilitation in texture segmentation behavior

The experiments showing the fingerprints in this subsection are part of a previous study (Zhaoping & May, 2007). Here, we illustrate the fingerprints with a simulation of V1's behavior using a previously developed V1 model (Li, 1999b, 2000) and compare the model's behavior with human data. Figure 7.3 shows this fingerprint. Figure 7.3A and 7.3B both have two orientation textures with a 90° contrast between them. The texture borders pop out automatically, as the saliency of such texture borders increases with the orientation contrast at the border (Nothdurft, 1992). A texture border bar is salient since it has fewer iso-orientation neighbors than do the texture bars away from the border, and hence the neuron responding to it experiences weaker iso-orientation suppression. However, in Figure 7.3B, the vertical texture border bars in addition enjoy full collinear facilitation, since each has more collinear neighbors than do other texture border bars in either Figure 7.3A or Figure 7.3B. The vertical texture border bars are thus more salient than other border bars. In general, given an orientation contrast at a texture border, the border bars parallel to the texture border are predicted to be more salient than other border bars (Li, 1999b, 2000), and we call these border bars collectively a collinear border.

We hence predict that the border in Figure 7.3A takes longer to locate than the border in Figure 7.3B. This is tested in an experiment with such stimuli in which the texture border is sufficiently far from the display center to bring performance away from ceiling. We asked human subjects to press a left or right button as soon as possible after stimulus onset to indicate whether the border is in the left or the right half of the display. Our prediction is indeed confirmed (Figure 7.3G). Higher saliency of a collinear border is probably the reason why Wolfson and Landy (1995) observed that it is easier to discriminate the curvature of a texture border when it is collinear than otherwise.

Note that since both texture borders in Figure 7.3A and 7.3B are salient

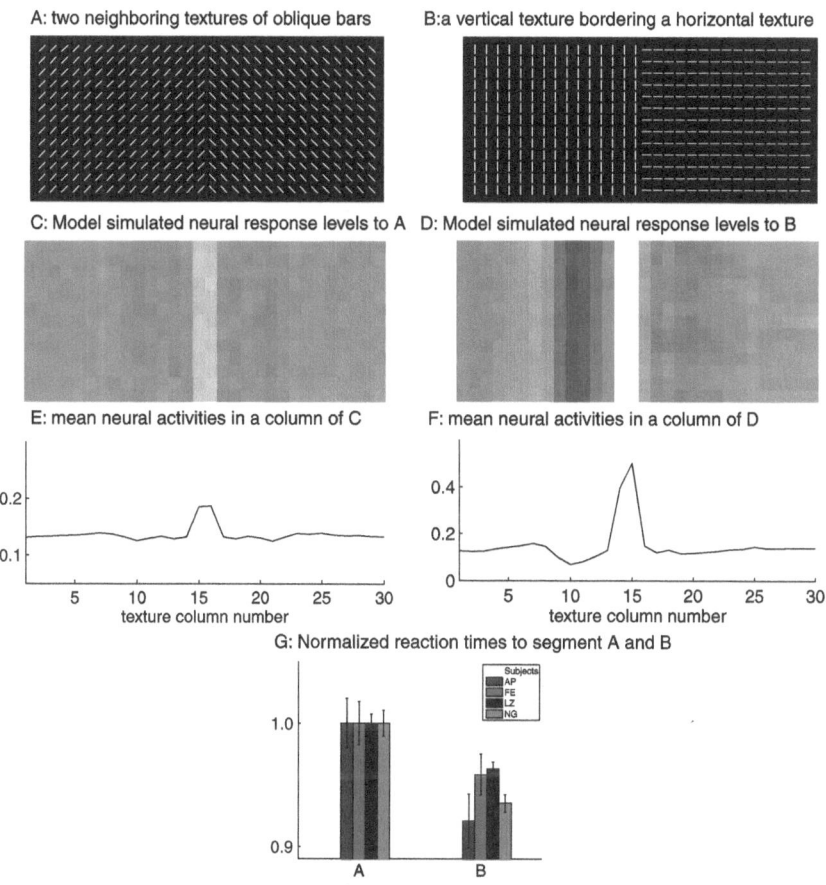

Figure 7.3 Fingerprint of the collinear facilitation in V1: a texture border with texture bars parallel to the border is more salient. (A, B) Stimulus patterns for
texture segmentation; each contains two neighboring orientation textures
with a 90° orientation contrast at the texture border. The texture border in
B appears more salient. (C, D) Simulation results from a V1 model (Li,
1999b, 2000 – used in all model simulations in this chapter) on the neural
activity levels in space for stimulus patterns A and B, respectively. Higher
activities are visualized by a lighter luminance at the corresponding image
location. (E, F) Neural activities in C and D, respectively, averaged in each
texture column. G: Normalized reaction times of human observers to
locate the texture border, in the units of the reaction time RT_A of the
subject for stimulus A. The RT_A for various subjects are respectively, 493,
465, 363, 351 ms. For each subject (same as in Figure 7.1D), it is statistically significant that $RT_A > RT_B$ ($p < .05$). [A color version of this figure
can be viewed at www.cognitiveneurosciencearena.com/brain-scans]

enough to require only short RTs, and since RTs cannot be shorter than a
certain minimum for each subject, a large difference in the degrees of border
highlights in our two stimuli can only give a small difference in their required

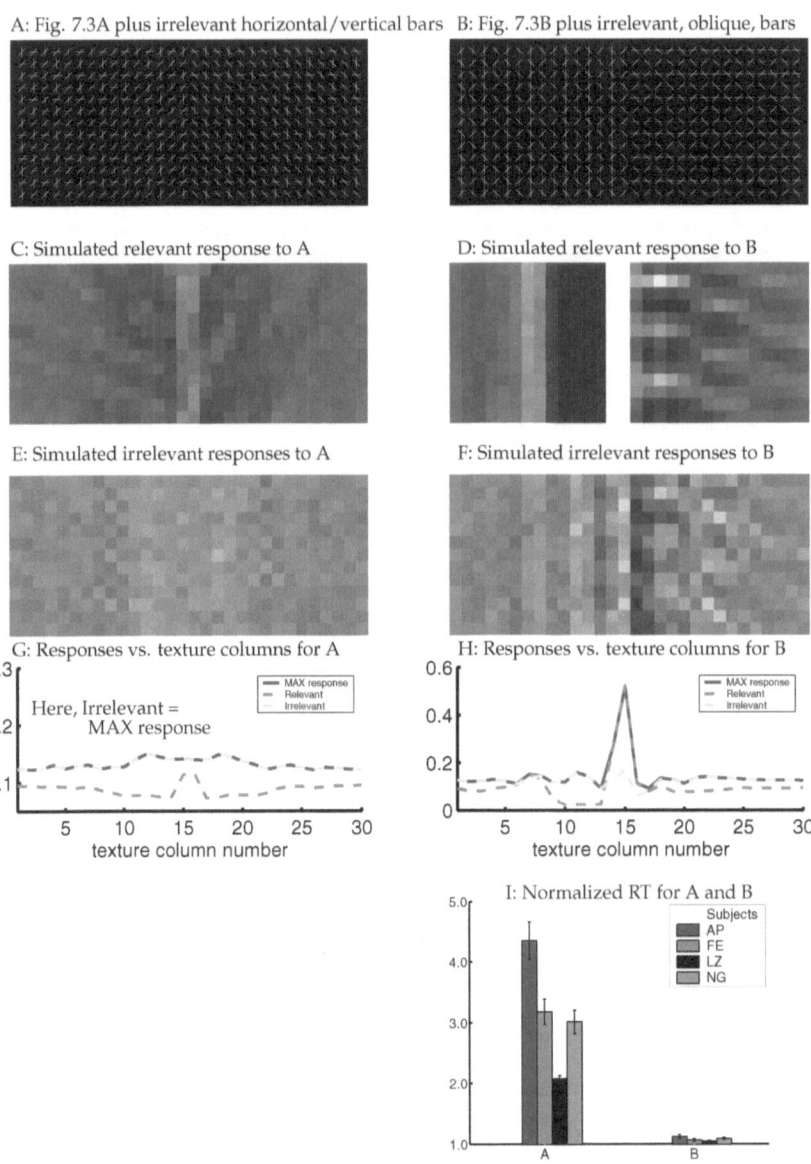

Figure 7.4 A more salient collinear texture border is less vulnerable to interference. Panels A and B are the stimuli in Figure 7.3A–B after superposing task-irrelevant bars, which form a checkerboard pattern. The simulated relevant responses are, respectively, in Panels C and D, and the irrelevant responses in Panels E and F, using the same format as Figure 7.3C–D. Panels G and H plot the responses vs. texture columns, for relevant, irrelevant, and the maximum of them – that is, saliency. Panel I gives the normalized reaction times to A and B. The subjects are the same as in Figure 7.1D. The

RTs. We can unveil this predicted large difference in V1 responses by *interference*, explained in Figure 7.4, thereby demonstrating another manifestation of the fingerprint in Figure 7.3. Figure 7.4A is made by superposing onto Figure 7.3A a checkerboard pattern of horizontal and vertical bars, similar to those in Figure 7.1; analogously, Figure 7.4B is made by superposing onto Figure 7.3B left-oblique and right-oblique bars. The superposed checkerboard patterns are irrelevant to the task of segmenting the textures. We refer to the responses to the task-relevant and task-irrelevant stimuli as "relevant" and "irrelevant" responses, respectively; similarly, the neuron populations tuned to the relevant and irrelevant orientations are referred to as "relevant" and "irrelevant" neuron populations, respectively. By the MAX rule in the V1 saliency hypothesis, the irrelevant responses compete with the relevant ones to dictate saliency at each location. If they win the competition at some locations, they can interfere with segmentation by misleading visual attention and thus prolong the RT. As illustrated in Figure 7.1, the irrelevant response level to any texture element location is comparable to that of the relevant response to the border, since an irrelevant bar has as few iso-orientation neighbors as a relevant texture border bar. Consequently, the maximum neural response at each texture element location is roughly the same across space, and the texture border highlight is now reduced or diminished. Indeed, RTs (Figure 7.4I) for the same texture segmentation task are much longer for the stimuli shown in Figure 7.4A and 7.4B than those for stimuli without irrelevant bars (Figure 7.3). Meanwhile, it is clear that the RT for Figure 7.4B is much shorter than the RT for Figure 7.4A, as the interference is much weaker in Figure 7.4B. The extra salient, collinear, vertical border bars evoke responses that are much higher than the irrelevant responses, and they are thus less vulnerable to being submerged by the higher background saliency levels, even though the relative border salience is somewhat reduced due to the raised background salience levels.

The arguments above are qualitative since we included only iso-orientation suppression and collinear facilitation in our argument and for simplicity have omitted the effect of general surround suppression, which, although weaker than the iso-orientation suppression, causes nearby neurons responding to different orientations to suppress each other and thus modulate the overall spatial patterns of the responses. To verify our qualitative arguments, we simulated the V1 responses using our previously developed V1 model (for details sufficient for the reproduction of the model behavior, see Li, 1998, 1999b), which includes all three forms of the contextual influences:

---

normalized RT for each subject is obtained by dividing the RT for A and B, respectively, by the RT of the subject for the corresponding stimulus without irrelevant bars (i.e., Figure 7.3A and 7.3B, respectively). The interference in B, even though significant (i.e., the normalized RT is significantly larger than 1 for each subject, with $p < .02$), is much less than in A. [A color version of this figure can be viewed at www.cognitiveneurosciencearena.com/brain-scans]

iso-orientation suppression, collinear facilitation, and general suppression. The model behavior, shown in Figure 7.3C–F and Figure 7.4C–F, confirmed our qualitative analysis. In viewing the model responses, note that the highest possible responses from the model neurons (at saturation) are set to 1 and that the model includes some levels of noise simulating input or intrinsic noise in the system. Also note that, without knowledge of quantitative details of the V1 mechanisms, the quantitative details of our model should be seen only as an approximation of the reality, to supplement our qualitative predictions. Nevertheless, as the model parameters were previously developed, fixed, and published, our predictions and simulation results were produced without model parameter tuning.[1]

Additional qualitative details, although not affecting our conclusions here, are also visible in the model behavior. For example, a local suppression of relevant responses near the texture border is due to the stronger iso-orientation suppression from the more salient (relevant) border bars. This local suppression is particularly strong next to the most salient vertical border bars (in Figure 7.3D and 7.3F). We call this local suppression region next to the border the *border suppression region* (Zhaoping, 2003).

Figure 7.5 demonstrates another fingerprint of the collinear facilitation. Figure 7.5A–H are analogous to Figure 7.4A–H. The task-relevant stimulus component is that of Figure 7.3A, while the task-irrelevant stimulus components are the horizontal bars in Figure 7.5A and vertical bars in Figure 7.5B. Without orientation contrast among the task-irrelevant bars, the irrelevant responses have a similar level to relevant responses in the background, since the level of iso-orientation suppression is about the same among the irrelevant bars as that among the relevant bars in the background. Based on the MAX rule, if there were no general surround suppression enabling interaction between differently oriented bars, there would be no interference to segmentation based on the relevant bars, which evoke a response highlight at the texture border. However, general surround suppression induces interactions between local relevant and irrelevant neurons. Thus, spatially inhomogeneous relevant responses induce inhomogeneity in the irrelevant responses, despite the spatial homogeneity of the irrelevant stimulus. In particular, because the relevant responses in the border suppression region generate weaker general suppression, the local irrelevant responses are slightly higher (or less suppressed). Hence, the irrelevant response as a function of the texture column number exhibits local peaks next to the texture

1 The methods for all the model simulations for this chapter are as follows. Each displayed model response area of 30 × 13 texture grid locations is, in fact, only a small central portion of a sufficiently large area of textures without the wrap-around or periodic boundary condition, in order to avoid the artifacts of the boundary conditions. The model inputs to each visual texture bar was set at a level $\hat{I} = 1.9$ (in notation used in Li, 1999b), corresponding to the intermediate contrast-level condition. For each input image, the model simulates the neural responses for a duration of at least 12 time constants. A model neuron's output was temporally averaged to get the actual outputs displayed in the figures.

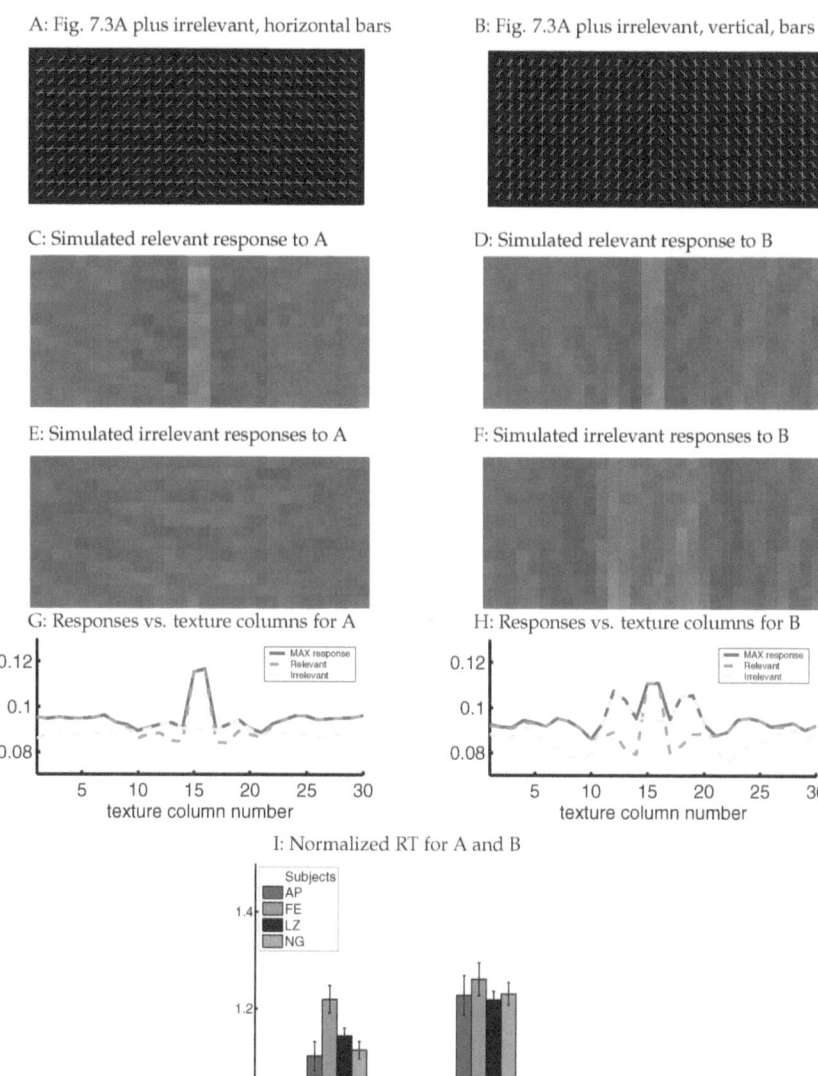

A: Fig. 7.3A plus irrelevant, horizontal bars

B: Fig. 7.3A plus irrelevant, vertical, bars

C: Simulated relevant response to A

D: Simulated relevant response to B

E: Simulated irrelevant responses to A

F: Simulated irrelevant responses to B

G: Responses vs. texture columns for A

H: Responses vs. texture columns for B

I: Normalized RT for A and B

Figure 7.5 Differential interference by irrelevant bars due to collinear facilitation. Panels A–H have the same format as Figure 7.4A–H. Their contents differ due to the change of stimuli in Panels A and B, which have Figure 7.3A as the relevant stimulus and uniformly horizontal (A) or vertical bars (B) as irrelevant stimuli. Panel I: Normalized reaction times to A and B – each is the RT divided by the RT of the same subject for Figure 7.3A (the stimulus without irrelevant bars). Subjects are the same as in Figure 7.1D. In three out of four subjects, RT_B for B is significantly longer than RT_A for A ($p <$ .01). By matched-sample t tests across subjects, $RT_B > RT_A$ significantly ($p <$ .01). For each subject, RTs for both A and B are significantly longer ($p <$ .0005) than that for Figure 7.3A (the stimulus without irrelevant bars). [A color version of this figure can be viewed at www.cognitiveneuro-sciencearena.com/brain-scans]

border, as is apparent in Figure 7.5G and 7.5H (and in Figure 7.4G and 7.4H). These irrelevant response peaks not only dictate the local saliencies, but also reduce the relative saliency of the texture border, thereby inducing interference. Figure 7.5A and 7.5B differ in the direction of the collinear facilitation among the irrelevant bars: it is in the direction across the border in Figure 7.5A and along the border in Figure 7.5B. Mutual facilitation between neurons tends to equalize their response levels – that is, smooth away the response peaks or variations in the direction along the collinear facilitation. Consequently, the irrelevant response peaks near the border are much weaker for Figure 7.5A (see Figure 7.5E and G) than for Figure 7.5B (see Figure 7.5F and H), predicting a stronger interference in Figure 7.5B than in Figure 7.5A. This is indeed confirmed in our data for the same segmentation task (Figure 7.5I).

2.3 Fingerprints of V1's conjunctive cells in bottom-up saliency

In Figure 7.6, among a background of purple–right-tilted bars, a unique green–left-tilted bar is salient due to its unique color *and* its unique orientation. We call such a singleton a double-feature singleton, and a singleton unique in only one feature is called a single-feature singleton. By measuring the reaction times to search for the singletons, one can measure the amount of the double-feature advantage – that is, how much more salient the double-feature singleton is compared to the corresponding single-feature singletons. We will explain below that the double-feature advantage depends in specific ways on the existence of conjunctive cells or neurons tuned conjunctively to features in both of the relevant feature dimensions (e.g., color and orientation). Since V1 has neurons tuned conjunctively to color (C) *and* orientation (O), or to orientation *and* motion direction (M), but none conjunctively to the color *and* motion direction, the V1 saliency hypothesis predicts specific double-feature advantages among various feature dimensions.

Take the example of a color and orientation double-feature (denoted as CO, and the corresponding single features as C and O, respectively). To each colored bar, let the neurons respond with outputs O_C, O_O, and O_{CO}, respectively, from neurons (or neural populations) tuned only to C, only to O, or conjunctively to CO. We use superscripts to denote the nature of the bar, so (O_C^C, O_O^C, O_{CO}^C) is the triplet of responses to a color singleton; (O_C^O, O_O^O, O_{CO}^O), that to an orientation singleton; $(O_C^{CO}, O_O^{CO}, O_{CO}^{CO})$ that to a double-feature singleton; and (O_C^B, O_O^B, O_{CO}^B) that to one of the many bars in the background.

For a neuron tuned only to color or orientation, its response should be independent of feature contrast in the other feature dimensions. Hence,

$$O_C^{CO} \approx O_C^C, \quad O_O^{CO} \approx O_O^O, \quad O_C^O \approx O_C^B, \quad O_O^C \approx O_O^B \tag{4}$$

Furthermore, iso-color and iso-orientation suppression implies

A: a portion of a stimulus example for double-feature CO singleton search

B: Normalized RT for double-feature singletons

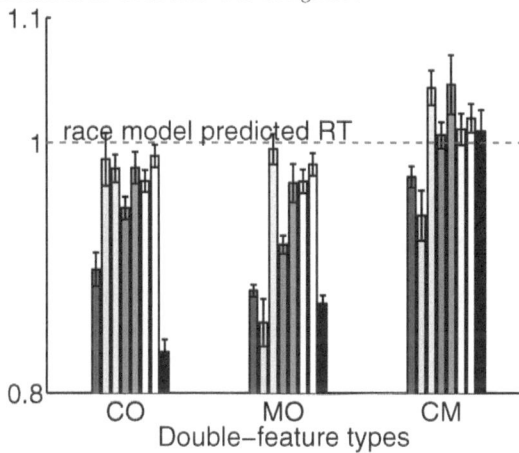

Figure 7.6 Fingerprint of the types of the conjunctive cells in V1. (A) A portion of an example stimulus to search for a CO singleton. (B) The normalized RTs (by the race-model predicted RTs, which are of order 500 ms) for the double-feature singletons for 7 subjects. Different subjects are denoted by the differently colored bars; only two subjects, denoted by blue and green colors (the first two subjects from the left in each double-feature group), are non-naive. Error bars denote standard errors of the mean. By matched-sample 2-tailed t tests, the observed RT^{CO} and RT^{MO} for the double-feature singletons CO and MO are significantly ($p = .03$ and $.009$, respectively) shorter than predicted by the race model, whereas the observed RT^{CM} for the double-feature singleton CM is not significantly different ($p = .62$) from the race-model prediction. More details are available in Koene and Zhaoping (2007). [A color version of this figure can be viewed at www.cognitiveneurosciencearena.com/brain-scans]

$$O_C^C > O_C^B \quad \text{and} \quad O_O^O > O_O^B \tag{5}$$

And generalizing iso-feature suppression to the conjunctive cells, we expect

$$O_{CO}^{CO} > O_{CO}^O, \quad O_{CO}^{CO} > O_{CO}^C, \quad O_{CO}^O > O_{CO}^B, \quad O_{CO}^C > O_{CO}^B \tag{6}$$

The MAX rule states that the maximum response $O^{\alpha}_{max} \equiv max(O^{\alpha}_C, O^{\alpha}_O, O^{\alpha}_{CO})$ determines the saliency of the bar for $\alpha = C, O, CO,$ or B. We denote O_{max} with and without the conjunctive cells by $O_{max}(conj)$ and $O_{max}(base)$, respectively; hence,

$$O^{\alpha}_{max}(base) = max[O^{\alpha}_C, O^{\alpha}_O] \qquad and$$

$$O^{\alpha}_{max}(conj) = max[O^{\alpha}_C, O^{\alpha}_O, O^{\alpha}_{CO}] \geq O^{\alpha}_{max}(base) \tag{7}$$

Since the singletons pop out, we have, with or without the conjunctive cells,

$$O^C_{max}, O^O_{max}, O^{CO}_{max} >> O^B_{max} \tag{8}$$

Without conjunctive cells, we note with Equation (4) that

$$O^B_{max}(base) = max(O^B_C, O^B_O) \approx max(O^O_C, O^O_O) \tag{9}$$

Then, combining equalities and inequalities (4), (5), (7), (8), and (9) gives

$$O^C_{max}(base) = O^C_C, \qquad O^O_{max}(base) = O^O_O \tag{10}$$

$$O^{CO}_{max}(base) = max[O^C_C, O^O_O] = max[O^C_{max}(base), O^O_{max}(base)] \tag{11}$$

So the double-feature singleton is no less salient than either single-feature singleton. With conjunctive cells, combining the equalities and inequalities (4)–(8),

$$O^{CO}_{max}(conj) = max[O^{CO}_C, O^{CO}_O, O^{CO}_{CO}]$$

$$= max[O^C_C, O^O_O, O^{CO}_{CO}]$$

$$= max[max(O^C_C, O^C_O), max(O^O_O, O^O_C), max(O^C_{CO}, O^O_{CO}, O^{CO}_{CO})]$$

The last equality arises from noting $O^C_C > O^O_C$, $O^O_O > O^C_O$, and $O^{CO}_{CO} > O^C_{CO}, O^O_{CO}$. Now re-arranging the variables in the various max(. . .) gives

$$O^{CO}_{max}(conj) = max[max(O^C_C, O^C_O, O^C_{CO}), max(O^O_C, O^O_O, O^O_{CO}), O^{CO}_{CO}]$$

$$= max[O^C_{max}(conj), O^O_{max}(conj), O^{CO}_{CO}]$$

$$\geq max[O^C_{max}(conj), O^O_{max}(conj)] \tag{12}$$

The double-feature singleton can be more salient than both the single-feature singletons if there are conjunctive cells whose response O^{CO}_{CO} has a non-zero chance of being the dictating response.

   Due to the variabilities in the neural responses, the actual neural output in a single trial may be seen as drawn randomly from probability distribution functions (PDFs). So O^C_{max}, O^O_{max}, and O^{CO}_{CO} are all random variables from their respective PDFs, making O^{CO}_{max} (which is the maximum of these three random variables) another random variable. As O^{α}_{max} determines RT by some mono-tonically decreasing function $RT(O^{\alpha}_{max})$ to detect the corresponding input item α, variabilities in neural responses give variabilities in RT^C, RT^O, or RT^{CO} to detect, respectively, the singleton unique in color, in orientation, or in both features. Hence, Equations (11) and (12) lead to

$$RT^{CO}(\text{base}) = \min(RT^C, RT^O) \tag{13}$$

$$RT^{CO}(\text{conj}) = \min[RT^C, RT^O, RT(O^{CO}_{CO})]$$
$$\leq \min(RT^C, RT^O) = RT^{CO}(\text{base}) \tag{14}$$

Hence, without conjunctive cells, RT^{CO} to detect a double-feature singleton can be predicted by a race model between two racers, O^C_{max} and O^O_{max}, with their respective racing times, RT^C and RT^O, as the RTs to detect the corresponding single-feature singletons. With conjunctive cells, RT^{CO} can be shorter than predicted by this race model. Averaged over trials, as long as the additional racer O^{CO}_{CO} has a non-zero chance of winning the race, the mean RT^{CO} should be shorter than predicted by the race model based only on the RTs for detecting the two single-feature singletons.

Hence, the fingerprints of V1's conjunctive cells are predicted as follows: compared to the RT predicted by the race model from the RTs for the corresponding single-feature singletons, RTs for the double-feature singleton should be shorter if the singleton is CO or OM, but should be the same as predicted if the singleton is CM.

We tested for these fingerprints in a visual search task for a singleton bar among 659 background bars (Koene & Zhaoping, 2007). Any bar – singleton or not – is about $1 \times 0.2°$ in visual angle, takes one of the two possible iso-luminant colors (green and purple), is tilted from the vertical to either the left or the right by a constant amount, and moves left or right at a constant speed. All the background bars are identical to each other by color, tilt, and motion direction, and the singleton pops out by unique color, tilt, or motion direction, or any combination of these. The singleton had a 10° eccentricity from the display center. The subjects had to press a button as soon as possible to indicate whether the singleton was in the left or right half of the display regardless of the singleton conditions, which were randomly interleaved and unpredictable by the subjects. To test the predictions, we compared the RTs – for example, RT^{CO}, for the double-feature singletons with the predictions from the race model – for example, $RT^{CO}(\text{base})$. The RTs predicted by the race model were calculated from the RTs for the single-feature singletons using Monte Carlo simulation methods with Equation (13), as follows. For instance, with features C, O, and CO, we randomly obtain one sample each from the collected data of RT^C and RT^O, respectively, and Equation (13) is then used to obtain a simulated sample of $RT^{CO}(\text{base})$. Sufficient number of samples can be generated by these Monte Carlo methods to obtain a histogram distribution of $RT^{CO}(\text{base})$ to compare with the human data RT^{CO} to test whether $RT^{CO} < RT^{CO}(\text{base})$.

Figure 7.6 plots the observed RTs normalized by the race-model predicted RTs for the double-feature singletons. The results confirm the predicted fingerprint. By matched sample 2-tailed *t* tests, the observed RT^{CO} and RT^{OM} for the double-feature singletons CO and OM are significantly shorter than predicted by the race model, whereas the observed RT^{CM} for the double-feature singleton CM is not significantly different from the race-model prediction. The normalized RT^{CO} and RT^{OM} are not significantly different from each other, but they are significantly shorter than the normalized RT^{CM}. Double-feature advantage for the CO singleton had also been observed previously (Krummenacher, Muller, & Heller, 2001). Nothdurft (2000) used a discrimination task, without requiring subjects to respond as soon as possible, and found no qualitative dependence of the double-feature advantages on the feature dimensions. We believe that RT tasks like ours are better suited for probing bottom-up selection, which by nature acts quickly and transiently (Jonides, 1981; Nakayama & Mackeben, 1989; van Zoest & Donk, 2004).

3 Summary and discussion

We modeled and derived the predictions of visual saliency behavior from neural properties known to be specific to V1 – namely, (1) the existence of cells tuned to the eye of origin of the inputs, (2) collinear facilitation between neurons, and (3) the existence of only certain types of conjunctively tuned neurons. Our predictions are consequences of combining (1) these V1 neural properties and the iso-feature suppression via intracortical interactions in V1, (2) the hypothesis that the receptive field location of the most responsive V1 neuron is the most likely to be selected in the bottom-up manner, and (3) the assumed shorter RTs for higher saliencies of the target locations in visual segmentation and search tasks. We presented experimental data confirming these predictions, thereby lending support to the V1 saliency hypothesis.

Previous frameworks for visual saliency and selection (Duncan & Humphreys, 1989; Itti & Koch, 2000; Koch & Ullman, 1985; Treisman & Gelade, 1980; Wolfe et al., 1989) have relied on the assumption that each feature dimension is independently processed in early vision. While it has been known that neural coding in early visual cortices is not independent, it has been assumed or hoped that at a functional level the feature independence would be achieved, or that neural properties specific to V1 would not be manifested so precisely in the visual selection behavior, such as the RT-based segmentation and search tasks. For example, some of these works (Itti & Koch, 2000; Koch & Ullman, 1985; Wolfe et al., 1989) have assumed that there are separate feature maps to process visual inputs of various features and that the activations from different feature maps are then summed, by the SUM rule, into a master saliency map to guide selection. Such a framework implies that the visual saliency map should be in higher cortical areas such as the lateral intraparietal (LIP) (Gottlieb et al., 1998).

In a previous work by two of us (Zhaoping & May, 2007), behavioral data on the interference by irrelevant features, like the ones in Figures 7.1 and 7.3, were presented to confirm the MAX rule, which, as shown in Introduction, arises directly from the V1 saliency hypothesis, implying that no separate feature maps, or any combination of them, are needed for bottom-up selection. However, the MAX rule in itself does not preclude another cortical area from being the locus for the visual saliency map. In fact, it does not even preclude the separate processings of different features, as long as the selection is done by attending to the RF location of the most activated feature unit regardless of its feature map origin. One could even, in principle, modify the previous saliency framework to suit the MAX rule without V1-specific neural properties, simply by replacing the SUM rule of the previous models by the MAX rule when combining activations of the separate feature maps to create a master saliency map.

It is therefore important, for our purpose, to affirm or refute the V1 saliency hypothesis by identifying the predicted fingerprints of V1 in bottom-up saliency behavior. In this chapter, the behavioral fingerprints specifically identify V1 since they rely on neural mechanisms, the existence of monocular cells, collinear facilitation, and particular types of conjunctive cells, specific to V1 only. In particular, automatic attraction to attention by ocular discontinuity excludes V2 and higher visual areas since V1 is the only visual cortical area with a substantial number of monocular cells, and our finding of zero double-feature advantage (over the race-model prediction) of the color and motion double-feature singleton cannot be explained by V2 mechanisms, since V2 and V3 contains neurons tuned conjunctively to color and motion direction (Gegenfurtner, Kiper, & Fenstemaker, 1996; Gegenfurtner, Kiper, & Levitt, 1997; Shipp, 2007, private communication; Tamura et al., 1996) and would create the double-feature advantage by our arguments in section 2.3.

It is likely that V1's saliency map is read by the superior colliculus, which receives input from V1 and directs gaze and thus attention. Indeed, microstimulation of the V1 cells can make monkey saccade to the RF location of the stimulated cell, presumably via V1's drive to superior collicus (Tehovnik et al., 2003). The selection of the receptive field of the most active V1 neuron could be done in principle in a single step by the superior colliculus. In practice, it is likely that this selection is partially or at a local level carried out by the deeper layers of V1 that receive inputs from Layer 2–3 V1 neurons (where intracortical interactions for contextual influences are implemented) and send outputs to the superior colliculus. Specifically, it is possible that some V1 neurons in Layer 5–6 carry out a local MAX rule to relay the local maximum responses (of the Layer 2–3 cells) to the superior colliculus, which carries out a global MAX rule to identify the selected location – this is an empirical question to be answered experimentally.

The V1 saliency hypothesis, however, does not preclude V1 from contributing to other functional goals such as object recognition and learning. Nor does it preclude higher cortical areas, such as V2, from contributing additionally to bottom-up saliency. Indeed, the superior colliculus receives inputs from many higher cortical areas (Shipp, 2004). It is likely that V1's contribution to bottom-up saliency is mainly dominant for the time duration immediately after exposure to visual inputs. Even though V2 and higher cortical areas should have the neural signals and information that would provide the double-feature advantage for the color-motion singleton, our finding of a lack of this advantage implies that the superior colliculus or some other brain area made the decision for attention shift without waiting for such information to arrive in our task and stimulus arrangement. This is not surprising, since being fast is presumably one of the priorities of bottom-up attentional shifts – as long as there is sufficient neural signal or information to arrive at a clear decision for a winner for attention, it is not imperative to ponder or dawdle for a refined decision. With a longer latency, especially for inputs when V1 signals alone are too equivocal to select the salient winner within that time duration, it is likely that the contribution from higher visual areas will increase relatively. These contributions from higher visual areas to bottom-up saliency are in addition to the top-down selection mechanisms that further involve mostly higher visual areas (Desimone & Duncan, 1995; Tsotsos, 1990; Yantis & Serences, 2003). Meanwhile, the bottom-up saliency signals observed in higher level visual areas, such as LIP (Gottlieb et al., 1998) and the frontal eye field (FEF) (Schall & Thompson, 1999), are likely relayed from lower visual areas, particularly V1, rather than computed or created within these higher areas.

The feature-blind nature of the bottom-up V1 selection also does not prevent top-down selection and attentional processing from being feature-selective (Chelazzi, Miller, Duncan, & Desimone, 1993; Treue & Martinez-Trujillo, 1999; Wolfe et al., 1989), so that, for example, the texture border in Figure 7.3A could be located through feature scrutiny or recognition rather than saliency. By exploring the potentials and limitions of the V1 mechanisms for bottom-up selection, it could position us better to understand the roles of the higher visual areas and top-down attention. After all, what V1 could not do must be carried out by higher visual areas, and the top-down attentional selection must work with or against the bottom-up selectional mechanisms in V1 (Zhaoping & Dayan, 2006; Zhaoping & Guyader, 2007; Zhaoping & May, 2007).

4 Acknowledgement

Work supported in part by the Gatsby Charitable Foundation and by the UK Research Council.

5 References

Allman, J., Miezin, F., & McGuinness, E. (1985). Stimulus specific responses from beyond the classical receptive field: Neurophysiological mechanisms for local–global comparisons in visual neurons. *Annual Review of Neuroscience*, 8, 407–430.

Bakin, J. S., Nakayama, K., & Gilbert, C. D. (2000) Visual responses in monkey areas V1 and V2 to three-dimensional surface configurations. *Journal of Neuroscience, 20*(21): 8188–8198.

Beck, D. M., & Kastner, S. (2005). Stimulus context modulates competition in human extra-striate cortex. *Nature Neuroscience, 8*(8), 1110–1116.

Burkhalter, A., & Van Essen, D. C. (1986). Processing of color, form, and disparity information in visual areas VP and V2 of ventral extrastriate cortex in the macaque monkey. *Journal of Neuroscience, 6*(8), 2327–2351.

Chelazzi, L., Miller, E. K, Duncan, J., & Desimone, R. (1993). A neural basis for visual search in inferior temporal cortex. *Nature, 363*(6427), 345–347.

Crick, F., & Koch C. (1995). Are we aware of neural activity in primary visual cortex? *Nature, 375*(6527), 121–123.

DeAngelis, G. C., Freeman, R. D., & Ohzawa, I. (1994). Length and width tuning of neurons in the cat's primary visual cortex. *Journal of Neurophysiology, 71*(1), 347–374.

Desimone, R., & Duncan, J. (1995). Neural mechanisms of selective visual attention. *Annual Review of Neuroscience, 18*, 193–222.

Duncan, J., & Humphreys, G. W. (1989). Visual search and stimulus similarity. *Psychological Review, 96*, 433–458.

Gegenfurtner, K. R., Kiper, D. C., & Fenstemaker, S. B. (1996). Processing of color, form, and motion in macaque area V2. *Visual Neuroscience, 13*(1), 161–172.

Gegenfurtner, K. R., Kiper, D. C., & Levitt, J. B. (1997). Functional properties of neurons in macaque area V3. *Journal of Neurophysiology, 77*(4), 1906–1923.

Gilbert, C. D., & Wiesel, T. N. (1983). Clustered intrinsic connections in cat visual cortex. *Journal of Neuroscience, 3*(5), 1116–1133.

Gottlieb, J. P., Kusunoki, M., & Goldberg, M. E. (1998). The representation of visual salience in monkey parietal cortex. *Nature, 391*(6666), 481–484.

Hegde, J., & Felleman, D. J. (2003). How selective are V1 cells for pop-out stimuli? *Journal of Neuroscience, 23*(31), 9968–9980.

Hirsch, J. A., & Gilbert, C. D. (1991). Synaptic physiology of horizontal connections in the cat's visual cortex. *Journal of Neuroscience, 11*(6), 1800–1809.

Hooge, I. T., & Erkelens, C. J. (1998). Adjustment of fixation duration in visual search. *Vision Research, 38*(9), 1295–1302.

Horwitz, G. D., & Albright, T. D. (2005). Paucity of chromatic linear motion detectors in macaque V1. *Journal of Vision, 5*(6), 525–533.

Huang, P. C., Hess, R. F., & Dakin, S. C. (2006). Flank facilitation and contour integration: Different sites. *Vision Research, 46*(21), 3699–3706.

Hubel, D. H, & Wiesel, T. N. (1959). Receptive fields of single neurones in the cat's striate cortex. *Journal of Physiology, 148*, 574–591.

Hubel, D. H., & Wiesel, T. N. (1968). Receptive fields and functional architecture of monkey striate cortex. *Journal of Physiology, 195*(1), 215–243.

Itti, L., & Koch, C. (2000). A saliency-based search mechanism for overt and covert shifts of visual attention. *Vision Research, 40*(10–12), 1489–1506.

Itti, L., & Koch, C. (2001). Computational modelling of visual attention. *Nature Reviews. Neuroscience, 2*(3), 194–203.

Jones, H. E., Grieve, K. L., Wang, W., & Sillito, A. M. (2001). Surround suppression in primate V1. *Journal of Neurophysiology, 86*(4), 2011–2028.

Jonides, J. (1981). Voluntary versus automatic control over the mind's eye's movement.

In J. B. Long & A. D. Baddeley (Eds.), *Attention and performance IX* (pp. 187–203). Hillsdale, NJ: Lawrence Erlbaum Associates.

Kapadia, M. K., Ito, M., Gilbert, C. D., & Westheimer, G. (1995). Improvement in visual sensitivity by changes in local context: Parallel studies in human observers and in V1 of alert monkeys. *Neuron, 15*(4), 843–856.

Knierim, J. J., & Van Essen, D. C. (1992). Neuronal responses to static texture patterns in area V1 of the alert macaque monkey. *Journal of Neurophysiology, 67*(4), 961–980.

Koch, C., & Ullman, S. (1985). Shifts in selective visual attention: Towards the underlying neural circuitry. *Human Neurobiology, 4*(4), 219–227.

Koene, A. R., & Zhaoping, L. (2007). Feature-specific interactions in salience from combined feature contrasts: Evidence for a bottom-up saliency map in V1. *Journal of Vision, 7*(7), 6, 1–14. Available at http://journalofvision.org/7/7/6/Koene-2007-jov-7-7-6.pdf

Kolb, F. C., & Braun, J. (1995). Blindsight in normal observers. *Nature, 377*(6547), 336–338.

Krummenacher, J., Muller, H. J., & Heller, D. (2001). Visual search for dimensionally redundant pop-out targets: Evidence for parallel-coactive processing of dimensions. *Perception & Psychophysics, 63*(5), 901–917.

Li, C. Y., & Li, W. (1994). Extensive integration field beyond the classical receptive field of cat's striate cortical neurons – classification and tuning properties. *Vision Research, 34*(18), 2337–2355.

Li, Z. (1998). A neural model of contour integration in the primary visual cortex. *Neural Computation, 10*, 903–940.

Li, Z. (1999a). Contextual influences in V1 as a basis for pop out and asymmetry in visual search. *Proceedings of the National Academy of Sciences, USA, 96*(18), 10530–10535.

Li, Z. (1999b). Visual segmentation by contextual influences via intracortical interactions in primary visual cortex. *Network: Computation in Neural Systems, 10*(2), 187–212.

Li, Z. (2000). Pre-attentive segmentation in the primary visual cortex. *Spatial Vision, 13*(1), 25–50.

Li, Z. (2002). A saliency map in primary visual cortex. *Trends in Cognitive Sciences, 6*(1), 9–16.

Livingstone, M. S., & Hubel, D. H. (1984). Anatomy and physiology of a color system in the primate visual cortex. *Journal of Neuroscience, 4*(1), 309–356.

Morgan, M. J., Mason, A. J., & Solomon, J. A. (1997). Blindsight in normal subjects? *Nature, 385*(6615), 401–402.

Nakayama, K., & Mackeben, M. (1989). Sustained and transient components of focal visual attention. *Visual Research, 29*, 1631–1647.

Nelson, J. I., & Frost, B. J. (1985). Intracortical facilitation among co-oriented, co-axially aligned simple cells in cat striate cortex. *Experimental Brain Research, 61*(1), 54–61.

Nothdurft, H. C. (1992). Feature analysis and the role of similarity in preattentive vision. *Perception & Psychophysics, 52*(4), 355–375.

Nothdurft, H. C. (2000). Salience from feature contrast: Additivity across dimensions. *Vision Research, 40*, 1183–1201.

Nothdurft, H. C., Gallant, J. L., & Van Essen, D. C. (1999). Response modulation by texture surround in primate area V1: Correlates of "popout" under anesthesia. *Visual Neuroscience, 16*, 15–34.

Nothdurft, H. C., Gallant, J. L., & Van Essen, D. C. (2000). Response profiles to texture border patterns in area V1. *Visual Neuroscience, 17*(3), 421–436.

Reynolds, J. H., & Desimone, R. (2003). Interacting roles of attention and visual salience in V4. *Neuron, 37*(5), 853–863.

Rockland, K. S., & Lund, J. S. (1983). Intrinsic laminar lattice connections in primate visual cortex. *Journal of Comparative Neurology, 216*(3), 303–318.

Schall, J. D., & Thompson, K. G. (1999). Neural selection and control of visually guided eye movements. *Annual Review of Neuroscience, 22*, 241–259.

Shipp, S. (2004). The brain circuitry of attention. *Trends in Cognitive Sciences, 8*(5), 223–230.

Sillito, A. M., Grieve, K. L., Jones, H. E., Cudeiro, J., & Davis, J. (1995). Visual cortical mechanisms detecting focal orientation discontinuities. *Nature, 378*, 492–496.

Super, H., Spekreijse, H., & Lamme, V. A. (2003). Figure–ground activity in primary visual cortex (V1) of the monkey matches the speed of behavioral response. *Neuroscience Letters, 344*(2), 75–78.

Tamura, H., Sato, H., Katsuyama, N., Hata, Y., & Tsumoto, T. (1996). Less segregated processing of visual information in V2 than in V1 of the monkey visual cortex. *European Journal of Neuroscience, 8*(2), 300–309.

Tehovnik, E. J., Slocum, W. M., & Schiller, P. H. (2003). Saccadic eye movements evoked by microstimulation of striate cortex. *European Journal of Neuroscience, 17*(4), 870–878.

Tong, F. (2003). Primary visual cortex and visual awareness. *Nature Reviews. Neuroscience, 4*(3), 219–229.

Treisman, A. M., & Gelade, G. (1980). A feature-integration theory of attention. *Cognitive Psychology, 12*(1), 97–136.

Treue, S., & Martinez-Trujillo, J. C. (1999). Feature-based attention influences motion processing gain in macaque visual cortex. *Nature, 399*, 575–579.

Ts'o, D. Y., & Gilbert, C. D. (1988). The organization of chromatic and spatial interactions in the primate striate cortex. *Journal of Neuroscience, 8*(5), 1712–1727.

Tsotsos, J. K. (1990). Analyzing vision at the complexity level. *Behavioral and Brain Sciences, 13*(3), 423–445.

von der Heydt, R., Peterhans, E., & Baumgartner, G. (1984). Illusory contours and cortical neuron responses. *Science, 224*(4654), 1260–1262.

van Zoest, W., & Donk, M. (2004). Bottom-up and top-down control in visual search *Perception, 33*(8), 927–937.

Wachtler, T., Sejnowski, T. J., & Albright, T. D. (2003). Representation of color stimuli in awake macaque primary visual cortex. *Neuron, 37*(4), 681–691.

Webb, B. S., Dhruv, N. T., Solomon, S. G., Tailby, C., Lennie, P. (2005). Early and late mechanisms of surround suppression in striate cortex of macaque. *Journal of Neuroscience, 25*(50), 11666–11675.

Wolfe, J. M., Cave, K. R., & Franzel, S. L. (1989). Guided search: An alternative to the feature integration model for visual search. *Journal of Experimental Psychology, 15*, 419–433.

Wolfe, J. M., & Franzel, S. L. (1988). Binocularity and visual search. *Perception & Psychophysics, 44*(1), 81–93.

Wolfson, S. S., & Landy, M. S. (1995). Discrimination of orientation-defined texture edges. *Vision Research, 35*(20), 2863–2877.

Yantis, S. (1998). Control of visual attention. In H. Pashler (Ed.), *Attention* (pp. 223–256). London: Psychology Press.

Yantis, S., & Serences, J. T. (2003). Cortical mechanisms of space-based and object-based attentional control. *Current Opinion in Neurobiology, 13*(2), 187–193.

Zhaoping, L. (2003). V1 mechanisms and some figure–ground and border effects. *Journal of Physiology, Paris, 97,* 503–515.

Zhaoping, L. (2006) Theoretical understanding of the early visual processes by data compression and data selection. *Network: Computation in Neural Systems, 17*(4), 301–334.

Zhaoping, L. (2007, November). *Popout by unique eye of origin: A fingerprint of the role of primary visual cortex in bottom-up saliency.* Paper presented at the annual meeting of the Society for Neuroscience, 2007, San Diego, CA. Program No. 717.8.

Zhaoping, L. (2008). Attention capture by eye of origin singletons even without awareness: A hallmark of a bottom-up saliency map in the primary visual cortex. *Journal of Vision, 8*(5), 1–18.

Zhaoping, L., & Dayan, P. (2006). Pre-attentive visual selection. *Neural Network, 19*(9), 1437–1439.

Zhaoping, L., & Guyader, N. (2007). Interference with bottom-up feature detection by higher-level object recognition. *Current Biology, 17,* 26–31.

Zhaoping, L., & May, K. (2007). Psychophysical tests of the hypothesis of a bottom up saliency map in primary visual cortex. *PLoS Computational Biology, 3*(4), e62.

Zhaoping, L., & Snowden, R. J. (2006). A theory of a salency map in primary visual cortex (V1) tested by psychophysics of color–orientation interference in texture segmentation. *Visual Cognition, 14*(4/5/6/7/8), 911–933.

8   Decision making and population decoding with strongly inhibitory neural field models

Thomas Trappenberg

This chapter discusses dynamic neural field models, which involve the mathematical quantification of cortical and subcortical maps. Several examples from different brain areas show the universality of these models that captures a principal information processing principle in the brain. The examples connect behavioural findings with physiological measurements in the brain. Such models are therefore important to close the gap.

1 Introduction

Understanding how the brain works is a scientific quest on many levels. This certainly includes areas such as the biophysics of ion channels or the developmental properties of the nervous system. But ultimately we would like to understand how specific brain mechanisms measured with neurophysiological methods give rise to cognitive mechanisms that can be measured on a behavioural level. While there has been some progress in recent years, the gap between neurophysiology and behaviour is often large. The aim of this book is to demonstrate how computational modelling approaches can help in bridging this gap.

In this contribution I discuss an example of a fairly general model of processing in neural tissues based on centre–surround lateral interactions. I refer to this model as a dynamic neural field (DNF) or centre–surround neural field (CDNF) model. It is called a field model because it describes a continuous (feature) space and incorporates continuous time dynamics. Describing brain functions with continuous models in space and time is not only mathematically useful and elegant, but it stresses the continuous character of cognitive processes. In contrast to many neural network models, it does not depend on the number of nodes and displays some characteristics that can be generalized to many brain functions.

I give several examples of how DNFs can elucidate neural response properties. For example, I review how tuning curves result from lateral interactions in centre–surround neural tissue. While classical tuning curves describe only the maximal response of neurons, I will show in addition that experimental results for the temporal evolution of neural responses are

consistent with such general models. Furthermore, important for this workshop is not only that single-cell data can be explained by such models, but
that these models, at the same time, explain behavioural findings. I will therefore give some examples of how the same model can explain physiological
single-cell recording data as well as psychophysical functions.

The examples of cell recording data in this chapter are taken from a variety
of neocortical areas, including primary visual cortex, temporal cortex and
parietal cortex, and a subcortical area related to eye movements, the superior
colliculus. This in itself should highlight some universality with respect to
the information processing principle discussed in this section. Besides tuning
curves, I discuss perceptual choice tasks and experimental data from the
recent literature on decision making. I also briefly discuss some issues in
population decoding. Some of these data and mechanisms have been
explained previously with similar or related models. However, the reason
for reiterating them here is to highlight the explanation of these data within
the same neural field framework and within the same parametrization for
all of the various application areas. Indeed, it is good news that various
researchers start to build converging consensus within computational principles rather than building highly specialized models of specific physiological
or behavioural findings.

There are also some new parts to the story presented here. In particular,
I concentrate to some extent on a regime in these networks with relatively
strong inhibition, which is not commonly the focus in the application of
such models. Furthermore, I mention new data from eye movement studies
that demonstrate how surprising nonlinearities in behavioural studies can be
explained with the neural field model of the superior colliculus.

2 Dynamic neural field model and cortical feature representation

2.1 The basic computational model

Centre–surround neural field models have been introduced and discussed at
least since the early 1970s (Grossberg, 1973; Wilson & Cowan, 1973). In the
following I discuss a neural field model of the formulation introduced
by Amari (1977) where the dynamics of an internal state variable $u(x)$ at
"location" x is given by

$$\tau \frac{\partial u(x,t)}{\partial t} = -u(x,t) + \int_y w(|x-y|)r(y,t)dy + I^{ext}(x,t). \tag{1}$$

In this formula, $I^{ext}(x,t)$ represents an external stimulus as received by the
neural field, and $r(x)$ is an external state variable. This rate is related to the
internal state variable $u(x)$ through a gain function $r = g(u)$. I use a sigmoidal

gain function $g(u) = 1/(1 + \exp(-\beta u))$ for the demonstrations in this chapter. Other gain functions are sometimes biologically more appropriate but will not change the principal findings in the examples presented here. Most of the discussions in this chapter are done with 1-dimensional neural fields, but all of the discussions can easily be generalized to higher dimensions where the 1-dimensional feature space is replaced by a higher dimensional feature map. An example of a 2-dimensional model is discussed in the last section when discussing eye movements and a model of the superior colliculus. Also, we always use period boundary conditions to minimize boundary effects. Thus the feature space is formally a ring in one dimension, where we frequently use a feature value of $(0, 2\pi)$, or more generally a torus in higher dimensions.

Equation (1) defines the dynamic evolution of the neural field, and it is important to note that this formulation is continuous in time and space. To evaluate such models on a computer, one needs to *discretize* both space and time. For example, one can define a small but finite time step Δt and spatial resolution Δx and write the differential equations as a difference or iterative equation for each *node* with index i as

$$u_i(t + \Delta t) = \left(1 - \frac{\Delta t}{t}\right)u_i(t) + \frac{\Delta t}{t}\left(\sum_j w_{ij}r_j(t)\Delta x + I_i^{ext}(t)\right) \tag{2}$$

This equation reduces to the widely used equation for recurrent attractor networks in the stationary limit or with time steps $\Delta t = \tau$. It is important to stress that the discretization is only a necessary step for computational purposes, but that the processes described by this model should approach continuality of time and (feature) space that is appropriate to describe cognitive processes. Thus, we have to make sure that numerical results do not change by altering the time step or including more nodes. Also, the interpretation of each node is not that of a simple neuron; rather, it stands for a collection (population) of neurons that receive similar input in a certain task (Gerstner, 2000; Wilson & Cowan, 1972) or that of a cortical minicolumn (Johansson & Lansner, 2007). However, we will see below some examples where such a population average describes quite well the trial-averaged activity of neurons.

The lateral interaction kernel $w(x,y)$, or its discrete version w_{ij}, is chosen to be a shifted Gaussian:

$$w_{ij} = A_w\left(\frac{1}{\sqrt{4\pi\sigma_r}}e^{-((i-j)*\Delta x)^2/4\sigma_r^2} - C\right). \tag{3}$$

Other popular choices are weight profiles in the form of a Mexican hat, but these functions result in similar behaviour when the extent of inhibition and excitation are matched in the periodic feature maps discussed here. Many studies assume this interaction structure from the start, and it is possible that such structures are genetically coded – in particular, if the feature space is topographic. However, such weights can also be learned from activity-dependent Hebbian learning on Gaussian patterns with width σ_r (Stringer, Trappenberg, Rolls, & Araujo, 2002; Trappenberg, 2002; Wu & Trappenberg, 2008). Such learning might be, for example, important in the formation of place fields in the hippocampus (Stringer et al., 2002).

Amari (1977) discussed such neural field models in detail and showed that there are several solutions to the dynamic equations. In particular, there is a regime of intermediary inhibition values in which an area of activity will persist even in the absence of continuous external input. Such an active area is sometimes called a bubble, a bump, or an activity packet, and this persistent activity has been implicated with physiological working memory (Wang, 1999, 2001). There are many variations of the basic model with different shapes of weights, various gain functions or slightly different formulations of the dynamic equation. However, as long as there is some balance between local excitation and long-distance inhibition, the principal findings are very similar.

Neural field models of lateral-inhibition type are in common use on various levels of description and application domains. For example, Wilson and Cowan (1973) introduced such a model from course-graining neuronal sheets of spiking neurons with excitatory and inhibitory neuronal pools. In this sense, the model can be viewed as physiologically based. On the other hand, Gregor Schöner (see for example, Schöner, 2008) has argued for many years that such models are useful descriptions of behavioural phenomena – in particular, since they describe the dynamics of many behavioural findings. In this chapter I want to describe specifically some examples where the model simultaneously describes physiological data of trial-averaged firing rates of single neurons and behavioural data such as reaction times and detection accuracies.

As a final comment on the neurophysiological plausibility of such models, the origins of long-range lateral inhibition may be subject to debate. Anatomically it seems that inhibitory neurons have a more local projection area than lateral excitatory connections (Hirsch & Gilbert, 1991; Kisvárday & Eysel, 1993; McDonald & Burkhalter, 1993). However, mechanistically there are several possible ways to accomplish extended inhibition. For example, Wilson and Cowan (1973, p. 62) cited experimental evidence that excitatory-to-inhibitory connections extend further than excitatory-to-excitatory connections; Taylor and Alavi (1993) argued that long-range inhibition could be mediated by cortico-thalamic interactions thorough the nucleus reticularis of the thalamus. In general, one could think about the long-range global inhibition as an inhibitory background field, often described as an inhibitory pool (Deco & Lee, 2002). Also, effective long-range inhibition can be seen experimentally (Grinvald, Lieke, Frostig, & Hildesheim, 1994; Nelson & Frost, 1978).

When thinking about the cortical implementation of this model, one is foremost bound to think about this model as a representation of a cortical organization of a hypercolumn. Indeed, the discussions in the next section illustrate this in some detail. Processing in hypercolumns is often though of as a winner-take-all process (Hubel & Wiesel, 1977), and the neural field models discussed here implement such a mechanism though lateral inter-actions. In this chapter, we also discuss competition between object represen-tations beyond the typical representation in hypercolumns of early sensory areas. There are two ways of thinking about this. First, functional structures of hypercolumns in higher cortical areas will, of course, represent much more abstract concepts and global object features compared to hypercol-umns in early sensory cortex. Second, competitive interactions can also be interpreted, in a more abstract way, as competition between hypercolumns or even whole objects as represented by a series of features represented by hypercolumns (Johansson & Lansner, 2007).

2.2 Competitive feature representations in cortical tissue

Hubel and Wiesel (1962) found that some neurons in the primary visual cortex of anaesthetized cats respond strongly to moving bars in their receptive field with a specific orientation and that the same neurons respond less when the orientation of the moving bar deviates from this preferred direc-tion. An example of such a tuning curve from experiments by Henry, Dreher, and Bishop (1974) is shown in Figure 8.1 by solid circles and an interpolated Gaussian curve. In this example, the preferred direction of this cell is plotted as zero degrees, and the rate diminishes with a bell-shaped curve with increas-ing distance from the preferred direction.

Such tuning curves can be modelled in different ways. For example, we could assume that the neuron receives input from cells that only fire to specific orientations and that the input weight of these connections is a bell-shaped function. However, since the orientation selectivity only arises in V1, we follow here the common idea that local arrangements of V1 cells provide a focal orientation selectivity as suggested by Hubel and Wiesel (1962; see also Nagano & Kurata, 1980) but that the broadness of the tuning curves is mediated by lateral interactions (Ben-Yishai, Bar-Or, & Sompolinsky, 1995; Bullier, 2006). This general idea is captured by the neural field model as illustrated in Figure 8.1. Each node in the model receives input from a certain area of the visual field (its receptive field) and is most responsive to a specific orientation (tuning curve). In the simulated experiment shown, I probed the response of the neural field by activating externally a very small region, essentially covering only one node in the simulations. The neurons for these specific stimuli are very active, and, most importantly, the activity is also "bleeding out" to neighbouring nodes. The maximal response for different orientations of input for a neuron with preferred orientation of zero degrees is plotted with open squares on top of the experimental tuning curve, demon-strating how the neural field model can reproduce the experimental findings.

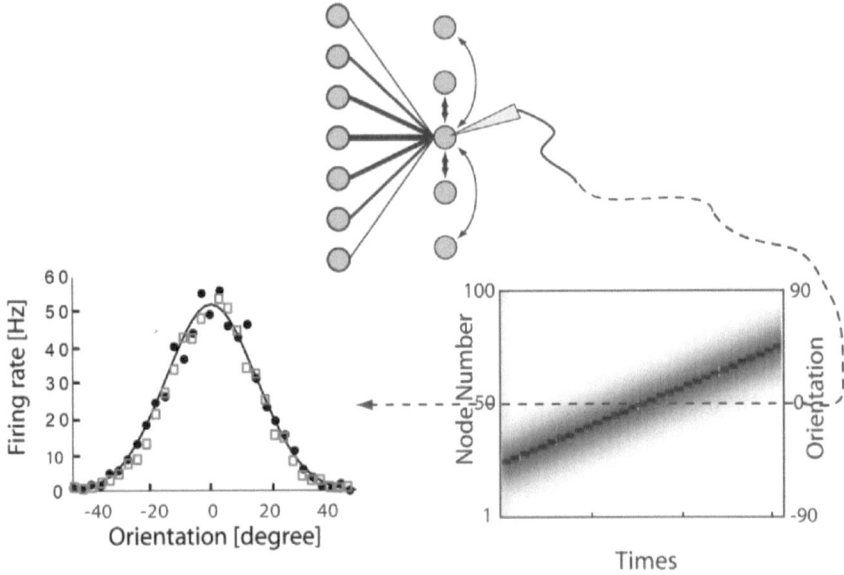

Figure 8.1 Experimental data (solid circles) from Henry et al. (1974) plot the firing rate of a neuron in cat primary visual cortex as a function of the orientation of moving line segments. Open squares correspond to the corresponding measurements in a neural field model. The evolution of the activity in the neural field is shown in the plot on the right.

In our interpretation of tuning curves we see that there is a wide response of neurons even when presenting a "pure" image. This is important as such states are likely to represent probabilities of specific events (Zemel, Dayan, & Pouget, 1998). Cortical neurons are also very noisy, which was combated with many trials in the experiments and by neglecting noise in the simulations. Decoding the information from a population becomes crucial for reliable information processing in the brain. I return to this point later in this chapter. The tuning curves discussed here only show the maximal response during each stimulus presentation, and I now turn to some temporal aspects that can illuminate other experimental findings.

3 Neural field models of perceptual choice and decision making

3.1 Temporal aspects of neural field responses

In the following experiments we have a closer look at the temporal developments in the neural field. While doing this, we will generalize the application of this model to other feature representations in cortex. Higher cortical areas represent increasingly complex features beyond orientation selectivity of V1 neurons, but we can still think of these representations in terms of neural fields. The only difference is that the represented feature value (orientation in V1) should be replaced by another feature value to which the other cortical area responds.

In this section I discuss cell recording data from monkey inferior temporal (IT) area associated with object recognition. Figure 8.2 shows data from recordings in this area by Chelazzi, Miller, Duncan, and Desimone (1993). In their experiment they showed different objects to monkeys and selected objects to which the recorded IT cell responded strongly (*good* objects) and weekly (*bad* objects). In the illustration of Experiment 1, the good object is represented with a square. The average firing rate of an IT cell to this good stimulus is shown in Figure 8.2C as a solid line. The period when the cue stimulus is presented is indicated by the left-hand grey bar in Figure 8.2C. The neuron responds strongly to this stimulus, but the trial-averaged rate of this neuron decays basically to the background level after the cue is removed.

The "bad" object is illustrated with a triangle in Figure 8.2, and the trial-averaged response of the same neuron recorded during Experiment 2 is shown as a dashed line in Figure 8.2C. Instead of increasing the rate during the cue presentation, this neuron seems to respond with a firing rate below the usual background rate.

Several aspects of this response are captured by the neural field model shown in Figures 8.2C and 8.2D. The solid line represents the activity of a

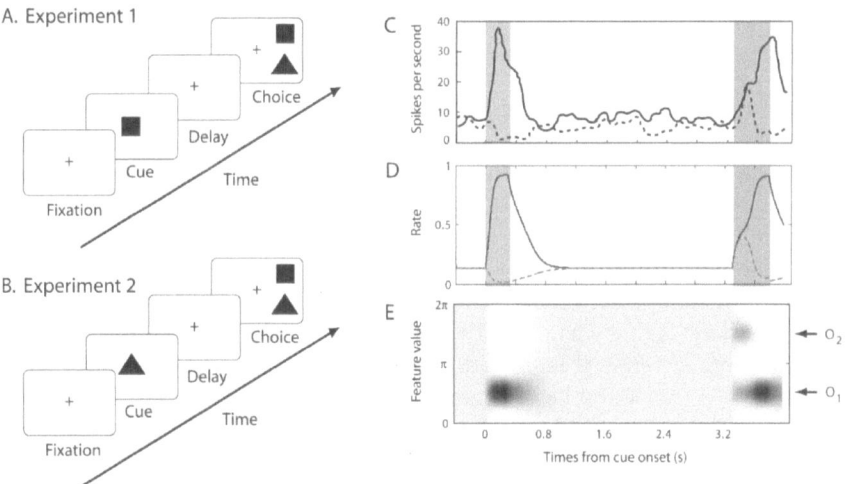

Figure 8.2 Experimental paradigm and data from Chelazzi et al. (1993) and their interpretation in the neural field model. A monkey is given a single visual cue and, following a delay, the monkey is shown two stimuli and must move its eyes to the remembered stimulus. The average response of an IT neuron is shown in Panel C, where the solid line corresponds to a "good" object and the dashed line to a "bad" object. The corresponding responses in the neural field model are shown in Panels D and E.

node within the response bubble of the neural field, such as a node recorded at position O_1 in Figure 8.2E. The activity declines after the external stimulus is removed. This was achieved by using a relatively strong inhibition constant of $c = 0.06$ in the simulations. A slightly smaller inhibition constant of $c = 0.05$ would result in a sustained-activity bubble, as used in working memory models. The activity of the dashed line corresponds to the activity of a node outside the activity bubble. There is a diminished activity through lateral inhibition during the stimulus presentations similar to the experimental data.

The details of the program and the chosen parameters are presented in the Appendix. It is worth noting that all the experiments in the different application areas are basically run with the same model and comparable parameters.

3.2 *Competitive aspects of neural field responses*

The second half of the experiments by Chelazzi et al. (1993) is even more interesting. After a delay, the monkey is shown the two objects, the "good" and the "bad" object, at the same time (the right-hand grey area in panels C and D in Figure 8.2). The monkey is instructed to saccade to the cued objects. Interestingly, the initial response of the IT neuron to the display is an increase in rate, regardless of which object was cued. However, after around 200 ms the responses in trials with good and bad object cues diverge. The activity increases further in trials where the cue is a good object. In contrast, further activity is suppressed in trials where the bad object is the cue.

The interpretation of these data with the neural field model is straightforward. Presenting two objects to the neural field means that two different regions are stimulated simultaneously. We take the (top-down) task instruction into account by making one of the inputs slightly stronger (only 1% in the shown simulations). Thus, the requirement to memorize the cue is not explicitly modelled in these effects; rather, only the effect of this memory is taken into account. Indeed, it is possible and common to use a similar model to store this short term memory which feeds into this IT map. This module could, for example, simulate working memory in prefrontal cortex or working memory loops between prefrontal and other cortical areas. The interesting point here is that we take only a small bias in the input to the neural field into account. At first, this small difference in input strength does not make a difference in the neural field response. However, after enough activity is generated in the neural field, the two active centres begin to interact through lateral inhibition, and the slight advantage of one centre leads to a winning active area.

I am not the first to offer this interpretation of these data. Usher and McClelland (2001) have recently argued that these data can be understood and described with competitive integrators that correspond to the neural field models discussed here. Indeed, their model represents a discrete implementation of the neural field model with strong inhibition. Usher and McClelland also discuss nicely how the neural field model unifies two historical models of decision making: accumulator models and diffusion models. With accumulator models, evidence for stimuli accumulates over time, and a decision is reached when evidence for a particular stimulus exceeds a decision threshold. With diffusion models, it is not the absolute evidence that is accumulated, but the relative difference between the evidence for each stimulus. The neural field model is thus a dynamic implementation of an intermediate model where there is some coupling between different areas of the neural field.

3.3 Decision making, psychophysics, and uncertainty

My discussion of perceptual choice has centred on electrophysiological data, but with crude behavioural resolution. To provide a cognitive perspective, behavioural measurements with higher resolution are required. Those are provided by motion discrimination experiments. In a typical motion discrimination task, a display of dots is presented on a computer screen, and some fraction of the dots are repeatedly displaced. The dots appear to move in the direction of displacement, typically to the left or to the right. Monkeys are trained to move their eyes in the direction of perceived movement, thus indicating their choice. Experimentalists manipulate the *coherence* of movement by controlling the percentage of displaced dots, allowing them to finely control the certainty of evidence. Recording response time and accuracy provides psychophysical measurements in response to the controlled parameter (coherence, in this case).

Psychophysical measurements from Roitman and Shadlen (2002) are shown as dots in Figures 8.3A and 8.3B. The figures show the *psychometric* and *chronometric* functions, respectively, depicting accuracy and reaction time as a function of the certainty of evidence. At low coherence levels, the monkey makes a lot of errors and has long reaction times. With increasing coherence, the monkey's accuracy increases and its reaction times are lower. Results from corresponding neural recordings in monkey lateral intraparietal area (LIP) are shown in Figure 8.3C. The data are similar to output from simulations, shown in Figure 8.3D. With increased coherence, the activity of neurons with a preferred response to the direction of motion increases (solid lines). The response of neurons with other preferred directions increases at first, before rapidly decreasing (dashed lines). Three different values of the coherence parameter are depicted in Figures 8.3C and 8.3D. Stronger input (greater certainty) more quickly "defeats" the weaker stimulus (dashed lines) leading to a quicker decision, as shown by the solid lines. Reaction times are equated with the time required for activity to reach a threshold level, and they decrease with increased coherence (Figure 8.3B). If the evidence has not reached threshold by 800 ms, the monkey responds according to the accumulated evidence, shown in Figure 8.3A. Model output is show with lines in Figure 8.3A and 8.3B.

The accuracy curves from the model are proportional to the evidence

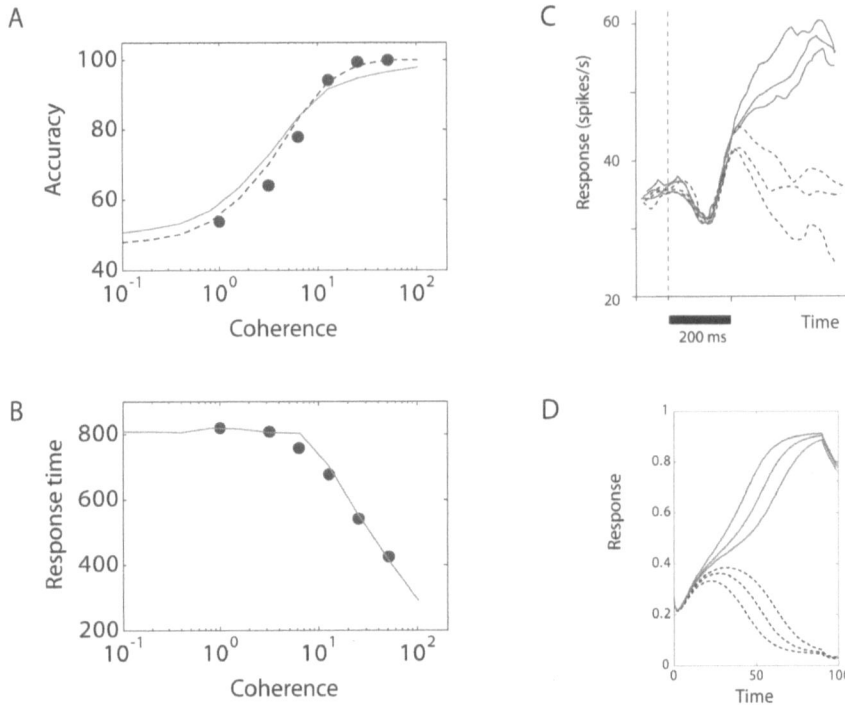

Figure 8.3 Data from Roitman and Shadlen (2002) as explained by the neural field
model. (A) The *psychometric* function plotting accuracy as a function of
certainty of evidence. Experimental data are shown as dots. The solid line
shows normalized output from the model. A linear transformation
weighted in favour of high-rate activity is shown by the dashed line.
(B) The *chronometric* function plotting response time as a function of certainty of evidence. Again, dots and lines correspond to data and output
from the model, respectively. (C) Neural response over time for different
levels of certainty of evidence. Solid and dashed lines show the spike rate
of neurons responsive to the chosen and unchosen stimuli, respectively.
(D) Output from the model in response to simulation of the task in
Panel C.

accumulated in LIP. In the simplest case, the evidence from the neural
population can be read out by calculating the normalized maximum response
for each perceptual choice, where the normalization provides a probabilistic
interpretation of the activity in LIP. This calculation is shown by the solid
line in Figure 8.3A. The curve describes the experimental data, providing
a mechanistic description of LIP activity in Roitman and Shadlen's
(2002) task.

Despite its ability to explain the data, the solid curve in Figure 8.3A does
not run through the dots. The monkey is perfectly accurate at even moderate

coherence levels, but the solid line approaches this limit smoothly. Thus, a decision based on LIP activity cannot be as simple as suggested by the probabilistic interpretation. In Figure 8.3A, the dotted line shows a linear transformation of the output of the model, where high-rate activity is given a greater weight than low-rate activity. This curve fits the data more accurately, but it does not tell us anything new. In this regard, it is sometimes easy to fit curves to experimental data with enough parameters, but a simpler model may be more scientifically useful, serving to identify information processing principles in the brain.

4 Population decoding with strong inhibition

As pointed out at the beginning of this chapter, the models discussed here describe instantaneous rates of populations of neurons. Single neurons in the cortex have typically low firing rates, and their firing is also very noisy. In the previous examples I averaged over such neuronal populations, and I also compared the model behaviour to a trial-based average of corresponding experiments. However, the question arises how a neural sheet receiving very noisy input can achieve reliable feature estimation from noisy population activity.

A possible answer to noisy population decoding comes from the recognition that centre–surround neural field models are ideal to solve this noisy decoding task. The theory of population decoding with neural fields has been developed recently in some detail (Deneve, Latham, & Pouget, 1999a, 2001; Deneve, Pouget, & Latham, 1999b; Pouget, Dayan, & Zemel, 2000; Pouget, Zhang, Deneve, & Latham, 1998; Wu, Amari, & Nakahara, 2002). It is common in these applications of neural field models to use a normalizing gain function that represents divisive inhibition so that the resulting activity packet has a Gaussian shape. It can then be shown (Pouget et al., 1998; Wu et al., 2002) that the activity pattern is equivalent to a maximum-likelihood estimate of Gaussian tuning curves.

Population decoding in the above-mentioned papers is usually done by applying input as initial states of the recurrent network. The decoded signal is then usually considered to be the asymptotic state of sustained-activity packets (bubbles). Here I summarize my recent study (Trappenberg, 2008) where I investigated the population decoding with DNFs with several additions and simplifications that add to their biological realism, and I also investigate which regime of the DNF models is best suited for population decoding. I thereby investigate these networks with sustained input over a period of time, either with fixed noisy input or with temporal changing input. All the simulations in this section are done without divisive inhibition that is commonly used in the related literature, showing that this form of global operation is not necessary for the population decoding. The major question I ask in this section is if there is an optimal value of global inhibition which controls the amount of competition in the network.

4.1 Population decoding with recurrent networks

I consider again the generic network model described above. As noted previously, this model exhibits several regimes characterized by different possible asymptotic states (Amari, 1977). If the inhibition C is low compared to excitation in the network, then the excitation will spread through the network, resulting in runaway activity. In contrast, if inhibition is dominating, then any activity in the field will decay without external reinforcement. In an intermediate regime, it is possible to have activity packets where localized activity is stable. I call this mode of the model the memory regime. The latter two regimes are depicted in Figure 8.4 in the context of population decoding. In these experiments, I supplied a static noisy input to the field over 20 time steps. This input was chosen as a Gaussian around the middle node ($x_0 = 50$) with additive white noise, η, of strength $n_h = 0.5$,

$$I_{ext} = I_0 + Ae^{-((i-j)*\Delta x)^2/2s^2} + n_h h, \tag{4}$$

where I_0 is a background field and η is a normal distributed random number. When an inhibition constant of $c = 0.05$ was used, the field developed into a clean bubble around the middle node after the input was removed at time $t = 20$, demonstrating perfect decoding. The sustained localized activity in the neural field without external input demonstrates the above-mentioned memory regime.

Traces of the noisy input are not apparent in the $c = 0.05$ case, and it seems one has to wait until the bubble forms to perform accurate decoding. This is different when running the same simulation with larger inhibition. Figure 8.4 shows the case for $c = 0.07$. Traces of the noisy input are now also visible during the time the external input is supplied, which is partly enhanced by the fact that a smaller range of values is depicted by the grey scale. The inhibition is now too large to sustain an activity packet after input is removed. However, the increased competition facilitates a cleaning of the signal even during the time when the signal is applied, so that some form of population decoding is supported. While this might be less accurate than in the previous case, an advantage would certainly be that the decoding can be achieved much earlier. This assertion is investigated in the next section.

4.2 Quality of decoding with varying inhibition over time

To assess the quality of decoding with time and different inhibition constants I ran decoding experiments over 100 trials in each condition. While I have used signals with static noise in Figure 8.4, I report now on the results when changing the noise after each unit of time, simulating ongoing fluctuations in the input signal over time. This represents a more plausible implementation of decoding conditions in the brain, although I found similar results in the static noise case.

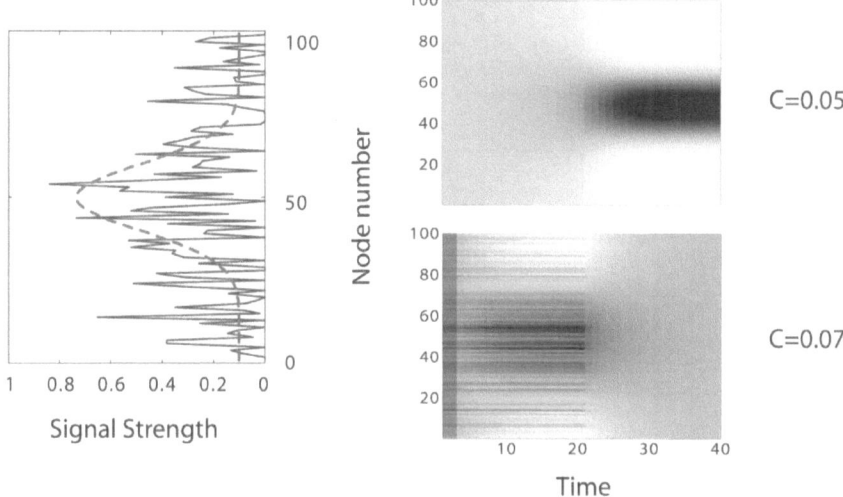

Figure 8.4 Noisy population decoding with weak and strong inhibition in neural fields. The noisy input (solid line in the left-hand graph), which is derived from the dashed line in the left-hand graph, is applied until $t = 20$ to the neural fields. The time evolution of the neural field is shown in the right-hand graphs, with moderate inhibition constant, $C = 0.05$, in the upper graph and larger inhibition constant, $C = 0.07$, in the lower graph. The activity of the neural field is thereby depicted on a grey scale with different ranges in the two figures. Therefore, the noisy input until $t = 20$ in the upper graph is not well visible since the activity over time is dominated by the sustained bubble at later times.

To assess the decoding quality I use a centre-of-mass scheme to determine the prediction of the feature value encoded in the neural field. This was done for the original input signal and at each time step during the dynamic decoding in the recurrent network. An example trace of the decoding error is shown in Figure 8.5A. For the shown strength of inhibition, the decoding error continuously decreases even while external input is supplied. The decoding error only improves slightly after the external input is removed at $t = 20$.

As can be expected, the decoding error increases with increasing noise level, as shown in Figure 8.5B, but the decoding error at $t = 20$ is always much smaller than the centre-of-mass decoding of the original signal. Finally, a major question motivating this study was to determine which network regime, in terms of inhibition strength, would be most suitable for decoding in such networks. The results of these studies are summarized in Figure 8.5C, which shows the decoding improvement at different time steps for different inhibition parameters. Early on there is little dependence of the results on the strength of inhibition, but later there is some advantage for inhibition

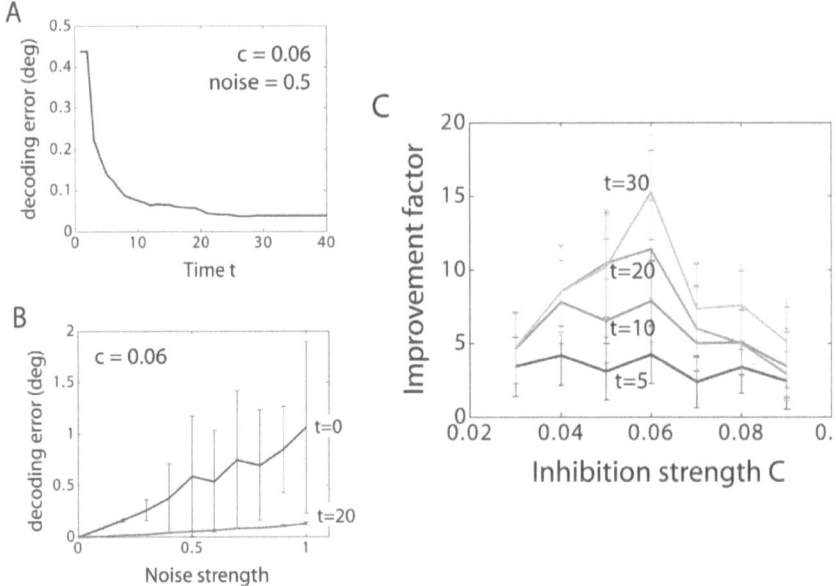

Figure 8.5 Temporal aspects and decoding quality with different levels of inhibition. (A) Time course of the decoding error in degrees (radians), with an inhibition constant of C = 0.06. The decoding error is decreasing even while the external input is supplied, until $t = 20$. The noise level in the shown example is $n_h = 0.5$. (B) The decoding error as a function of noise strength, n_n, when decoding is performed at $t = 0$, which corresponds to the original signal before modified by the recurrent network, and at $t = 20$, when the recurrent network was able to process the signal. (C) The improvement factor is calculated as the average decoding error at a specific time $t > 0$ compared to the average decoding error of the original signal, $t = 0$.

values around 0.06. Interestingly, this is close to the transition region between the domain of decaying input and the domain of sustained activity.

5 The interaction of bottom-up and top-down manipulations on saccade reaction times in monkeys

The last application of the neural field model that I want to mention here is recent work with Robert Marino and Doug Munoz (Marino, Trappenberg, & Munoz, 2006). Marino studied eye movements with monkeys in a gap paradigm where he manipulated two principal factors: the number of possible locations where a target could appear in each trial, and the intensity of a target. The resulting median saccade reaction times (SRT) are shown in Figure 8.6D. Reaction times are fairly long with faint targets, but SRTs become faster with increasing intensity. However, the surprising effect Marino discovered is that the rate of decrease slowed down and reaction times became even slightly longer for very strong intensities. Another surprising non-monotonicity also appeared when analysing the reaction times to the number of possible target locations shown in Figure 8.6F. While reaction times first increase with increasing number of possible target locations, saccade onsets are again faster for 8 possible target locations.

These results are stunning in themselves, but they are also not that obvious to reconcile with the model of the intermediate layer of the superior colliculus (SC) that we devised some time ago (Trappenberg, Dorris, Klein, & Munoz, 2001). The SC is thought to be a topographically organized motor map that drives fast eye movements – so-called saccades. We found that a large variety of effects of saccadic eye movements can be explained when considering that this motor map integrates a variety of endogenous and exogenous signals with a competitive mechanism described by the DNF

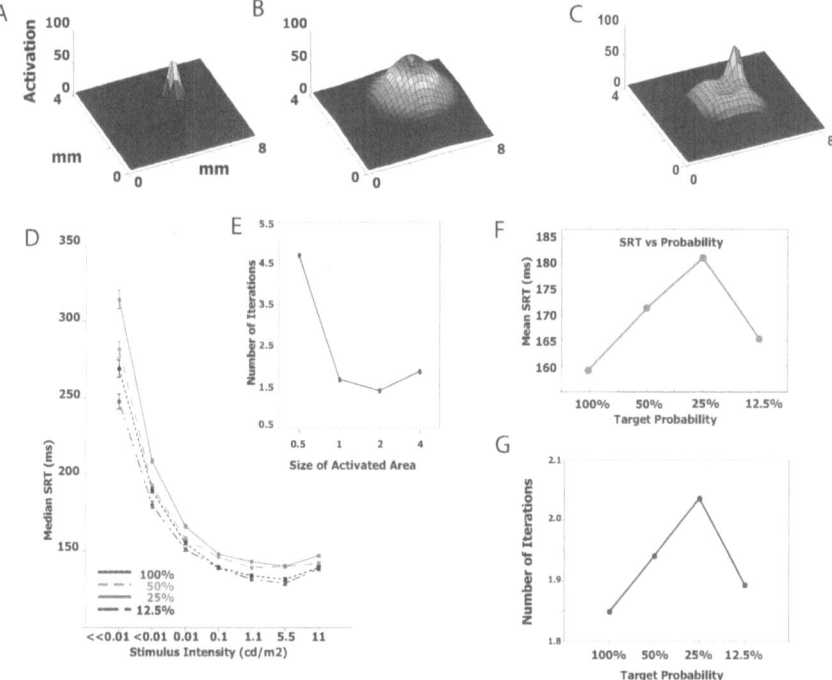

Figure 8.6 Data of behavioural experiments and 2-dimensional DNF simulations of saccadic eye movements. The upper row shows the 2-dimensional DNF neural sheet activity with a narrow (A) and wider (B) input. Neural sheet activity (C) with 4 endogenous and 1 exogenous inputs. Behavioural monkey data (D) and DNF simulation data (E) with different target intensities. Larger target intensity was thereby simulated with wider input in the DNF simulations. Monkey data (F) and DNF simulation data (G) of experiments with a varying number of possible target locations. [A colour version of this figure can be viewed at www.cognitiveneurosciencearena. com/brain-scans]

models considered in this chapter. Indeed, we even found direct evidence of the centre–surround interaction structure (Dorris, Olivier, & Munoz, 2007; Trappenberg et al., 2001).

When applying this model to the new experimental conditions with increased levels of external input to the neural field simulating increased target intensities, the buildup of SC activity become faster with increasing intensity and there is no reason that reaction times become slower again. In the case of an increasing number of possible target locations, we expect an increasing number of inputs representing the expectation of a target at different locations, which increases the inhibition within the neural field. This in turn should increase the reaction times with an increasing number of target locations. However, the decrease in reaction time with 8 possible target locations is stunning.

The puzzle can be resolved by considering area effects. If intensities will not only increase the strength of input but also mainly alter the area of the stimulated region within the SC, as illustrated in Figures 8.6A and 8.6B, then we get a reaction-time curve shown in Figure 8.6E that resembles the nonlinearity of the experimental results. The reason for the slowdown in reaction times with the largest intensities comes thereby from the fact that the activated area extends to the inhibitory surround in the neural field. As long as the input is small, increasing input will add to the buildup of SC activity and thus will increase SRTs. However, when the activated area becomes comparable with the extent of the centre–surround scale, then SRTs will become longer again.

A corresponding explanation can be given to the effect of altering the number of possible target locations. At first there is an increase in areas that are stimulated by endogenous input (top-down input that represents expectation), as shown in Figure 8.6C, but a further input will not alter much the inhibitory input from remote areas but will help the winning location to build up faster. Thus, the model predicts an increase in the area of activation with increased intensity. Marino is currently running experiments to investigate this prediction.

6 Conclusion

Measuring behavioural responses in specific circumstances is a major tool to study cognition. Results of such studies are valuable in themselves. For example, knowing the specific parameters of a speed–accuracy tradeoff can help to design appropriate systems for human–computer interactions. On the other side of neuroscience, neurophysiological studies are a major tool in understanding brain mechanisms. In particular, single-cell recordings in behaving monkeys and functional brain imaging in behaving humans have started to map brain activities in specific tasks. Such studies are in themselves valuable. For example, such studies can illuminate the effects of neurodegenerative diseases. But to understand how the brain thinks, we have ultimately to close the gap between neurophysiology and behaviour, and computational modelling approaches are a central tool in this domain.

In this chapter, I have given some examples where models are able to explain the principal trends in cell recordings and psychophysical curves. Making contact with both levels is important. This allows us to embed behavioural findings into physiological mechanisms, on the one hand, and to specify the consequences of neuronal activity on behaviour, on the other. It also provides us with a functional explanation of behavioural findings. For example, cell recordings in the perceptual choice task have shown that there is first a common build up of activity before they split and, on a behavioural level, that there is an accuracy–speed tradeoff in responses to such perceptual choices. The model quantifies the hypothesis that the perceptual choice is facilitated by competitive mechanisms in the neural representation of the stimuli and that these mechanisms will directly influence behaviour. Using computationally specific implementation of such models will even allow us to specify these relations quantitatively.

The model studied here is fairly simple. It consists of a neural sheet with centre–surround lateral interactions. The continuous formulation of this model is useful, as this allows mathematically elegant treatments and also makes it somewhat universal. While the model is simple, I find it interesting to see that it can be applied to so many different areas in neuroscience and cognitive science. I believe that this demonstrates that this model captures principal brain mechanisms that are important for brain-style information processing. The fact that this model can also explain so many effects measured in behavioural studies and neurophysiological studies shows that the experimentally measured effects are understood. Thus, the simplicity of the model is not a drawback; indeed, it should be a major attraction of this model. This fact also highlights that, while the experimental effects look at a first glance exciting, it might be worthwhile to investigate new questions where the behaviour cannot be explained with such simple models and might therefore highlight some novel information processing mechanisms in the brain. Models should indeed guide us more in discovering such novel mechanisms by making concrete predictions that can be verified experimentally.

7 Appendix: the basic program

The following is a complete listing of the program to simulate the experiment by Chelazzi et al. (1993). The other experiments use basically the same program.

The following listing is the main program with the specific experimental settings. After defining some constants, Hebbian learning is used to set up the centre–surround connectivity matrix that drives the experiments. Training on Gaussian training patterns leads to a Gaussian connectivity matrix.

After training, we define some input patterns and then call a numerical differential equation solver to integrate with the neural field over specific times with the various inputs and plot the results at the end of the program. The four short functions called in the main program are listed further below.

```
%%%%%%%%%%%%%%%%%%%%%%%%%%%%%%%%%%%%%%%%%%%%%%%%%%%%%%%%%%%%%%%%
% 1-d Continuous Attractor Neural Network with Hebbian learning%
% two gaussian signal: decision network                       %
%%%%%%%%%%%%%%%%%%%%%%%%%%%%%%%%%%%%%%%%%%%%%%%%%%%%%%%%%%%%%%%%

clear; close all;
nn = 100; dx=2*pi/nn; % number of nodes and resolution in deg

%weight matrices
    sig = 2*pi/20;
    w_sym=hebb(nn,sig,dx);
    w_inh=0.07;%use 0.04, 7,6,3; 3*(sqrt(2*pi)*sig)^2/nn;
    w=500*(w_sym-w_inh);

%inputs
    perc=0.01; Is=11;
    Ia=(1+0.5*perc)*Is;
    Ib=(1-0.5*perc)*Is;


%%%%%%%%%%%%%%%%%%%%%%%%%%%%%%%%%%%%%%%%%%%%%%%%%%%%%%%%%%%%%%%%
%%%%%%%%%%%%%%%%%%%    Experiment    %%%%%%%%%%%%%%%%%%%%%%%%%%%
%%%%%%%%%%%%%%%%%%%%%%%%%%%%%%%%%%%%%%%%%%%%%%%%%%%%%%%%%%%%%%%%

    param=0;
%%%% no external input
    u0 = zeros(nn,1)-10;
    I_ext=zeros(nn,1);
    tspan=[0,40]
    [t,u]=ode45('rnn_ode_u',tspan,u0,[],nn,dx,w,I_ext);
    r=f1(u);

%%%% external cue
    u0 = u(size(t,1),:);
    I_ext=zeros(nn,1);
    loc1=pi/2;%+pi/16;
    loc2=3*pi/2;%-pi/16;
    I_ext=I_ext+in_signal_pbc(loc1,Is,sqrt(2)*sig,nn,dx);
    tspan=[40 70];
    [t2,u]=ode45('rnn_ode_u',tspan,u0,[],nn,dx,w,I_ext);
    r=[r;f1(u)];
    t=[t;t2];

%%%% no external input
    u0 = u(size(t2,1),:);
    I_ext=zeros(nn,1);
    param=0;
    tspan=[70,370];
    [t2,u]=ode45('rnn_ode_u',tspan,u0,[],nn,dx,w,I_ext);
    r=[r;f1(u)];
    t=[t;t2];

%%%% both external inputs
    u0 = u(size(t2,1),:);
    I_ext=zeros(nn,1);
    loc1=pi/2;%+pi/16;
    loc2=3*pi/2;%-pi/16;
    I_ext=I_ext+in_signal_pbc(loc1,Ia,sqrt(2)*sig,nn,dx);
    I_ext=I_ext+in_signal_pbc(loc2,Ib,sqrt(2)*sig,nn,dx);
    tspan=[370 420];
    [t2,u]=ode45('rnn_ode_u',tspan,u0,[],nn,dx,w,I_ext);
    r=[r;f1(u)];
    t=[t;t2];
```

```
%%%% no external input
    u0 = u(size(t2,1),:);
    I_ext=zeros(nn,1);
    param=0;
    tspan=[420,440];
    [t2,u]=ode45('rnn_ode_u',tspan,u0,[],nn,dx,w,I_ext);
    r=[r;f1(u)];
    t=[t;t2];

    %%% plot results

    surf(t',1:nn,r','linestyle','none'); view(0,90);
    figure
    plot(t,r(:,round(loc1/dx))); hold on;
    plot(t,r(:,round(loc2/dx)),'r')
```

The numerical ODE solver calls the file rnn_ode_u.m that contains the differential equation:

```
function udot=rnn(t,u,flag,nn,dx,w,I_ext)
% odefile for recurrent network
    tau_inv = 1./2;    % inverse of membrane time constant
    r=f1(u);
    sum=w*r*dx;
    udot=tau_inv*(-u+sum+I_ext);
return
```

The function f1.m simply contains the sigmoidal gain function, which can easily be replaced by other gain functions:

```
function f1=rnn(u)
% gain function: logistic
    beta =.1;
     alpha=.0;%-100*(u>0);
    f1=1./(1+exp(-beta.*(u-alpha)));
return
```

The function in_signal_pbc.m produces a Gaussian signal with periodic boundary conditions:

```
function y = in_signal_pbc(loc,ampl,sig,nn,dx)
    y=zeros(nn,1);
    for i=1:nn;
        di=min(abs(i*dx-loc),2*pi-abs(i*dx-loc));
        y(i)=ampl/(sqrt(2*pi)*sig)*exp(-di^2/(2*sig^2));
    end
return
```

Finally, we also included Hebbian learning to choose the weight matrix:

```
function w = hebb(nn,sig,dx)
% self organization of symmetric cann interactions
    lrate=1;    % learning rate
    w=zeros(nn);
```

```
%%%%%%% learning session
for epoch=1:1
    for loc=1:nn;
        r=in_signal_pbc(loc*dx,1,sig,nn,dx);
        dw=lrate.*r*r';
        w=w+dw;
    end
end
w=w/nn;
return
```

8 References

Amari, S. (1977). Dynamics of pattern formation in lateral-inhibition type neural fields. *Biological Cybernetics, 27*, 77–87.

Ben-Yishai, R., Bar-Or, R. L., & Sompolinsky, H. (1995). Theory of orientation tuning in visual cortex. *Proceedings of the National Academy of Sciences, USA, 92*(9), 3844–3848.

Bullier, J. (2006). What is the role of top-down connections. In J. L. van Hemmen & T. Sejnowski (Eds.), *23 problems in system neuroscience* (pp. 103–132). New York: Oxford University Press.

Chelazzi, L., Miller, E. K., Duncan, J., & Desimone, R. (1993). A neural basis for visual search in inferior temporal cortex. *Nature, 363*, 345–347.

Deco, G., & Lee, T. (2002). A unified model of spatial and object attention based on inter-cortical biased competition. *Neurocomputing, 44–46*, 775–781.

Deneve, S., Latham, P. E., & Pouget, A. (1999a). Reading population codes: A neural implementation of the ideal observer. *Nature Neuroscience, 2*, 740–745.

Deneve, S., Latham, P. E., & Pouget, A. (2001). Efficient computation and cue integration with noisy population codes. *Nature Neuroscience, 4*, 826–831.

Deneve, S., Pouget, A., & Latham, P. (1999b). Divisive normalization, line attractor networks and ideal observers. In M. S. Kearns, S. A. Solla, & D. A. Cohn (Eds.), *Advances in neural information processing systems* (Vol. 11). Cambridge, MA: MIT Press.

Dorris, M. C., Olivier, E., & Munoz, D. P. (2007). Competitive integration of visual and preparatory signals in the superior colliculus during saccadic programming. *Journal of Neuroscience, 27*(1), 5053–5062.

Gerstner, W. (2000). Population dynamics of spiking neurons: Fast transients, asynchronous states, and locking. *Neural Computation, 12*(1), 43–89.

Grinvald, A., Lieke, E., Frostig, R., & Hildesheim, R. (1994). Cortical point-spread function and long-range lateral interactions revealed by real-time optical imaging of macaque monkey primary visual cortex. *Journal of Neuroscience, 14*, 2545–2568.

Grossberg, S. (1973). Contour enhancement, short-term memory, and constancies in reverberating neural networks. *Studies in Applied Mathematics, 52*, 217–257.

Hansel, D., & Sampolinsky, H. (2005). Continuous attractor neural networks. In L. N. de Castro & F. J. V. Zuben (Eds.), *Recent developments in biologically inspired computing* (pp. 398–425). Hershey, PA: IDEA Group.

Henry, G. H., Dreher, B., & Bishop, P. O. (1974). Orientation specificity of cells in cat striate cortex. *Journal of Neurophysiology, 37*, 1394–1409.

Hirsch, J., & Gilbert, C. (1991). Synaptic physiology of horizontal connections in the cat's visual cortex. *Journal of Neuroscience, 11*, 1800–1809.

Hubel, D., & Wiesel, T. (1962). Receptive fields, binocular interaction and functional architecture in the cat's visual cortex. *Journal of Physiology, London, 160*, 106–154.

Hubel, D., & Wiesel, T. (1977). Ferrier lecture: Functional architecture of macaque monkey visual cortex. *Proceedings of the Royal Society London, Series B. Biological Sciences, 198*, 1–59.

Johansson, C., & Lansner, A. (2007). Towards cortex sized artificial neural systems. *Neural Networks, 20*, 48–61.

Kisvárday, Z. F., & Eysel, U. T. (1993). Functional and structural topography of horizontal inhibitory connections in cat visual cortex. *European Journal of Neuroscience, 5*(12), 1558–1572.

Marino, R. A., Trappenberg, T. P., & Munoz, D. P. (2006). Modeling the bottom-up and top-down effects on saccadic reaction times in monkeys using artificial neural network models of the superior colliculus. In *28th international symposium Montreal: Computational neuroscience*. Montreal: GRSNC (available online at http://www.grsnc.umontreal.ca/XXVIIIs/doc/XXVIII-Programme.pdf).

McDonald, C. T., & Burkhalter, A. (1993). Organization of long-range inhibitory connections within rat visual cortex. *Journal of Neuroscience, 13*(2), 768–781.

Nagano, T., & Kurata, K. (1980). A model of the complex cell based on recent neurophysiological findings. *Biological Cybernetics, 33*(2), 103–105.

Nelson, J., & Frost, B. (1978). Orientation-selective inhibition from beyond the classical visual receptive field. *Brain Research, 139*(2), 357–365.

Pouget, A., Dayan, P., & Zemel, R. (2000). Information processing with population codes. *Nature Reviews. Neuroscience, 1*, 125–132.

Pouget, A., Zhang, K., Deneve, S., & Latham, P. (1998). Statistically efficient estimation using population coding. *Neural Computation, 10*(2), 373–401.

Roitman, J. D., & Shadlen, M. N. (2002). Response of neurons in the lateral intraparietal area during a combined visual discrimination reaction time task. *Journal of Neuroscience, 22*(21), 9475–9489.

Schöner, G. (2008). Dynamical systems approaches to cognition. In R. Sun (Ed.), *The Cambridge handbook of computational psychology*. New York: Cambridge University Press.

Stringer, S., Trappenberg, T., Rolls, E., & Araujo, I. (2002). Self-organising continuous attractor networks and path integration: One-dimensional models of head direction cells. *Network: Computation in Neural Systems, 13*, 217–242.

Taylor, J., & Alavi, F. (1993). A global competitive network for attention. *Neural Network World, 5*, 477–502.

Trappenberg, T. (2002). *Fundamentals of computational neuroscience*. Oxford: Oxford University Press.

Trappenberg, T. (2008). Tracking population densities using dynamic neural fields with moderately strong inhibition. *Cognitive Neurodynamics, 2*, 171–177.

Trappenberg, T., Dorris, M., Klein, R., & Munoz, D. (2001). A model of saccade initiation based on the competitive integration of exogenous and endogenous signals in the superior colliculus. *Journal of Cognitive Neuroscience, 13*, 256–271.

Usher, M., & McClelland, J. L. (2001). On the time course of perceptual choice: The leaky competing accumulator model. *Psychological Review, 108*, 550–592.

Wang, X.-J. (1999). Synaptic basis of cortical persistent activity: The importance of NMDA receptors to working memory. *Journal of Neuroscience, 19*, 9587–9603.

Wang, X.-J. (2001). Synaptic reverberation underlying mnemonic persistent activity. *Trends in Neuroscience, 24*, 455–463.

Wilson, H., & Cowan, J. (1972). Excitatory and inhibitory interactions in localized populations of model neurons. *Biophysical Journal, 12,* 1–24.

Wilson, H., & Cowan, J. (1973). A mathematical theory of the functional dynamics of cortical and thalamic nervous tissue. *Kybernetik, 13,* 55–80.

Wu, S., Amari, S., & Nakahara, H. (2002). Population coding and decoding in a neural field: A computational study. *Neural Computation, 14,* 999–1026.

Wu, S., & Trappenberg, T. (2008). In R. Wang, F. Gu, & E. Shen (Eds.), *Advances in cognitive neurodynamics ICCN 2007*. Berlin: Springer.

Zemel, R., Dayan, P., & Pouget, A. (1998). Probabilistic interpretation of population code. *Neural Computation, 10,* 403–430.

9   The importance of neurophysiological constraints for modelling the emergence of modularity

John A. Bullinaria

This chapter is concerned with the large-scale structure of the human brain, how that relates to human behaviour, and how certain types of modularity could have emerged through evolution as a result of a combination of fitness advantages and neurophysiological constraints. It is argued that computational modelling involving evolutionary neural network simulations is the best way to explore such issues, and a series of simulation results are presented and discussed. The incorporation of known neurophysiological constraints into the models is seen to be of crucial importance.

1 Introduction

Cognitive neuropsychology and fMRI studies have provided a great deal of information about large-scale brain structure. Numerous neuropsychological studies of brain-damaged patients have examined patterns of deficits and observed double dissociations (Teuber, 1955) that have been taken to provide evidence of modularity (Shallice, 1988). More recently, the modular structures inferred in that way have been further refined by fMRI studies (e.g., Binder, Medler, Desai, Conant, & Liebenthal, 2005; Huettel, Song, & McCarthy, 2004), and many large-scale "modules" are now thought to consist of collections of lower level modules, some of which are actually used more widely (e.g., Bookheimer, 2002; Duncan, 2001; Duncan & Owen, 2000; Marcus, 2004). Despite this large body of research, there remain many controversies concerning modules and modularity.

First, there is disagreement over how to define modules. The hard-wired, innate, and informationally encapsulated modules of Fodor (1983) are, for example, rather different from the modules in the cognitive models of Coltheart, Curtis, Atkins, and Haller (1993). What Bullinaria (2005a) would call a module, Plaut (1995) would claim is not a module. The range of different definitions has been discussed by Seok (2006), and it is becoming clear that it is quite appropriate to use different definitions across different applications and different levels of description (e.g., Geary & Huffman, 2002; Seok, 2006).

Next, there is disagreement whether particular data implies modularity or not. For example, doubt has been cast on the inference of modularity from double dissociation (e.g., Bullinaria, 2005a; Dunn & Kirsner, 1988; Plunkett & Bandelow, 2006; Van Orden, Pennington, & Stone, 2001), while other earlier claims have been shown to be mere artifacts of the models involved (Bullinaria & Chater, 1995).

Finally, even when it is agreed that certain types of modularity exist, there remains disagreement over the extent to which it is innate rather than learned during the individual's lifetime and what cost–benefit trade-offs affect that distinction (e.g., Elman, Bates, Johnson, Karmiloff-Smith, Parisi, & Plunkett, 1996; Jacobs, 1999; Geary & Huffman, 2002; O'Leary, 1989). It is known that the costs of learning can drive the evolution of behaviours that reduce that cost and eventually result in the learned behaviours being assimilated into the genotype – a process often referred to as the Baldwin Effect (Baldwin, 1896; Bullinaria, 2003). It may well be that modular processing is first learned and only later becomes assimilated, or partially assimilated, into the genotype by such a process. On the other hand, if the tasks to be performed change too quickly for evolution to track them, or if the necessary neural structures are too complex for easy genetic encoding, then the need to learn them will persist.

One promising approach for making progress on the issue of brain modularity, while avoiding much of the existing controversy, is to consider from a theoretical point of view what are the possible advantages and disadvantages of modularity. This may then inform why and how and when it may have emerged in brains as a result of evolution. A reliable approach for testing such ideas is to build computational (neural network) models that perform simplified behavioural tasks and to study the various trade-offs inherent in them. Such models can be built "by hand", but simulating the evolution of such systems is often a more productive approach (Bullinaria, 2005b). Natural selection will presumably have resulted in the most advantageous structures for biological systems, and what emerges from the evolutionary models can be compared with what is believed to be known about real brains. Of course, the evolution of brains is constrained by various biological practicalities (Allman, 1998; Striedter, 2005), and realistic models need to reflect this. Interestingly, evolutionary simulation results show that imposing known physical/neurophysiological constraints on the models can have a drastic effect on what emerges from them. Moreover, such models can also begin to explore the interaction of learning and evolution and thus explicitly address the nature/nurture debate with regard to brain structures.

The remainder of this chapter is structured as follows. In the next section, the main advantages of neural modularity are reviewed. Then an approach is described for simulating the evolution of neural networks that must carry out pairs of simple tasks, and it is shown how this can be applied to studying the evolution of modularity in such systems, with particular reference to the accommodation of known neurophysiological constraints on the emergent structures. Empirical results are presented from a series of computational experiments that are designed to explore the circumstances under which modularity will emerge. The chapter ends with some discussion and conclusions.

2 Advantages of modularity

The literature already suggests numerous (overlapping) reasons why modularity may be advantageous for information processing systems such as brains. In fact, it can be regarded as a general design principle for both vertebrates and invertebrates (e.g., Leise, 1990). Perhaps the most obvious advantage of modularity corresponds to the familiar idea of breaking a problem into identifiable sub-tasks and making effective use of common processing subsystems across multiple problems (e.g., Kashtan & Alon, 2005; Marcus, 2004; Reisinger, Stanley, & Miikkulainen, 2004). This is certainly standard practice when writing complex computer programs, and Chang (2002) has presented a specific cognitive model in which modularity is shown to allow improved generalization in complex multi-component language processing tasks. In fact, numerous different types of modular neural networks and their associated successful applications were reviewed some time ago by Caelli, Guan, and Wen (1999). Such forms of modularity might have evolved in brains specifically to deal with particular existing tasks. It is also possible that new ways were discovered to use existing modules that gave individuals abilities they did not have before and that these provided such an evolutionary advantage that they were quickly adopted into the whole population. In fact, it has been argued that modularity will prove particularly advantageous when adaptation to changing environments is required (e.g., Kashtan & Alon, 2005; Lipson, Pollack & Suh, 2002), especially since the opportunity to employ useful modules from a rich existing source will invariably be more efficient than generating a new ability from scratch. That modularity can act to improve robustness and evolvability is certainly not a new idea (e.g., Wagner, 1996).

Another way of thinking about brain modularity is in terms of low-level neural processes. It seems intuitively obvious that attempting to carry out two distinct tasks using the same set of neurons and connections will be less efficient than having separate modules for the two tasks. This intuition was explored in a series of simple neural network simulations by Rueckl, Cave, and Kosslyn (1989), in which standard multi-layer perceptron neural networks were trained using gradient descent to perform simplified visual object recognition. This involved carrying out separate "what" and "where" classification tasks on images presented as a 5 × 5 input grid, as shown in Figure 9.1. They found that disruptive interference between connection weight updates for the two tasks did result in poorer learning performance for a single network compared to a network with separate modules (blocks of hidden units) for the two tasks. The problem with such a simulation approach, however, is that the learning algorithms used were far from

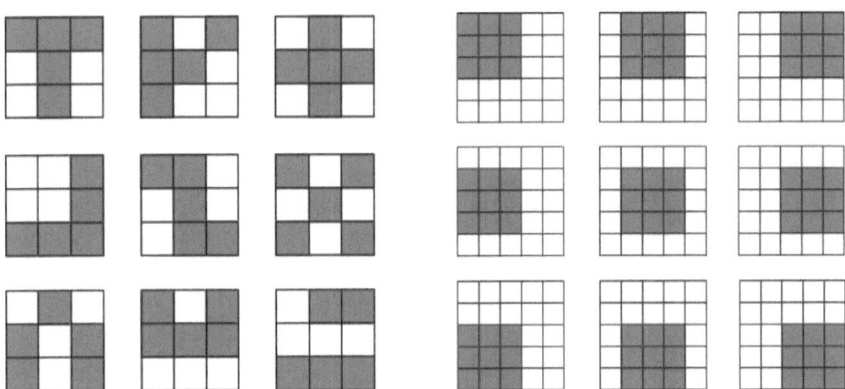

Figure 9.1 The simplified What–Where training data studied by Ruekl et al. (1989) and most subsequent studies. There are nine distinct 3 × 3 images (What) that may appear in any of nine positions (Where) in the 5 × 5 input grid.

biologically plausible, and it is possible that brains have evolved better learning algorithms that do not suffer such disruptive interference. In fact, it was later shown that a slightly different learning algorithm (equally biologically implausible) did perform significantly better, did not suffer such disruptive interference, and performed better when the neural architecture was non-modular (Bullinaria, 2001). This naturally cast doubt on the whole idea that low-level disruptive interference was relevant for modularity.

In an attempt to clarify this matter, an extensive follow-up study was carried out to explore further the trade-off between minimizing interference and restricting the allowed neural architecture (Bullinaria, 2007b). First, by using simulated evolution to optimize the neural architecture for various fixed gradient descent learning algorithms, for the same simplified What–Where task studied by Rueckl et al. (1989), it was confirmed that learning algorithms prone to significant learning interference between independent tasks tended to lead to modular architectures, whereas learning algorithms that did not suffer such interference led to non-modular architectures emerging. Then, if the learning algorithm was allowed to evolve alongside the architecture, the learning algorithm least prone to cross-task interference consistently emerged, along with non-modular architectures, and it was shown that this set-up did indeed provide the best performance on the chosen tasks.

The problem remaining was that the simplified What–Where learning task, used in virtually all the earlier studies, was far removed from the kinds of tasks that are faced by real biological individuals. Rather than learning a small set of input–output mappings by repeated exposure to them, they typically experience steady streams of input patterns drawn from continuous distributions, and they need to learn classification boundaries that allow them to generalize, in order to respond correctly to inputs they have never seen before. To explore more realistic learning scenarios of this type, a series of artificial data sets were generated based on various decision boundaries in a two-dimensional input space normalized to a unit square. Some examples are shown in Figure 9.2. Such inputs might correspond to observable features of other animals (shapes, sizes, etc.), and the output classes could represent important properties of them (dangerous, edible, etc.). The neural networks experience a stream of random data points from the input space, and their fitness corresponds to how well they perform the right classifications *before* learning from that experience. Evolving the neural network learning algorithms to perform well on such generalization tasks invariably led to the better (less interference prone) learning algorithm emerging, but the associated architectures were now more variable (Bullinaria, 2007b). There was a trade-off between leaving the learning algorithm more freedom to use all the available network connections versus minimizing the task interference by placing restrictions on how the connections were used. Whether that

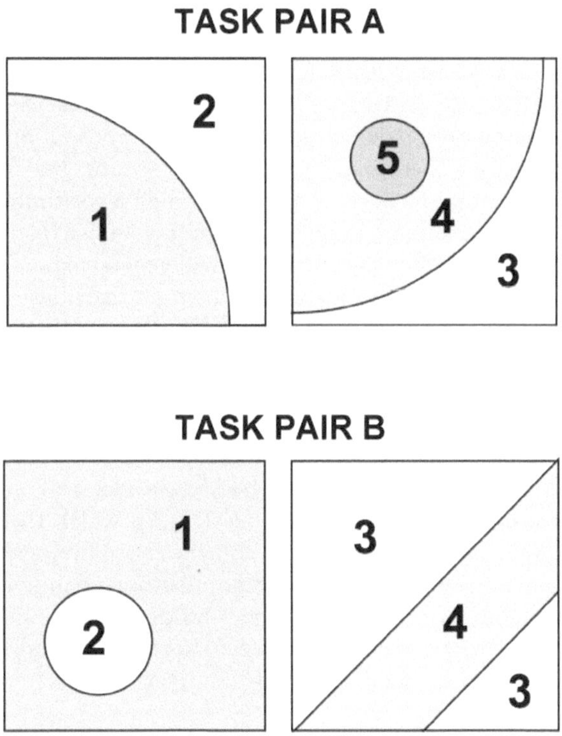

Figure 9.2 Two representative pairs of classification generalization tasks based on a two-dimensional input space. The upper pair (A) has one 2-class task and one 3-class task, both with circular boundaries. The lower pair (B) has two 2-class tasks, one with circular boundary and one with linear boundary.

trade-off favoured modularity or non-modularity proved to be rather task-dependent, affected by numerous factors such as the number of classes that need to be learned, the complexity of the various classification boundaries, and the relative difficulties of the two tasks. For example, Task Pair A in Figure 9.2 leads to non-modular architectures, while Task Pair B usually results in modular architectures.

It seemed, then, that the emergence of modularity depended both on the power of the learning algorithm and on the properties of the tasks, but the simulations did provide an explanation of why low-level neural modules should evolve. The third factor explored by Bullinaria (2007b), after the learning algorithm and task type, related to the fact that there are numerous neurophysiological constraints placed upon biological brains that were absent in the previous models. The remainder of this chapter describes the approach that was used to investigate these factors and expands upon the analysis and discussion of what emerged from those simulations.

3 Simulating the evolution of modularity

The idea of applying simulated evolution by natural selection to neural networks is now well established (e.g., Cantû-Paz & Kamath, 2005; Yao, 1999). In principle, all aspects of the neural networks can be evolved (the architecture, the properties of the neurons, the connection weights, the learning algorithm, the data preprocessing, and so on), but for modelling brains it is most natural to have an innate learning algorithm to adjust the connection weights during the individual's lifetime, rather than having them evolved (Elman et al., 1996). The evolution will generate appropriate neural architectures, learning algorithms, and initial weight distributions that allow the learning to take place most effectively. There are still many different ways this can be done, as discussed by Bullinaria (2007a), all involving a population of neural networks, each specified by an innate genotype that represents the various evolvable parameters. However, the key results concerning modularity appear to be quite robust with respect to the details of the evolutionary process, as demonstrated by the identical outcomes from the generational approach of Bullinaria (2007b), in which the populations are simulated one generation at a time, and the more biologically inspired approach of Bullinaria (2001), involving populations of individuals of different ages with competition, procreation, and deaths each simulated year. Since it seems not to be crucial which evolutionary process is used, this chapter will adopt the computationally simpler approach of Bullinaria (2007b).

The neural networks used are standard multi-layer perceptrons (MLPs) with a single hidden layer, sigmoidal processing units, and trained using a gradient descent learning algorithm with a cost function E that is an evolvable linear combination of sum-squared error (SSE) and cross entropy (CE). This allows the evolutionary process to choose between SSE, as used in the study of Rueckl et al. (1989), and CE, which is now known to be more appropriate for classification tasks (Bishop, 1995). The weight update equation for the connection weight w_{ij} between hidden unit i with activation h_i and output unit j with activation o_j and target t_j then takes the form

$$\Delta w_{ij} = -\eta_{HO} \frac{\partial E}{\partial w_{ij}} = \eta_{HO} h_i (t_j - o_j)[(1 - \mu)(1 - o_j)o_j + \mu]$$

in which $\mu \in [0, 1]$ specifies the learning cost function, with $\mu = 0$ being pure SSE, and $\mu = 1$ being pure CE. The intermediate value of $\mu \sim 0.1$ is equivalent to using SSE with a sigmoid prime offset, which was suggested by Fahlman (1988) as a way of resolving some of the learning difficulties inherent in using pure SSE. Note that it has already been established that having independent evolvable learning rates η_L for the distinct network components L (input to hidden weights IH, hidden unit biases HB, hidden to output weights HO, and output biases OB) can lead to massive improvements in performance over a single learning rate across the whole network (Bullinaria, 2005b). Similarly, it proves advantageous to evolve different uniform ranges $[-l_L, +u_L]$ from which to draw the random initial weights for each component L. In total, then, the network initialization and learning process is specified by 13 evolvable parameters (four η_L, four l_L, four u_L, and μ), and the evolutionary process is expected to optimize these to give the best possible learning performance on the given tasks.

In this context, modularity can be defined in terms of the pattern of connectivity in the network, ranging from fully modular, with separate sets of hidden units for each task, to fully distributed, with all hidden units used for all tasks. For two tasks, this can conveniently be parameterized as shown in Figure 9.3, with the total number of hidden units Nhid partitioned into

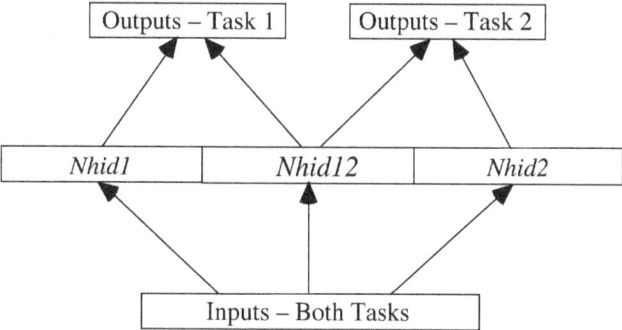

Figure 9.3 The neural network architecture designed to study the evolution of modularity. Rectangles represent sets of neurons and arrows represent full connectivity. There is a common set of input units, a set of output units for each task, and a set of hidden units partitioned according to which output set the neurons connect to.

Nhid1 that only participate in Task 1, *Nhid2* that only participate in Task 2, and *Nhid12* that participate in both. If *Nhid* is fixed, that leaves 2 parameters that need to be evolved to specify the neural architecture, in addition to the 13 learning parameters.

Having specified each neural network by its 15 innate parameters, a fitness measure for that network must be determined, to which the simulated natural selection process can be applied. For current purposes, fitness can be conveniently measured in terms of the number of training epochs required to reach the first full epoch of perfect performance (Bullinaria, 2007a, 2007b). For the What–Where task (of Figure 9.1), an epoch corresponds to the full set of 81 training items, and performance is measured on that training set. For the online generalization tasks (of Figure 9.2), an epoch is defined as a block of 400 training items, and performance is measured on those items *before* training on them. In both cases, the random sequence of training patterns and online learning will introduce a stochastic element into the fitness measurement. This will combine with the various other sources of randomness (such as the choice of initial populations, the selection of mates for reproduction, the crossover and mutation processes, and the initial weights of each new individual) to give a certain variability across runs, as will be seen in the simulation results plotted later.

Each simulated evolution run starts with a population of 100 networks with randomly allocated innate parameters. The learning and initial weight parameters are taken from ranges that span those values generally used in hand-crafted networks – namely, [0, 4] – and the learning algorithm and architecture parameters are taken from their full valid ranges. Each network starts with random weights drawn from its innately specified ranges, and it trains until it reaches its first epoch of perfect performance, which determines its fitness. The next generation is then made up of the fittest half of the population, plus their children. These children inherit innate parameters taken randomly from the ranges spanned by their two parents (representing crossover), with random adjustments applied from a Gaussian distribution (representing mutation). The whole new generation is then initialized and the learning and selection process repeated. Over many generations the innate parameters are optimized, and the individual performances improve (Bullinaria, 2007a, 2007b).

As noted above, the neural architectures that emerge from simulated evolution of this type depend on the pair of tasks that need to be performed (Bullinaria, 2007b). The aim of this chapter is to consider further how known neurophysiological constraints will affect what evolves. There are two obvious factors that need to be considered: first, the fact that real brain sizes are constrained by numerous factors such as growth costs, energy/ oxygen consumption, heat dissipation, and such like; second, that space and other requirements preclude full connectivity within the brain. The next two sections present results from a series of simulations that explore these issues.

4 The effect of brain size

In the artificial neural networks studied here, it is the number of hidden units that corresponds to brain, or brain region, size. If all the learning parameters are optimized, the more hidden units employed, the fewer epochs of training are generally required. However, most simulations have the hidden units processed serially on conventional computers, rather than in parallel as in real brains, so it usually works out more computationally efficient to keep their numbers as low as possible. The relevant lower bound on the number of hidden units will depend mainly on the complexity of the task at hand. Brains have rather different processing overheads, with their sizes constrained by various physical factors, such as growth costs, energy/ oxygen consumption, heat dissipation, and such like. The various "design problems" that are encountered as brains get bigger or smaller have been discussed by Kaas (2000), and it is not surprising that brain sizes should vary so much across species when the range of demands placed on them by different animals and environments are taken into account. Striedter (2005, chapters 4 & 5) provides a good review of evolutionary changes in the sizes of brains and brain regions.

The problem that faces brain modellers here is the enormous difficulty in reliably matching the scales of the simplified neurons, representations, learning algorithms, tasks, and so on in the models against the corresponding components found in real brains. What can, *and should*, be done, however, is to run the models many times, with a wide range of numbers of hidden units, to determine how the precise numbers affect the results. In fact, doing this in previous neuropsychological modelling studies has revealed serious modelling artifacts, with neural networks behaving rather differently when they have barely enough hidden units to carry out the given tasks, compared to when they have brain-like magnitudes of resources (e.g., Bullinaria, 2005a; Bullinaria & Chater, 1995). Moreover, it has also been suggested that restrictions on the number of available neurons can lead to advantages for modularity for representing complex high-dimensional spaces (Ballard, 1986), and that is obviously of direct relevance to the current study.

The way to explore this issue is clearly to run the simulations described in the previous section with a wide range of different numbers of hidden units. Figure 9.4 shows the results obtained for the What–Where problem, with Task 1 being What and Task 2 being Where. In the top-left graph it is seen that the more efficient CE learning algorithm emerges across the full range of values. However, the evolved learning rates do vary considerably with the number of hidden units, as seen in the top-right graph. This confirms the importance of allowing the learning rates for the various network components to vary independently and of evolving them rather than attempting to set them by hand. The bottom-left graph shows that the evolved architectures remain totally non-modular from near the minimal network size required to perform the given task (9 hidden units) right up to over a

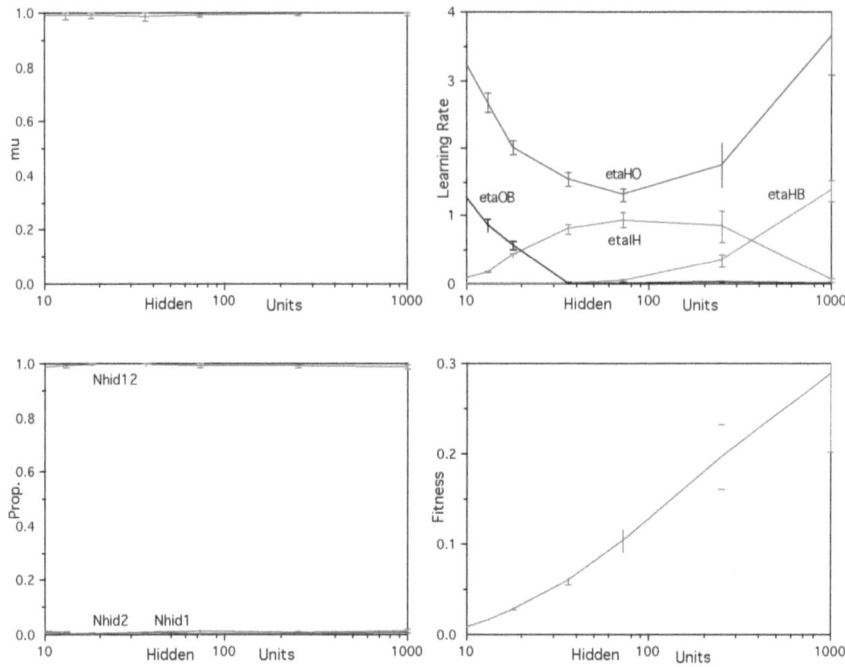

Figure 9.4 Dependence of the evolved What–Where networks on their number of hidden units. *Top left:* the learning cost function parameter μ; *top right:* learning rates η_L; *bottom left:* architecture parameters; *bottom right:* average fitness. [A colour version of this figure can be viewed at www.cognitive neurosciencearena.com/brain-scans]

hundred times that size (1000 units). Finally, the average evolved network fitness (inverse number of training epochs) is seen in the bottom-right graph to increase approximately logarithmically with the number of hidden units. This is why the total number of hidden units itself has not been allowed to evolve in the simulations, because if it were, it would just keep on increasing to whatever limit was imposed on it. From a practical point of view, this would slow down the simulations unnecessarily (in real time on a non-parallel processor) because of the increased computations required, rather than having the process settle down as the optimal configuration emerges.

For the two generalization tasks of Figure 9.2, the corresponding dependences of the evolved network architectures on the number of hidden units are shown in Figure 9.5. The results here are much noisier than for the What–Where task, particularly for the very small networks. As noted before, Task Pair A leads to non-modular architectures evolving, and there is broad independence of the total number of hidden units. For Task Pair B, modular architectures evolve reliably for large numbers of hidden units (200 or more), but for smaller numbers there is an increasing tendency for varying degrees of non-modularity to occur.

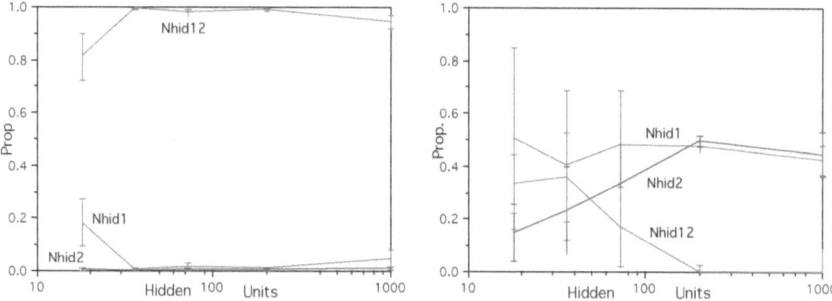

Figure 9.5 Dependence of the evolved generalization task network architectures on their total number of hidden units. *Left*: Task Pair A; *right*: Task Pair B. [A colour version of this figure can be viewed at www.cognitiveneuroscience arena.com/brain-scans]

It is possible to check the quality of the evolved architectures by generating contour plots of average fitness as a function of architecture, using the evolved learning parameters (Bullinaria, 2001). Figure 9.6 shows such plots for Task Pairs A and B for 18 hidden units in total. The apex of each triangle corresponds to a fully distributed architecture, with *Nhid12 ~ Nhid*, while the base represents full modularity, with *Nhid12 ~ 0*. The contours represent epochs of training, running from the lowest value up to 1.3 times that lowest value, with darker shading indicating higher fitness. Even though the evolutionary runs result in variable results for such low numbers of hidden units, these plots show clear non-modularity for Task Pair A and clear modularity for Task Pair B, as is found in the evolutionary runs for higher numbers of hidden units. As has already been observed in other studies (e.g., Bullinaria, 2005a; Bullinaria & Chater, 1995), to achieve reliable results it is best to avoid using networks that have close to the minimal number of hidden units required to perform the given tasks.

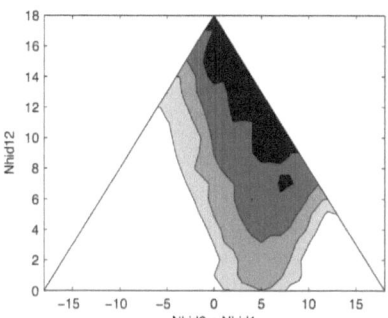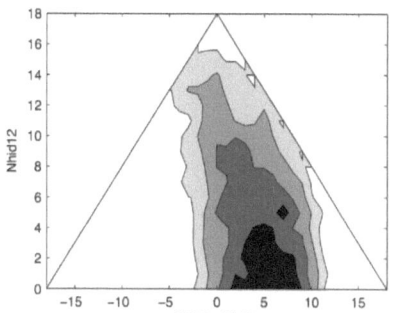

Figure 9.6 Contour plots of network learning performance as a function of architecture for 18 hidden units in total, with higher fitness shown darker. *Left*: Task Pair A; *right*: Task Pair B.

The fact that in each case, for reasonably large networks, the emergent architectures are independent of network size means that it is possible to set the number of hidden units at some convenient value and not be too concerned that there remains considerable uncertainty in how exactly a network size maps onto a real brain region size.

5 The effect of neural connectivity levels

An important factor related to the size of the brain, or brain region, is that of the neural connectivity levels within in it. As the number of neurons grows, so does the number of potential connections between them, but those connections would take up valuable space in the brain, corresponding to the volume that would need to be occupied by workable axons and dendrites. Moreover, the longer range connections required to fully connect large brains would also reduce the speed of neural computation. Beyond a certain size, full connectivity becomes an inefficient use of resources, as Ringo (1991) has demonstrated with an explicit model. Such biological factors and their consequences for brain structure have also been discussed in some detail by Stevens (1989), Kaas (2000), Changizi (2001), and Karbowski (2003), with the general conclusion that the degree of connectedness should fall as the brain size increases. The ways by which evolution might have optimized neural circuits in the light of these issues (i.e., to minimize conduction delays, signal attenuation, connection volumes, and so on) has been considered by Chklovskii, Schikorski, and Stevens (2002), and they concluded that approximately 60% of the space should be taken up by the wiring (i.e., axons and dendrites), which is close to the proportion actually found. It is clear that the issue of neural connectivity imposes important constraints on real brain structures and should therefore be taken into account in any reliable models of their emergence.

The most obvious way to minimize the volume of connections and increase the rate of information flow would be to keep the connections as short as possible. Jacobs and Jordan (1992) considered how such a bias towards short connections in neural network models would affect the neural architecture, and they found that it led to a tendency to decompose tasks into sub-tasks, which is clearly of direct relevance to the issue of modularity. Similar architecture issues were studied in the computational embryogeny approach of Bowers and Bullinaria (2005), in which processes were evolved that allowed neurons to grow from stem cells to connect up input and output neurons and learn to perform simple mappings between them. The problem with such models is that there is always a danger of identifying factors as physical constraints when they are actually unconstrained physical solutions to other problems, and this could introduce a bias into what the models subsequently tell us. For example, it could be that short connections emerged as the best way to implement modularity that has evolved because of its computational advantages, rather than as a consequence of the physical space taken up by the connections, in which case short connections should not be built into models as a constraint on what can evolve.

The safest way to proceed is to assume nothing about the physical brain structures other than the fact that full neural connectivity is not possible. The idea is to see what architectural changes will emerge simply by restricting the *proportion* of connections, without regard to any physical properties like the neuron locations and connection lengths. The natural assumption is that, for any constrained level of connectivity, evolution will result in the various physical costs being minimized by a suitable neuronal layout. Various aspects of this assumption have already been studied in some detail: Mitchison (1991) considered how different branching patterns affect the minimization of cortical wiring. Cherniak (1995) looked at neural component placement with a similar aim. Zhang and Sejnowski (2000) developed models that explained the empirical universal scaling law between grey and white matter. Chklovskii (2004) argued that under certain biologically plausible constraints the optimal neuronal layout problem has an exact solution that is in broad agreement with the layouts found in real brains. Striedter (2005, chapter 7) provides a good overview of the current beliefs concerning the evolution of neuronal connectivity.

It is reasonably straightforward to extend the evolutionary neural network approach described above to test the effect of restricted neural connectivity levels. The only extra feature needed is a restriction on the degree of connectivity between layers to some fraction f of full connectivity, where full connectivity means that each node of each layer is connected to each node of the next layer. One way this reduced connectivity can be achieved is by randomly removing a fraction $1 - f$ of connections from the fully connected network. Within the block structure of Figure 9.3, full connectivity means $Nhid12 = Nhid$ and $Nhid1 = Nhid2 = 0$, where $Nhid$ is the total number of hidden units. Here, the total connectivity levels can also be reduced more systematically by increasing the module sizes $Nhid1$ and/or $Nhid2$ at the expense of $Nhid12$, or by reducing the total number of connected neurons $Nhid1 + Nhid2 + Nhid12$ below the maximum allowed number $Nhid$. The evolutionary process will determine the best way to reduce the connectivity level, by being free to adjust the innate degree of connectivity f_{HO} between the blocks of hidden and output units (achieved by randomly removing a fraction $1 - f_{HO}$ of allowed connections) as well as the existing architecture parameters $Nhid1$, $Nhid2$, and $Nhid12$. The total proportion of connectivity between the hidden and output layers is then easily calculated to be

$$f = \frac{(Nhid1 + Nhid12) \cdot Nout1 + (Nhid2 + Nhid12) \cdot Nout2}{Nhid \cdot (Nout1 + Nout2)} f_{HO}$$

The idea is to run a number of simulations to explore how the outcomes vary for different values of f. In each simulation, f is set to be a particular fixed value, and the architecture parameters are allowed to evolve as before, but now the equation needs to be inverted to compute the value of f_{HO} that gives the right total connectivity level f.

   The results of doing this for the What–Where task with 72 hidden units are shown in Figure 9.7. The learning algorithm parameter μ and the learning rates η_L remain fairly steady as the level of connectivity is reduced, at least until around $f \sim 0.25$ where there remain so few connections left that the network is unable to perform the task reliably. The fitness, measured as the inverse of the number of epochs of training required, falls steadily as the connectivity is reduced. The interesting discovery is that the number of hidden units shared by both output tasks, $Nhid12$, falls almost linearly with f until it reaches one-half, when it stays close to zero for all lower levels of connectivity. Modules emerge for each of the two output tasks, with their relative sizes in proportion to the difficulty of the tasks, with a slight floor effect as the connectivity level gets close to the minimum required for the tasks. That smaller modules are needed for the linearly separable Where task, compared to the non-linearly separable What task, was found in the earlier studies too (Bullinaria, 2001; Rueckl et al., 1989). The minimum workable connectivity level can be decreased by increasing the total number of hidden

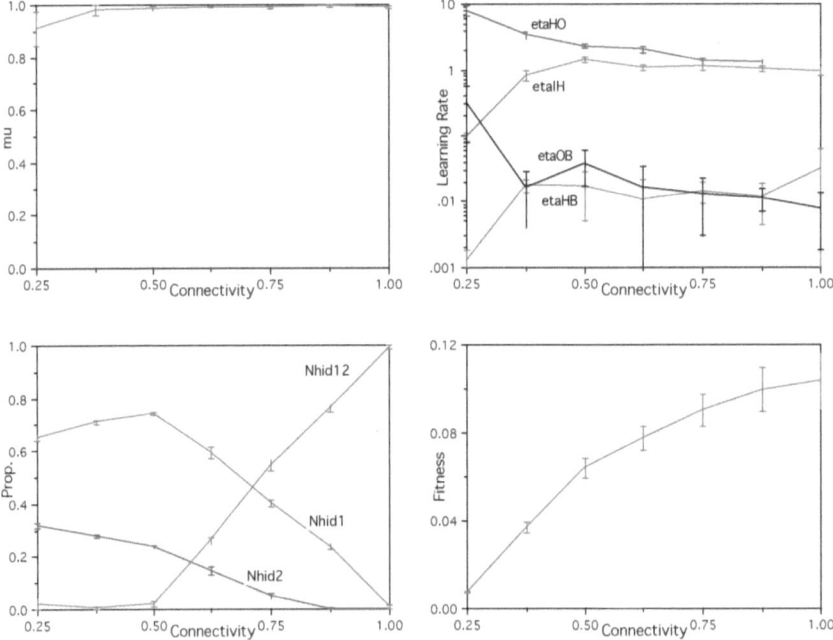

Figure 9.7 Dependence of the evolved What–Where networks on the fraction f of full neural connectivity. *Top left:* the learning cost function parameter μ; *top right:* learning rates η_L; *bottom left:* architecture parameters; *bottom right:* average fitness. [A colour version of this figure can be viewed at www.cognitiveneurosciencearena.com/brain-scans]

units, but modularity still emerges as the preferred architecture for all connectivity fractions below one-half. When connections need to be removed to achieve a lower value of f, the choice is effectively between randomly removing them from anywhere versus systematically removing them from hidden units that contribute to both tasks to increase the modularity, and the simulated evolution shows that increasing the modularity is the better option. Connectivity levels in human brains are certainly considerably below one-half, and even in much simpler organisms they are likely to be below that level (Chklovskii et al., 2002). It seems, then, that a reason has been found why modularity should emerge in simple neural structures performing multiple tasks.

Naturally, it would not be wise to base such a conclusion on one particular problem. The robustness of the result needs to be tested on other pairs of tasks. Figure 9.8 shows the corresponding results for the generalization Task Pair A shown in Figure 9.2, for 200 hidden units. For full connectivity, a purely non-modular architecture evolves, as for the What–Where tasks, and the result of restricting the connectivity level follows a similar pattern. The learning parameters are less tightly constrained in this case, indicating that they have less influence on the performance. The magnitudes of the evolved

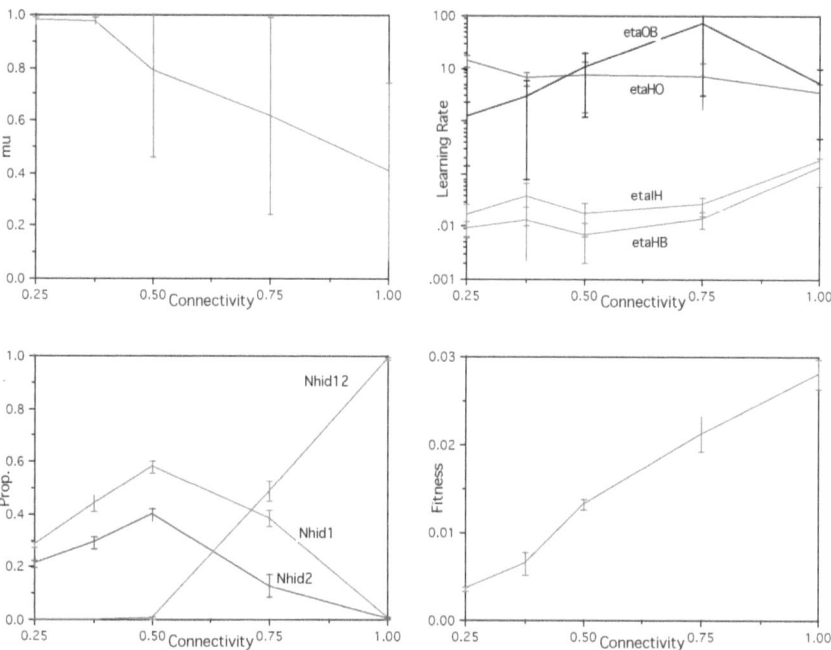

Figure 9.8 Dependence of the evolved generalization Task Pair A networks on the fraction f of full neural connectivity. *Top left:* the learning cost function parameter μ; *top right:* learning rates η_L; *bottom left:* architecture parameters; *bottom right:* average fitness. [A colour version of this figure can be viewed at www.cognitiveneurosciencearena.com/brain-scans]

learning rates are also rather different for this task, as one might expect given the different numbers of input and output units and the training data distributions, but they are again fairly steady across the range of connectivity levels. Most importantly, the degree of modularity again falls almost linearly from full to half connectivity, and the architecture remains purely modular for all lower connectivity levels.

It can be seen more clearly how the performance on this task depends on the network architecture in the fitness versus architecture contour plots of Figure 9.9, generated in the same way as those of Figure 9.6. Even though the fitness levels are averaged over 500 individuals each, the plots are still rather noisy, but the patterns are easy to see, and the evolutionary pressure is clearly enough for the preferred architectures to emerge. The top-left graph is for the evolved networks with full connectivity, with the best performance clearly around the fully non-modular region. The top-right graph is for networks with a connectivity level of $f = 0.75$, and the peak performance area has now shifted to the intermediate regions near the centre of the triangle. The bottom-left graph is for networks with a connectivity level of $f = 0.5$, and the

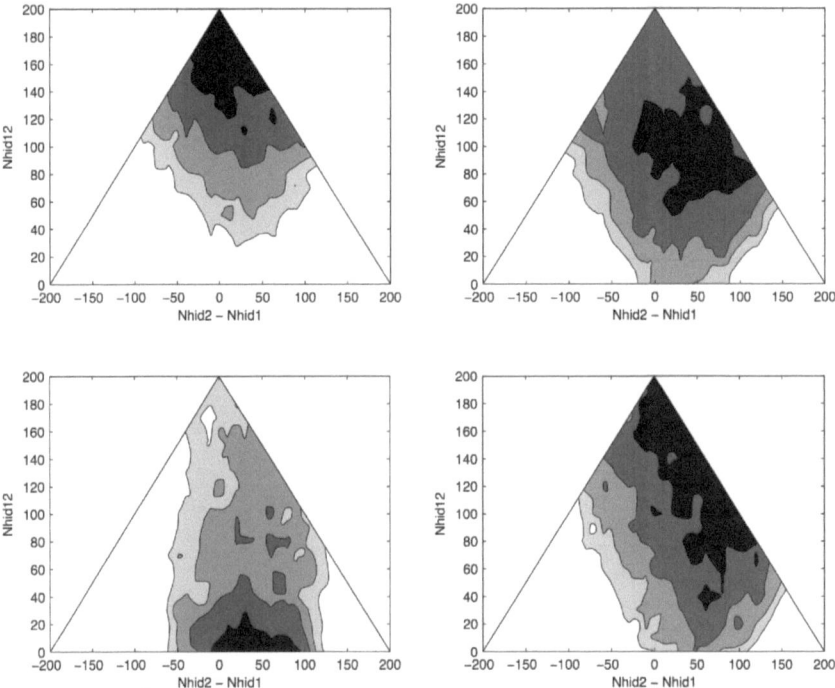

Figure 9.9 Contour plots of network learning performance as a function of architecture, for Task Pair A and 200 hidden units, with higher fitness shown darker. *Top left:* The evolved full connectivity networks; *top right:* 0.75 connectivity networks; *bottom left:* 0.5 connectivity networks; *bottom right:* 0.5 connectivity networks without the connectivity level enforced.

peak performance area is now clearly in the modular region. As a consistency check, the bottom-right graph shows the performance of the evolved $f = 0.5$ networks when f_{HO} is left unconstrained at 1 for all architectures. In this case, even though all the other network parameters have been optimized to deal with restricted connectivity, as soon as the restriction to keep f at 0.5 for all values of $Nhid12$ is removed, the advantage of non-modularity returns. It seems that the increase of modularity as the connectivity level reduces is a robust effect.

Finally, for completeness, it is worth checking how reduced connectivity levels affect networks evolved to deal with task pairs for which modularity consistently emerges even when full connectivity is possible. Figure 9.10 shows such results for the generalization Task Pair B shown in Figure 9.2, for 200 hidden units. Again, the learning parameters are fairly stable across connectivity levels and the fitness increases with connectivity. In this case, the architecture simply remains modular throughout.

Interestingly, below connectivity levels of 0.5, the networks evolved to perform the generalization tasks employ a different connection-reduction strategy from those for the What–Where learning task. Rather than removing random connections by reducing f_{HO}, they instead use the architecture

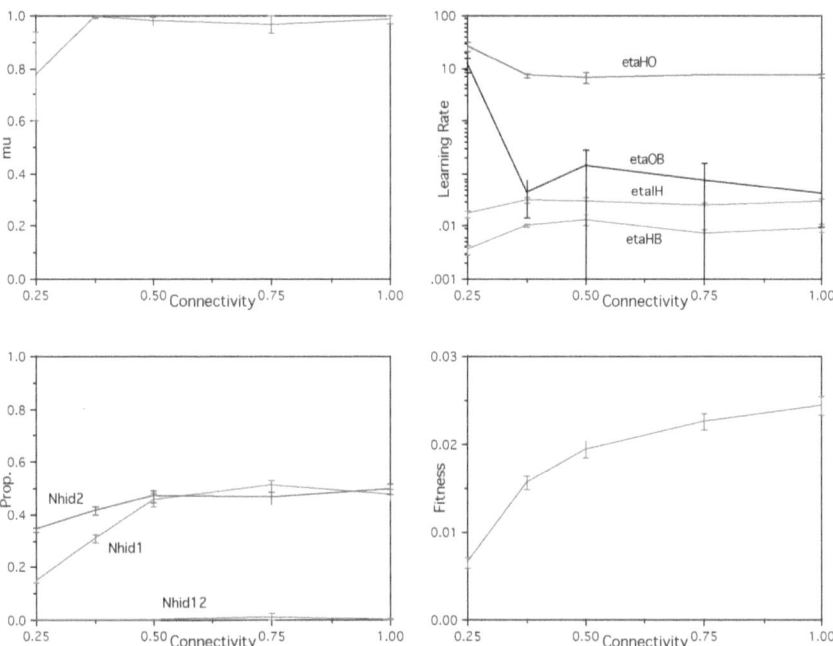

Figure 9.10 Dependence of the evolved generalization Task Pair B networks on the fraction f of full neural connectivity. *Top left:* the learning cost function parameter μ; *top right:* learning rates η_L; *bottom left:* architecture parameters; *bottom right:* average fitness. [A colour version of this figure can be viewed at www.cognitiveneurosciencearena.com/brain-scans]

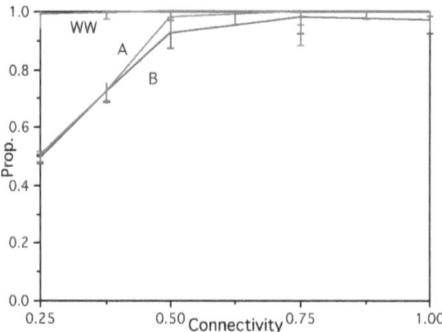

Figure 9.11 Proportional usage of the available hidden units as the connectivity level is restricted. The evolved What–Where task networks (WW) are seen to behave rather differently from those evolved for the generalization task pairs (A and B). [A colour version of this figure can be viewed at www.cognitiveneurosciencearena.com/brain-scans]

parameters to reduce their usage of the available pool of hidden units by having some of them not connected to any output units. This is seen more clearly in Figure 9.11. Given the small number of output units driven by each module hidden unit in these tasks, this strategy makes good sense. That the simulated evolution automatically determines such sensible strategies, which may easily have been missed by human modellers, is another reason why the evolutionary approach shows such good promise in this area.

6 Discussion and conclusions

This chapter has shown how simulations of the evolution of simple neural networks, which must learn to perform pairs of simple tasks, can be used to explore the factors that might lead to the emergence of modularity in brains. It noted that particular attention needs to be paid to two important high-level physical constraints that affect real brains – namely, the restrictions placed on brain size and the associated levels of neural connectivity. However, to understand the reasons for particular brain structures, it is always important to distinguish between the physical properties that really do constrain those structures and the computations they perform, and those physical properties that could potentially be different if the associated advantages were sufficient to cause them to evolve. For this reason, models across the spectrum were studied, ranging from the kind of unconstrained systems often used when building the best possible artificial systems, to those with levels of constraint approaching those found in biological systems.

It is clear from the plots of fitness in Figures 9.4, 9.7, 9.8, and 9.10 that there is a fitness advantage for larger networks and higher levels of connectivity. However, as discussed above, there are strong biological reasons that make that infeasible. Once those biological factors are incorporated into the models, modular architectures emerge automatically as the best approach for dealing with those constraints. This confirms empirically earlier suggestions that this would happen (e.g., Kaas, 2000).

The tasks and neural networks discussed in this chapter are, of course, far removed from the levels of complexity seen in the higher organisms alive today, but they may be more directly relevant to the low-level processing requirements of simpler organisms at earlier stages of evolutionary history. There remains a good deal of controversy in the field of early nervous system evolution (e.g., Holland, 2003), but having demonstrated how simple low-level modules can emerge, it is reasonable to conjecture that evolution will find ways of using them as building blocks for creating higher level modules and, eventually, whole brains as they exist today (e.g., Alon, 2003). Certainly, the kinds of "distinct output tasks" used in the simulations presented in this chapter will almost certainly correspond to separate components or sub-tasks of more complex higher level systems. Exploring how the prior existence of simple low-level modules of the type considered here will facilitate (or not facilitate) the emergence of more complex modules (with their higher level advantages) would clearly be a fruitful area of future research.

Having established that modularity emerges as the result of restricted connectivity levels, the next question to ask is how those restricted connectivity levels are actually implemented in biological brains. Chklovskii (2004) has considered in some detail the issue of optimal neuronal layout. Even highly modular architectures will still need long-range connections between modules. It seems most probable that something like "small-world" or scale-free networks (Watts & Strogatz, 1998) will prove to be the most efficient organizational structure, as has been discussed by Sporns, Tononi, and Edelman (2000), Buzsáki, Geisler, Henze, and Wang (2004), and Sporns, Chialvo, Kaiser, and Hilgetag (2004). Moreover, it has been suggested that the trade-off between computational requirements and physical constraints may be responsible for the diversity of interneurons in the mammalian cortex (Buzsáki et al., 2004; Chklovskii et al., 2002). Such neural structures, of course, go way beyond the simple architecture of Figure 9.3, and more complex neural representation schemes will be required to extend the evolutionary neural network approach in that direction.

Perhaps the best way to make further progress in this area would be to allow increasingly general neural structures to evolve. Bowers and Bullinaria (2005) started looking at this by evolving systems that grew neural structures from stem cells, with connections growing along chemical gradients. Approaches such as this will obviously require significantly more than the two evolvable parameters needed to specify the architecture in Figure 9.3, and that will make the evolutionary process considerably more demanding, but it will allow an almost unlimited range of neural structures to emerge. It will also allow appropriate structures to emerge in response to the training data, rather than being fixed innately, and this could begin to address questions about what degree of modularity should be innate and how much should be learned within a lifetime. Taking the evolving neural network approach in this direction will be difficult, but the results presented in this chapter suggest that it could be an extremely profitable way to make further progress in understanding brain structures.

7 References

Allman, J. M. (1998). *Evolving brains*. New York: W. H. Freeman.

Alon, U. (2003). Biological networks: The tinkerer as an engineer. *Science, 301*, 1866–1867.

Baldwin, J. M. (1896). A new factor in evolution. *The American Naturalist, 30*, 441–451.

Ballard, D. H. (1986). Cortical connections and parallel processing: Structure and function. *Behavioral and Brain Sciences, 9*, 67–120.

Binder, J. R., Medler, D. A., Desai, R., Conant, L. L., & Liebenthal, E. (2005). Some neurophysiological constraints on models of word naming. *NeuroImage, 27*, 677–693.

Bishop, C. M. (1995). *Neural networks for pattern recognition*. Oxford, UK: Oxford University Press.

Bookheimer, S. (2002). Functional MRI of language: New approaches to understanding the cortical organization of semantic processing. *Annual Review of Neuroscience, 25*, 151–188.

Bowers, C. P., & Bullinaria, J. A. (2005). Embryological modelling of the evolution of neural architecture. In A. Cangelosi, G. Bugmann, & R. Borisyuk (Eds), *Modeling language, cognition and action* (pp. 375–384). Singapore: World Scientific.

Bullinaria, J. A. (2001). Simulating the evolution of modular neural systems. In *Proceedings of the twenty-third annual conference of the Cognitive Science Society* (pp. 146–151). Mahwah, NJ: Lawrence Erlbaum Associates.

Bullinaria, J. A. (2003). From biological models to the evolution of robot control systems. *Philosophical Transactions of the Royal Society of London, A, 361*, 2145–2164.

Bullinaria, J. A. (2005a). Connectionist neuropsychology. In G. Houghton (Ed.), *Connectionist models in cognitive psychology* (pp. 83–111). Hove, UK: Psychology Press.

Bullinaria, J. A. (2005b). Evolving neural networks: Is it really worth the effort? In *Proceedings of the European symposium on artificial neural networks* (pp. 267–272). Evere, Belgium: d-side.

Bullinaria, J. A. (2007a). Using evolution to improve neural network learning: Pitfalls and solutions. *Neural Computing & Applications, 16*, 209–226.

Bullinaria, J. A. (2007b). Understanding the emergence of modularity in neural systems. *Cognitive Science, 31*, 673–695.

Bullinaria, J. A., & Chater, N. (1995). Connectionist modelling: Implications for cognitive neuropsychology. *Language and Cognitive Processes, 10*, 227–264.

Buzsáki, G., Geisler, C., Henze, D. A., & Wang, X.-J. (2004). Interneuron diversity series: Circuit complexity and axon wiring economy of cortical interneurons, *Trends in Neurosciences, 27*, 186–193.

Caelli, T., Guan, L., & Wen, W. (1999). Modularity in neural computing. *Proceedings of the IEEE, 87*, 1497–1518.

Cantû-Paz, E., & Kamath, C. (2005). An empirical comparison of combinations of evolutionary algorithms and neural networks for classification problems. *IEEE Transactions on Systems, Man, and Cybernetics—Part B: Cybernetics, 35,* 915–927.

Chang, F. (2002). Symbolically speaking: A connectionist model of sentence production. *Cognitive Science, 26,* 609–651.

Changizi, M. A. (2001). Principles underlying mammalian neocortical scaling. *Biological Cybernetics, 84,* 207–215.

Cherniak, C. (1995). Neural component placement. *Trends in Neurosciences, 18,* 522–527.

Chklovskii, D. B. (2004). Exact solution for the optimal neuronal layout problem. *Neural Computation, 16,* 2067–2078.

Chklovskii, D. B., Schikorski, T., & Stevens, C. F. (2002). Wiring optimization in cortical circuits. *Neuron, 34,* 341–347.

Coltheart, M., Curtis, B., Atkins, P., & Haller, M. (1993). Models of reading aloud: Dual-route and parallel-distributed-processing approaches. *Psychological Review, 100,* 589–608.

Duncan, J. (2001). An adaptive coding model of neural function in prefrontal cortex. *Nature Reviews. Neuroscience, 2,* 820–829.

Duncan, J., & Owen, A. M. (2000). Common regions of the human frontal lobe recruited by diverse cognitive demands. *Trends in Neurosciences, 23,* 475–483.

Dunn, J. C., & Kirsner, K. (1988). Discovering functionally independent mental processes: The principle of reversed association. *Psychological Review, 95,* 91–101.

Elman, J. L., Bates, E. A., Johnson, M. H., Karmiloff-Smith, A., Parisi, D., & Plunkett, K. (1996). *Rethinking innateness: A connectionist perspective on development.* Cambridge, MA: MIT Press.

Fahlman, S. E. (1988). Faster learning variations of back propagation: An empirical study. In D. Touretzky, G. E. Hinton, & T. J. Sejnowski (Eds.), *Proceedings of the 1988 Connectionist Models Summer School* (pp. 38–51). San Mateo, CA: Morgan Kaufmann.

Fodor, J. A. (1983). *The modularity of the mind.* Cambridge, MA: MIT Press.

Geary, D. C., & Huffman, K. J. (2002). Brain and cognitive evolution: Forms of modularity and functions of mind. *Psychological Bulletin, 128,* 667–698.

Holland, N. D. (2003). Early central nervous system evolution: An era of skin brains? *Nature Reviews. Neuroscience, 4,* 617–627.

Huettel, S. A., Song, A. W., & McCarthy, G. (2004). *Functional magnetic resonance imaging.* Sunderland, MA: Sinauer Associates.

Jacobs, R. A. (1999). Computational studies of the development of functionally specialized neural modules. *Trends in Cognitive Science, 3,* 31–38.

Jacobs, R. A., & Jordan, M. I. (1992). Computational consequences of a bias toward short connections. *Journal of Cognitive Neuroscience, 4,* 323–336.

Kaas, J. H. (2000). Why is brain size so important: Design problems and solutions as neo-cortex gets bigger or smaller. *Brain and Mind, 1,* 7–23.

Karbowski, J. (2003). How does connectivity between cortical areas depend on brain size? Implications for efficient computation. *Journal of Computational Neuroscience, 15,* 347–356.

Kashtan, N., & Alon, U. (2005). Spontaneous evolution of modularity and network motifs. *Proceedings of the National Academy of Sciences, 102,* 13773–13778.

Leise, E. (1990). Modular construction of nervous systems: A basic principle of design for invertebrates and vertebrates. *Brain Research Review, 15,* 1–23.

Lipson, H., Pollack, J. B., & Suh, N. P. (2002). On the origin of modular variation. *Evolution, 56*, 1549–1556.

Marcus, G. (2004). *The birth of the mind*. New York: Basic Books.

Mitchison, G. (1991). Neuronal branching patterns and the economy of cortical wiring. *Philosophical Transactions of the Royal Society of London, B, 245*, 151–158.

O'Leary, D. D. M. (1989). Do cortical areas emerge from a protocortex? *Trends in Neurosciences, 12*, 400–406.

Plaut, D. C. (1995). Double dissociation without modularity: Evidence from connectionist neuropsychology. *Journal of Clinical and Experimental Neuropsychology, 17*, 291–321.

Plunkett, K., & Bandelow, S. (2006). Stochastic approaches to understanding dissociations in inflectional morphology. *Brain and Language, 98*, 194–209.

Reisinger, J., Stanley, K. O., & Miikkulainen, R. (2004). Evolving reusable neural modules. In *Proceedings of the Genetic and Evolutionary Computation conference (GECCO 2004)* (pp. 69–81). New York: Springer-Verlag.

Ringo, J. L. (1991). Neuronal interconnection as a function of brain size. *Brain Behavior and Evolution, 38*, 1–6.

Rueckl, J. G., Cave, K. R., & Kosslyn, S. M. (1989). Why are "what" and "where" processed by separate cortical visual systems? A computational investigation. *Journal of Cognitive Neuroscience, 1*, 171–186.

Seok, B. (2006). Diversity and unity of modularity. *Cognitive Science, 30*, 347–380.

Shallice, T. (1988). *From neuropsychology to mental structure*. Cambridge, UK: Cambridge University Press.

Sporns, O., Chialvo, D. R., Kaiser, M., & Hilgetag, C. C. (2004). Organization, development and function of complex brain networks. *Trends in Cognitive Sciences, 8*, 418–425.

Sporns, O., Tononi, G., & Edelman, G. M. (2000). Theoretical neuroanatomy: Relating anatomical and functional connectivity in graphs and cortical connection matrices. *Cerebral Cortex, 10*, 127–141.

Stevens, C. F. (1989). How cortical interconnectedness varies with network size. *Neural Computation, 1*, 473–479.

Striedter, G. F. (2005). *Principles of brain evolution*. Sunderland, MA: Sinauer Associates.

Teuber, H. L. (1955). Physiological psychology. *Annual Review of Psychology, 6*, 267–296.

Van Orden, G. C., Pennington, B. F., & Stone, G. O. (2001). What do double dissociations prove? *Cognitive Science, 25*, 111–172.

Wagner, G. P. (1996). Homologues, natural kinds, and the evolution of modularity. *American Zoologist, 36*, 36–43.

Watts, D. J., & Strogatz, S. H. (1998). Collective dynamics of "small world" networks. *Nature, 393*, 440–442.

Yao, X. (1999). Evolving artificial neural networks. *Proceedings of the IEEE, 87*, 1423–1447.

Zhang, K., & Sejnowski, T. J. (2000). A universal scaling law between gray matter and white matter of cerebral cortex. *Proceedings of the National Academy of Sciences, 97*, 5621–5626.

10 Selective attention in linked, minimally cognitive agents

Robert Ward and Ronnie Ward

We investigate minimally cognitive agents for attention and control of action. A visual agent (VA) can attend to one target while ignoring another and can then reallocate processing. Here we review some of our previous results with VA, showing how an analysis of a minimally cognitive agent can both support previous ideas of selective attention and action, as well as suggest new possibilities. We investigate a simple case in which multiple agents are combined to solve tasks that no single agent can complete on its own.

1 Introduction

Cognitive agents and their environment interact within a continuous perception–action cycle: perceptions guide actions, actions alter the environment, and the altered environment in turn generates new perceptions (Gibson, 1979). However, computational models in psychology have tended *not* to focus on the dynamic nature of the perception–action loop. Even models of perception–action linkage are usually static in nature: processing begins with the presentation of static stimuli and ends with the selection of an action that has no effect on the stimulus environment (Cohen, Dunbar, & McClelland, 1990; Schneider, 1995; Ward, 1999). In many cases, action is not modeled seriously, perhaps consisting of an activated output node representing one of several discrete alternative responses. In this respect, the models largely follow the available empirical work. Although much is known about the capacity, time-course, and brain regions important for selective attention and action, in the large majority of studies the responses measured have little to do with the dynamic nature of the perception–action cycle. Usually, participants make some discrete response, such as a keypress, and the trial ends with no impact on the agent's environment.

Theories and models generated for static situations and discrete responses may be inadequate for more interesting tasks and realistic agent–environment interaction. We want to push psychological models towards more complex forms of action and to encourage thinking about how selective attention and action work within the perception–action cycle. In this respect, a useful class of models, sometimes called embodied, situated, and dynamical (ESD) agents, stress what Clark (1999) calls "the unexpected intimacy between the brain, body, and world". The ESD approach emphasizes the contextually bound nature of solutions to cognitive problems, and it therefore considers specific tasks in their entirety, rather than isolated but general-purpose attention, perception, or motor subsystems. In ESD agents, control of action does not arise through general-purpose representation; rather, it emerges from a distributed system involving the brain, body, and world.

An important form of the ESD approach for our purposes is the "minimally cognitive" agent, developed by Beer (1996). "Minimally cognitive" agents are intended to be tractable models that are embodied, situated, and dynamic and that can do interesting cognitive tasks. For example, Figure 10.1 shows the visual agent (VA) developed by Slocum, Downey, and Beer (2000), evolved by genetic algorithm to selectively attend and respond to targets in its 2D environment. The agent was equipped with an array of proximity sensors, left and right motor units to control its movement, and a small hidden layer of eight units. The targets "fell" from the top of the 2D environment, and VA was meant to "catch" them by moving under them

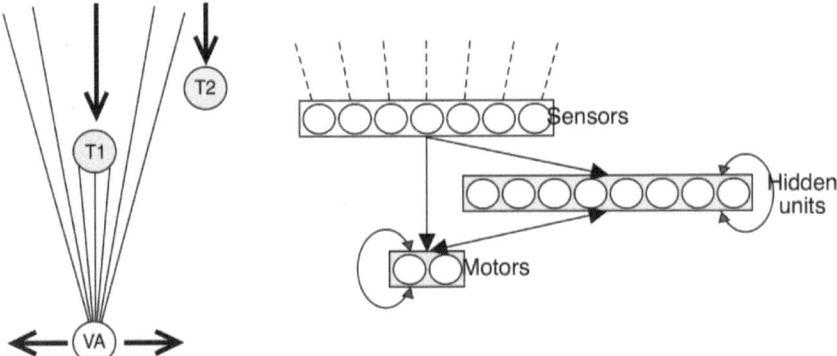

Figure 10.1 The visual agent (VA) originally developed by Slocum et al. (2000). *Left:* Schematic of the agent and environment. The agent is evolved to catch two targets (T1 and T2), falling from the top of the environment. VA has a left and a right motor, enabling it to move along the bottom of the environment. It is evolved by genetic algorithm to catch the targets as close to center as possible. The rays emanating from VA represent proximity sensors. *Right:* The continuous-time recurrent neural network (CTRNN) that controls VA. The seven input units each pass on activation from an associated proximity sensor. The activation of the input units reflects where the proximity sensor is intersected. Each input unit is connected to the two motor units and to each of the eight hidden units. Each of the hidden units receives connections from every unit in the network (including itself) and sends connections to every hidden and motor unit. The two motor units receive inputs from the input and hidden layer. These units also send connections back to the hidden layer, and both are self-connected and connected to each other.

before impact. In our simulations, the two targets fell straight down and with constant velocity, the first target (T1) by definition having greater velocity than the second target (T2). Following Slocum et al. (2000), the agent was controlled by a continuous-time recurrent neural network (CTRNN), with a reasonably generic network architecture.

Although catching one target is trivial, catching two targets is definitely not. Catching two targets requires numerous cognitive operations for success, including: (1) correctly prioritizing one of the two targets for immediate action; (2) before impact, tying motor systems to the movements of the selected target and insulating them from the effects of the other target; and (3) after catching the first target, the second target, which was previously insulated from the motor system, must now be allowed to control the motor system. These processes of selection, response control, and reconfiguration following a change of targets are all important themes in current selective attention research (for an overview, see Driver, 2001). Not only does the task require interesting cognitive abilities, but it is also a good example of an embodied agent, embedded within a genuine perception–action loop: action involves an extended sequence of motor commands in response to sensory inputs; the senses guide action, and action generates new sensory inputs.

VA therefore seems like a useful starting place for looking at a new class of models that extend beyond the limits of static situations, free of any meaningful agent–environment interaction. We have then recently undertaken *psychological* analyses of this agent. Our aim was to understand the bases for selective attention and control of action within the agent. Our approach might be fairly described as "artificial comparative psychology". We are seeking both to understand the psychological principles that allow VA to control its attention and then to compare these principles to what is found in the human attention literature.

We reasoned that, as long as VA is doing a *genuine* selective attention task, two outcomes are possible, both of which are interesting. First, it might be that VA operates according to new principles not yet seen in the human literature. In this case, VA acts as a model to new experimentation and theory. Second, it might be that the agent operates according to models already developed in the human literature. In this case, VA serves as a tractable, embodied model system. However, the validity of either outcome depends on the validity of the selective attention, an issue we address in greater detail elsewhere (Ward & Ward, 2008).

2 The psychology of attention in a minimally cognitive agent

In fact, our analyses of VA uncovered both established and novel concepts in the human attention literature. Here we briefly review our analyses of VA, describing its psychological mechanisms for attention and control; however, complete details are available elsewhere (Ward & Ward, 2006, 2008).

2.1 Reactive inhibition

It can seem effortless to switch attention from one object to the next, but this is probably an illusion. Robust costs are found whenever people have to switch their attention, particularly when reallocating attention to an object that was previously being ignored. These costs are large enough and reliable enough that the phenomenon has been given its own label. *Negative priming* refers to costs observed when a previously ignored object must be subsequently attended and reported. Models of *reactive inhibition* (Houghton & Tipper, 1994) propose that negative priming is a direct reflection of the competitive strength of the object for attention. According to reactive inhibition accounts, a target is selected by inhibiting the nontarget items. The more salient the nontarget – that is, the more powerfully it tends to attract attention – the greater the inhibition needed to prevent the object from gaining control of behavior. The flipside is that if attention needs to be subsequently allocated to this inhibited item, it will take time to overcome this inhibition. Thus, more negative priming is observed for highly salient nontargets than for less salient ones (Grison & Strayer, 2001). The otherwise unintuitive result – that the *more* salient an ignored item, the *harder* it is to switch attention to it – follows in a straightforward way from reactive inhibition.

We found that VA attended to the first target (T1) through reactive inhibition of the second (T2), as demonstrated through a variety of analyses. One example is the time required to reallocate processing from T1 to T2. After T1 impacted, it was cleared from the environment, so that it could no longer activate VA's sensor array. After catching T1, we noticed a clear hesitation before VA moved again to catch T2. We measured this hesitation as a function of the salience of T2. On some trials, the speed and spatial separation of the falling targets meant that a good catch of T1 would leave T2 outside the sensors. We called these out-of-view (OOV) trials. On OOV trials, after catching T1, VA would be looking at a simply blank sensor array. For efficient performance on OOV trials, VA would need memory for the position of T2. In fact, memory for T2 position was well above chance. VA moved directly towards the location of the unseen target on about 85% of novel test trials.

However, what was interesting from the perspective of attention and control of action was that hesitation varied according to the visibility of the T2 item, as shown in Figure 10.2. On a sample of 500 novel trials, hesitation on OOV trials was reduced compared to trials in which T2 was in view, $F(1, 498)$ = 95.9, $p < .0005$. This finding follows cleanly from the idea of reactive inhibition. The more salient in-view T2s required more inhibition than the OOV ones. Therefore, when it was time to reallocate attention to T2, there was more inhibition to overcome.

Overall, we found four lines of evidence for reactive inhibition in VA. First, as already mentioned, VA was slower to respond to more salient,

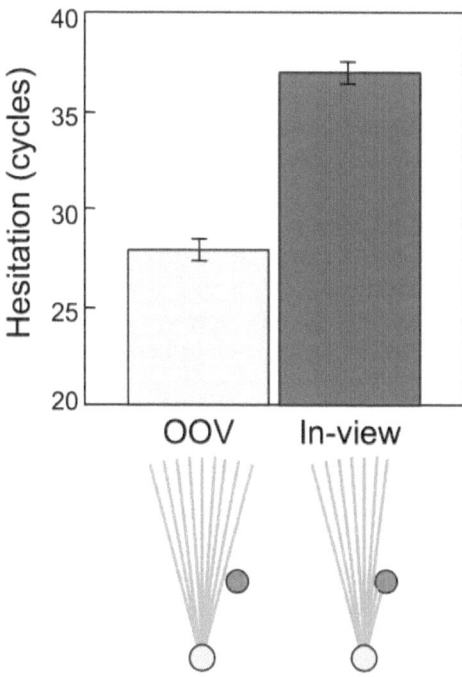

Figure 10.2 Hesitation after T1 catch. After catching the first target (T1), VA
hesitates before moving again to catch the second target (T2). Hesitation is a function of T2 salience and, in particular, whether T2 is in-
view or out-of-view (OOV) following T1 catch. On in-view trials, T2
is within the sensor array following T1 catch, but not on OOV trials
(as indicated by the icons below the graph). VA is faster to respond
to T2 when it is OOV compared to in-view. This is consistent with a
reactive inhibition account, as described in the text.

compared to less salient, T2s. Second, we isolated a signal within VA's
hidden layer that inhibited responses towards salient T2s. Third, we found
that inhibition of T2 depended upon the degree of conflicting motor activations within VA. Fourth, we observed the opponent-process loops within
VA's evolved network. These results all pointed to an internal, inhibitory
signal based on T2 salience (details available in Ward & Ward, 2008).

We also found some similar results in experiments with people. We presented observers with a simulation of the two falling targets, using stimulus
parameters essentially identical to what we used with VA. Before the trial
began, the two balls appeared on the screen. After the trial was initiated, the
balls began to "approach" the observer, increasing in size. Participants could
move the viewing window left and right by pressing the arrow keys, like a
first-person shooter game, and they were instructed to "catch" the targets as

accurately as possible by centering the viewing window on each target at the time of impact.

After catching T1, there was a clear period of hesitation, approximately 120 ms, before the next movement towards T2. Human participants, just like VA, were slower to respond towards in-view compared to OOV T2s. We also found that people were also slow to respond to T2s made salient by other means, including color contrast (Ward & Ward, 2008). In this case, one target was nearly the same color as the background, and the other was a very salient bright red. T2 could either be relatively salient (T1 = background color; T2 = red), or unsalient (T1 = red; T2 = background color). In this experiment, the stimulus parameters were set so that there were no OOV trials. The simple question was how would hesitation vary as a function of T2 salience. As predicted, hesitation to respond to T2 increased when T2 was relatively salient. Again, this otherwise counter-intuitive result follows straightforwardly from a reactive inhibition account.

2.2 Distributed, not localized, conflict monitoring

One way to think about the effect of the two targets on VA's behavior is in terms of *cognitive conflict*. Cognitive conflict occurs when neural pathways associated with different concurrent processes interfere with each other (Botvinick, Braver, Barch, Carter, & Cohen, 2001). In psychology, the Stroop task is perhaps the best-studied example of cognitive conflict. In this case, the identity of a colored word can interfere with color naming (Mari-Beffa, Estevez, & Danziger, 2000). The two-target catching task is another example of a conflict task. VA is most clearly in a state of cognitive conflict when it is located between the two targets. In these circumstances, T1 will be signaling a move in one direction, and, at the same time, T2 will be signaling a move in the opposite direction.

So how is cognitive conflict managed? Botvinick et al. (2001) proposed a two-part "evaluate–regulate" approach. The top-down conflict-monitoring system first detects conflict in underlying neural structures. Control mechanisms are then invoked that regulate processing in some appropriate way to reduce conflict. Botvinick et al. suggest that conflict monitoring is localized to a dedicated module, argued to be the anterior cingulate cortex (ACC) in humans. In support of this claim, Botvinick and colleagues (Botvinick, Cohen, & Carter, 2004; Botvinick et al., 2001) reviewed a variety of evidence showing ACC activation during conditions producing response conflict.

We therefore examined VA for evidence of a localized conflict-monitoring mechanism. For VA, we did not observe any set of units dedicated to monitoring or regulation. Instead, we found a system that used competitive balancing of response biases and distributed conflict management (Ward & Ward, 2006). This does not mean that other types of systems might not use a dedicated module for conflict detection, and, in fact, in the next section we use just such a dedicated module. However, the current simulation does demonstrate that a model system operating within a genuine perception–action cycle and facing real problems of response conflict (and solving them!) does not require this kind of dedicated module. It is worth remembering that there was no module for either conflict detection or regulation built in to the VA architecture. Instead, the computational properties of the catching task shaped the solution.

Further investigation is still needed to understand how conflict is managed by VA. Yet these results might still help us understand current data regarding the ACC. It is well established that ACC is activated during conflict tasks, but the significance of this activation is currently being debated. For example, neurological patients with damage to ACC can be unimpaired in response conflict tasks (Fellows & Farah, 2005). ACC activation might therefore be a correlate rather than causal factor of performance in response conflict tasks, such as arousal (Critchley et al., 2003) or effort (Mulert, Menzinger, Leicht, Pogarell, & Hegerl, 2005). Our results with VA, and its success in resolving cognitive conflict without explicit conflict monitoring, lead us to likewise question the causal link between ACC activation and the resolution of response conflict (Ward & Ward, 2006). The ACC may be one of many sites in which there is competition such as we observed in VA: biased activation for different actions.

3 Emergent abilities in combined agents

Let us assume for exposition's sake that the human brain has on the order of 100 billion neurons – about 20 billion cortical neurons (Pakkenberg & Gundersen, 1997) and about 100 billion cerebellar neurons (Andersen, Korbo, & Pakkenberg, 1992) – each one of which has far more sophisticated processing capabilities (Graham & van Ooyen, 2004) than the 17 simple CTRNN units we used with VA. Furthermore, there are an estimated 240 *trillion* synapses in the cortex alone (Koch, 1999), compared to 170 synapses for VA. This is a big gap. So, how do we scale up from a minimally cognitive agent to a brain? Which direction do we go? With more VAs, or with bigger VAs? In other words, is the human brain's cognitive architecture best thought of as a relatively small number of large multi- or general-purpose networks or as a collection of billions of small special-purpose agents? Here we describe a simple simulation pointing towards the second possibility, linking together two different VAs by a third VA.

3.1 Linked VAs and emergent attention

We began with two copies of the VA network discussed above, complete with reactive inhibition and distributed conflict monitoring. In every way these two copies of VA were identical, with one exception – their sensors detected different kinds of objects. In our previous simulations, target color was not an issue, but here we introduced targets in two colors, red and green.

One copy of VA – the green VA – had sensors that could only detect green targets; red targets were effectively invisible and would not activate the sensors in any way. The other VA copy – the red VA – could only detect red targets, and not green ones. We will call the red and green copies of VA the *component* VAs.

The two component VAs were linked to a single shared set of motor units (see Figure 10.3). The sensor arrays of the component VAs were exactly aligned with each other, so that both component VAs were always looking at the identical region of space. So now consider what happens if we drop two green targets to this combination of VAs. The green targets are invisible to the red VA, and so the red agent provides no task-relevant activation to the motor units. The green VA would therefore have sole control of the motor system, and it would proceed to catch the two targets, on its own, as it had already been evolved to do. Of course, the converse would be true if two red targets were dropped. For same-color targets, then, there is no cognitive conflict between the component VAs.

But consider the case in which a red and a green target are dropped. Now each of the component VAs will attempt to catch the target it can see. In this case of mixed-color targets, the two VAs are in a state of cognitive conflict, one pulling towards the red and the other towards the green target. Because there is no way to regulate the conflict between agents, we would expect interference from T2 on T1 catch in these conditions. Thus, while each of the

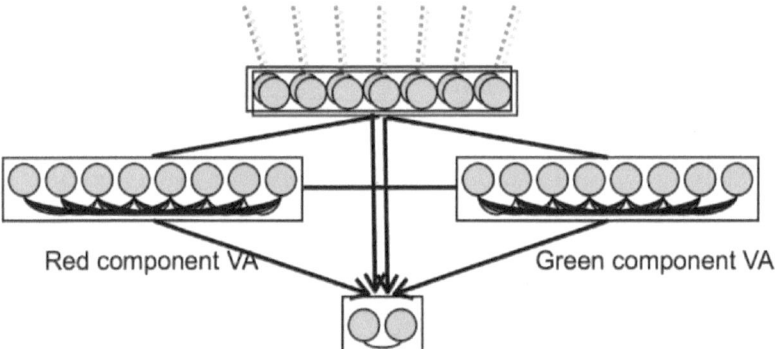

Figure 10.3 The linked visual agent (LVA). LVA combines three agents. There are two *component* agents, which have identical network structure and parameters to VA: the red (darker dots) and green (lighter dots) VAs. Each of the component agents can only see targets in one color. The input and hidden layers of each component project to a shared motor layer. Thus, the component VAs can be considered to compete for control of motor activity. Note that the proximity sensors of the red and green component VAs are exactly aligned so that they receive input from the identical region of space. Communication between the agents is added by a third "conflict-monitor" agent, consisting solely of weights between the hidden layers of the two component agents. [A color version of this figure can be viewed at www.cognitiveneurosciencearena.com/brain-scans]

component VAs is a sophisticated information processor with memory and reactive inhibition, by linking them to the same set of motors we have turned the ensemble into something that might show little or no ability for selective action.

To resolve this conflict, we added a third "agent" that linked each of the hidden units in one component agent to each of the hidden units in the other. This third agent consisted only of this weight matrix. We refer to the collection of agents (the two component VAs and the third "conflict monitoring" agent) as the linked visual agent, or LVA. The weights and network parameters of the component VAs were fixed, and we evolved only the weights of the conflict monitor.

This small experiment was successful. LVA was able to process the mixed-color targets with high accuracy, a task that none of its components alone would be capable of. T1 and T2 catch accuracy were over 99%. Some results of this simulation are shown in Figure 10.4, which shows hesitation times for 500 novel same-color trials and 500 equivalent mixed-color trials. Recall that for same-color trials, there is no necessary conflict between the component VAs, while for mixed-color trials, there will be conflict. LVA showed an overall cost for reallocating resources from T1 to T2, a pattern very similar to VA. After catching T1, LVA hesitated for a significant period before moving again to catch T2. Figure 10.4 illustrates hesitation to respond to T2 as a function of same- versus mixed-color targets, and whether T2 was in-view or OOV after T1 catch. First let us consider LVA performance on same-color trials (i.e., two red or two green targets). As can be seen in Figure 10.4, LVA's

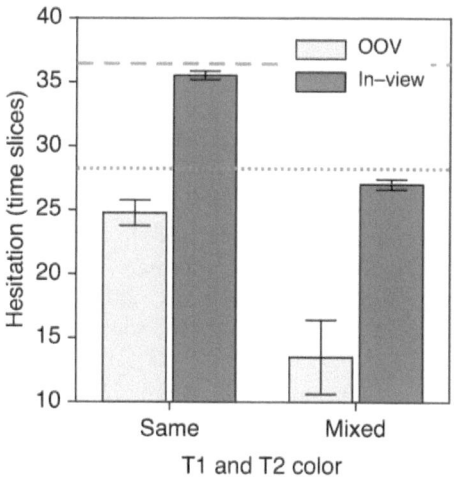

Figure 10.4 Hesitation in LVA. Hesitation after T1 catch is shown for same-color (T1 and T2 are both red or both green), and mixed-color (one red target, one green target) trials. For comparison, hesitation of VA is shown for OOV (dotted line) and in-view (dashed line) trials.

performance on same-color targets was almost identical to the component VAs. In particular, hesitation was longer when T2 was in-view compared to OOV, $F(1, 498) = 137.8$, $p < .0005$, suggesting that inhibition is proportional to target salience. This result is expected, and it simply shows that the addition of the third agent – the "conflict monitor" – did not fundamentally change the operation of the component VAs. The component VAs still used reactive inhibition to selectively respond to T1, and so they were slower to reallocate processing towards salient T2s.

It is interesting to see a similar pattern even for the mixed-color trials. Again, when reallocating from T1 to T2 there was a significant hesitation, and this hesitation was reduced for the less salient OOV targets, $F(1, 498) = 78.5$, $p < .0005$. Here, then, we also see evidence for reactive inhibition and increased inhibition for salient T2s. In this way, selective attention and action in the LVA is coordinated using mechanisms very similar to those in the component VAs.

There are several other interesting aspects to these results. First, while hesitation for LVA with same-color targets was very similar to those of the component VAs, hesitation was in fact slightly reduced for LVA, $F(1, 498) = 7.52$, $p = .006$. Evidently LVA was able to reallocate more efficiently than a component VA. Recall that LVA consists of the unaltered component VAs plus the third "conflict monitor" agent, consisting of weights between the hidden layers of the component VAs. The better performance for LVA must then be due to the use of units in the "other" component VA, mediated by the conflict monitor weights. That is, on a trial with two green targets, LVA could reallocate more effectively than the component green VA could on its own. This suggests that LVA was using units in the red component VA to assist performance on trials with two green targets.

Further analysis suggested that this benefit was due primarily to more efficient processing of OOV T2s. The benefit for OOV over in-view T2s was larger for LVA than the component VA, as evidenced by the two-way interaction of Agent type (LVA or VA) and Visibility (in-view or OOV), $F(1, 498) = 9.62$, $p = .002$. Hesitation for the in-view T2s was very similar (36.4 to 35.3 timeslices for VA and LVA, respectively), with a larger difference for OOV T2s (28.3 to 24.8 timeslices). Although clearly this effect must be mediated by the conflict-monitor weights, additional analysis would be needed to determine how the "other" component VA facilitates reallocation to an OOV T2.

Another interesting result evident in Figure 10.4 is that reallocation was more efficient in the mixed- than same-color case, $F(1, 98) = 320$, $p < .0005$. That is, it appears that cognitive conflict produced by the two targets was greater when both targets were loading on the same component VA. This suggests that T2 attracted less inhibition in the mixed- than the same-color case. This is to be expected since targets in the mixed case were not competing for the resources of the same component VA. Less competition means both less need for inhibition and faster subsequent reallocation.

Finally, we have described the third VA here as the "conflict monitor". We can be sure that this conflict monitor was responsible for the coordination of component VAs in the mixed-color case, since it is only these weights that allow the component VAs to communicate. Further analysis of these connections would therefore be very interesting – for example, with respect to the specific proposals of Botvinick et al. (2001). Perhaps these connections monitor the level of conflict between activation of motor units by the component VAs, or perhaps they use some other method for conflict evaluation and regulation.

3.2 Scaling up: more agents or bigger agents?

The minimally cognitive agent was intended as a research tool rather than as a research hypothesis. We have seen the value of this tool in our previous studies, which show how understanding a minimally cognitive agent can help understand human performance. However, we would now like to undertake some wild speculation. We have seen how even a very small number of very simple processing units can exhibit sophisticated behavior. What if we take seriously the hypothesis that small agents could act as the basis of a cognitive architecture?

The idea of intelligent behavior emerging from large collections of simple processors is an old one. Godfrey Thompson (1916) argued against the idea of Spearman's general factor (g), suggesting instead a collection of many specific mental abilities, or "bonds". The pros and cons of attempting complex cognition through a Minsky-style "Society of Mind" (Minsky, 1986), or collection of many small agents, are reasonably well understood in computer science and robotics (e.g., subsumption architecture; Brooks, 1986).

3.2.1 Monolithic systems

However, the question of more, or of more powerful, agents is not often asked in neuroimaging and neuropsychology. Currently in psychology and neuroscience there is overwhelming agreement about the structure of the human brain's cognitive architecture, at least if we move to a high-enough level of abstraction: the brain consists of a reasonably limited number of sophisticated, somewhat specialized processing systems. We will come back later to what is meant by a "reasonably limited number". In any case, the underlying assumption is that this limited number of cognitive modules are organized so that even simple tasks activate very large numbers of neurons. For example, it is typical to see reference to a small number of broad and powerful systems when interpreting brain imaging results, such as attention, memory, executive control, and so on.

We suggest that the general assumptions of the standard view (we will call it the "monolithic view") play out something as follows. What do we as psychologists and cognitive neuroscientists study? We study cognitive functions, specified at a fairly abstract level – for example, attention, or executive control. By way of comparison, most experimental psychologists would not say (admit?) that they study a specific task. Any single task, however well studied, is meant to give insights into more general cognitive mechanisms. The operating assumption is that the results from different tasks will converge to give a picture of these high-level mechanisms, abstracted away from specifics of stimuli and specific presentation parameters. For example, visual search across space and across time might be assumed (or perhaps hypothesized is more accurate) to tap into a common set of attentional processes.

Similarly, it is assumed or hypothesized that there are component brain systems recruited for these different functions. A complex task like visual search might involve multiple such brain systems, including, for example, executive and visual memory systems. Each of these systems is meant to participate in many tasks.

So now let us return to the issue of what is a "reasonably limited number" of component cognitive systems? It would be pointless to give a specific number, or even a specific order of magnitude. The important assumption of the monolithic view is surely that the number of component cognitive systems is much smaller than the number of tasks a person is capable of performing.

There are certainly reasons to support the majority view – that the monolithic view is appropriate. For one thing, brain imaging studies show considerable overlap in the brain areas activated during related tasks. For example, dorsolateral prefrontal cortex is active in a wide variety of demanding cognitive tasks (Duncan & Owen, 2000). One idea is that this area of cortex is analogous to a general-purpose computing facility (Duncan, 2001). Instead of a single, well-defined and specific tuning, many neurons in the frontal lobe have firing properties that vary based on task demands (e.g., Rao, Rainer, & Miller, 1997). Further support of the monolithic view also comes from brain lesion studies. For example, damage to the temperoparietal junction of the right hemisphere frequently results in unilateral spatial neglect, in which the left side of space is ignored. The neglect syndrome might itself be consistent with the possibility that a "spatial module" has been damaged, a module that provides input into many diverse kinds of spatial tasks (Halligan, Fink, Marshall, & Vallar, 2003).

3.2.2 "Society of Mind"

However, there are also reasons to suspect that the monolithic view is not correct and that, instead, cognition might result from numerous, more specific processors. One thing that makes brain imaging interesting is that differences as well as similarities can be found even between even highly related tasks. Likewise, the effects of brain lesion are frequently very hard to predict, and there can be intriguing and surprising patterns of preserved and lost function. For example, in neglect there are frequent double dissociations between tasks like cancellation and line bisection (Ferber & Karnath, 2001).

What about the evidence of general-purpose frontal structures, as evidenced by the fact that a single neuron might encode different kinds of information depending on the current task, as in the Rao et al. (1997) study? This kind of adaptability in coding is certainly required for a programmable, monolithic structure. However, this adaptability is also perfectly consistent with a Society of Mind, and Minsky (1986) includes the notion of agents carrying references to other agents, in a task-dependent manner (e.g., "frames").

Most psychologists will also be familiar with the fractionation that seems to inevitably occur with increased study of a specific function. Take visual-object recognition for example. Current thinking seems opposed to the idea of a general-purpose visual recognition system. Instead, there appear to be multiple, more specific subsystems for dealing with specific kinds of recognition problems. For example, there appear to be specialized face-identification systems, areas for places, body parts, viewpoint-invariant and viewpoint-specific recognition, holistic versus componential recognition, and so on. Likewise, the concept of memory has fractionated into a variety of different forms of memory: short-term, long-term, episodic, autobiographical, visual working memory, acoustic loop, and so on. A general trend seems to be that increased study of a cognitive function leads to evidence of more and more diverse forms of that function, rather than to grand unified theories.

In any case, our intention here is not to present a knock-down argument against a monolithic view, but to raise awareness about an alternative possibility and speculate about how results with VA might inform a view more like the Society of Mind. Again, it is worth considering the orders of complexity separating VA and the human brain – VA: 17 simple "interate and fire" units; human brain: on the order of 100,000,000,000 neurons, each a sophisticated information processing device. In principle, it appears that the human brain would have capacity for something on the order of billions of networks of VA's complexity. In such a collection of very large numbers of agents, the organizing principle would be small assemblies of units, each assembly capable of performing a very specific function. Everyday tasks, like talking on the phone, would require large numbers of agents, for parsing the words being spoken, anticipating the flows of conversation, activating appropriate schemas, modeling the speaker's mental state, planning and executing the next speech act, keeping the phone in the correct position, maintaining body posture, monitoring concurrent stimulation, and so on. Each of these small agents would possess its own selective processing system as needed, its own task-relevant memory system, task-specific object-recognition systems, and so on. Such a brain architecture would imply a hyper-distributed attention system, in which targets of action emerge from the combined processing of many simple agents, perhaps on the order of thousands to millions.

There are some well-known difficulties with the Society of Mind view. Perhaps most worrying is related to action selection. How is it determined which of perhaps billions of agents will control behavior? What is to prevent an incoherent paralysis resulting from the activation of many conflicting agents? There are no easy answers, but certainly in other domains it is evident that collections of many "agents" can produce purposeful behavior. For example, it is not clear how cells maintain their specialization and organization to produce coherent structure and function at the level of the organism, but somehow they do. We do not want to trivialize these issues. Instead, we want to suggest that the cognitive power of small assemblies, as demonstrated by VA and many other examples, may be well appreciated in the computer science community (e.g., Singh, 2003), but perhaps not in other domains of cognitive science. A cognitive architecture described by massive numbers of these assemblies may be a reasonable hypothesis for psychologists and neuroscientists.

4 Conclusions

We have examined psychological principles underlying selective attention and control in artificial, dynamic, "embodied" agents. Can these analyses aid our understanding of human selective attention and control? We think so. Given that VA is doing a genuine attention task, only two outcomes are possible: either VA is using attention mechanisms already explored in the human literature, or it is using novel, undescribed mechanisms. Both of these outcomes are interesting, and we have reviewed examples of each: "reactive inhibition" for the control of action, similar to previous descriptions in people (Ward & Ward, 2008), and a novel system of distributed control of cognitive conflict (Ward & Ward, 2006).

Here we have considered a case in which multiple agents coordinated their processing to achieve goals that neither could achieve separately. In this case, we combined two component agents, one that could only "see" red and an (otherwise identical) agent that could only see green. We linked the motor systems and neural networks of these agents with an explicit module for the control of cognitive conflict, in the form of a third agent. Again, we observed a system of reactive inhibition develop. The "conflict-monitor" agent switched processing flexibly between the component red and green agents, using reactive inhibition to reallocate processing appropriately. We have previously seen reactive inhibition in the most primitive agents and environment (VA), in the most complex (people), and again now in the case of coordinated agents. This result provides further support for the argument that reactive inhibition may be a fundamental mechanism for response control, in some ways perhaps analogous to the way lateral inhibition is a fundamental mechanism for contrast enhancement.

Finally, we have discussed an issue that will be old news to computer scientists but may not be so well understood in psychology and neuroscience:

very small networks can be surprisingly powerful. It may be worth considering seriously, as a neuroscientific hypothesis, the idea that human cognition emerges from the interaction of massive numbers of connected small networks. Computing and information sciences have defined well-known advantages and disadvantages to such Society of Mind approaches, which may usefully guide and constrain neuroscientific study.

5 References

Andersen, B. B., Korbo, L., & Pakkenberg, B. (1992). A quantitative study of the human cerebellum with unbiased stereological techniques. *Journal of Comparative Neurology*, *326*(4), 549–560.

Beer, R. D. (1996). Toward the evolution of dynamical neural networks for minimally cognitive behavior. In P. Maes, M. Mataric, J. Meyer, J. Polack, & S. Wilson (Eds.), *From animals to animats 4: Proceedings of the fourth international conference on simulation of adaptive behavior*. Cambridge, MA: MIT Press.

Botvinick, M. M., Braver, T. S., Barch, D. M., Carter, C. S., & Cohen, J. D. (2001). Conflict monitoring and cognitive control. *Psychological Review*, *108*(3), 624–652.

Botvinick, M. M., Cohen, J. D., & Carter, C. S. (2004). Conflict monitoring and anterior cingulate cortex: An update. *Trends in Cognitive Science*, *8*(12), 539–546.

Brooks, R. (1986). A robust layered control system for a mobile robot. *IEEE Journal of Robotics and Automation*, *2*(1), 14–23.

Clark, A. (1999). An embodied cognitive science? *Trends in Cognitive Science*, *3*(9), 345–351.

Cohen, J. D., Dunbar, K., & McClelland, J. L. (1990). On the control of automatic processes: A parallel distributed processing account of the Stroop effect. *Psychological Review*, *97*(3), 332–361.

Critchley, H. D., Mathias, C. J., Josephs, O., O'Doherty, J., Zanini, S., Dewar, B. K., et al. (2003). Human cingulate cortex and autonomic control: Converging neuroimaging and clinical evidence. *Brain*, *126*(10), 2139–2152.

Driver, J. (2001). A selective review of selective attention research from the past century. *British Journal of Psychology*, *92*(1), 53–78.

Duncan, J. (2001). An adaptive coding model of neural function in prefrontal cortex. *Nature Reviews. Neuroscience*, *2*(11), 820–829.

Duncan, J., & Owen, A. M. (2000). Common regions of the human frontal lobe recruited by diverse cognitive demands. *Trends in Neurosciences*, *23*(10), 475–483.

Fellows, L. K., & Farah, M. J. (2005). Is anterior cingulate cortex necessary for cognitive control? *Brain*, *128*(4), 788–796.

Ferber, S., & Karnath, H. O. (2001). How to assess spatial neglect – line bisection or cancellation tasks? *Journal of Clinical and Experimental Neuropsychology*, *23*(5), 599–607.

Gibson, J. J. (1979). *The ecological approach to visual perception*. Boston, MA: Houghton Mifflin.

Graham, B. P., & van Ooyen, A. (2004). Transport limited effects in a model of dendritic branching. *Journal of Theoretical Biology*, *230*(3), 421–432.

Grison, S., & Strayer, D. L. (2001). Negative priming and perceptual fluency: More than what meets the eye. *Perception & Psychophysics*, *63*(6), 1063–1071.

Halligan, P. W., Fink, G. R., Marshall, J. C., & Vallar, G. (2003). Spatial cognition: Evidence from visual neglect. *Trends in Cognitive Science, 7*(3), 125–133.

Houghton, G., & Tipper, S. P. (1994). A model of inhibitory mechanisms in selective attention. In D. Dagenbach & T. Carr (Eds.), *Inhibitory mechanisms in attention, memory, and language.* San Diego, CA: Academic Press.

Koch, C. (1999). *Biophysics of computation: Information processing in single neurons.* New York: Oxford University Press.

Mari-Beffa, P., Estevez, A. F., & Danziger, S. (2000). Stroop interference and negative priming: Problems with inferences from null results. *Psychonomic Bulletin & Review, 7*(3), 499–503.

Minsky, M. (1986). *The society of mind.* New York: Simon & Schuster.

Mulert, C., Menzinger, E., Leicht, G., Pogarell, O., & Hegerl, U. (2005). Evidence for a close relationship between conscious effort and anterior cingulate cortex activity. *International Journal of Psychophysiology, 56*(1), 65–80.

Pakkenberg, B., & Gundersen, H. J. (1997). Neocortical neuron number in humans: Effect of sex and age. *Journal of Comparative Neurology, 384*(2), 312–320.

Rao, S. C., Rainer, G., & Miller, E. K. (1997). Integration of what and where in the primate prefrontal cortex. *Science, 276*(5313), 821–824.

Schneider, W. X. (1995). VAM: A neuro-cognitive model for visual attention control of segmentation, object recognition, and space-based motor action. *Visual Cognition, 2,* 331–375.

Singh, P. (2003). Examining the society of mind. *Computing and Informatics, 22,* 521–543.

Slocum, A. C., Downey, D. C., & Beer, R. D. (2000). Further experiments in the evolution of minimally cognitive behavior: From perceiving affordances to selective attention. In J.-A. Meyer, A. Berthoz, D. Floreano, H. L. Roitblat, & S. W. Wilson (Eds.), *From animals to animats 6: Proceedings of the sixth international conference on simulation of adaptive behavior.* Cambridge, MA: MIT Press.

Thomson, G. H. (1916). A hierarchy without a general factor. *British Journal of Psychology, 8,* 271–281.

Ward, R. (1999). Interactions between perception and action systems: A model for selective action. In G. W. Humphreys, J. Duncan, & A. Treisman (Eds.), *Attention, space and action: Studies in cognitive neuroscience.* Oxford, UK: Oxford University Press.

Ward, R., & Ward, R. (2006). Cognitive conflict without explicit conflict-monitoring in a dynamical agent. *Neural Networks, 19*(9), 1430–1436.

Ward, R., & Ward, R. (2008). Selective attention and control of action: Comparative psychology of an artificial, evolved agent and people. *Journal of Experimental Psychology: Human Perception and Performance, 34,* 1165–1182.

11 Full solution for the storage of correlated memories in an autoassociative memory

Emilio Kropff

We complement our previous work (Kropff & Treves, 2007) with the full (nondiluted) solution describing the stable states of an attractor network that stores correlated patterns of activity. The new solution provides a good fit of simulations of a model network storing the feature norms of McRae, Cree, Seidenberg, and McNorgan (2005), experimentally obtained combinations of features representing concepts in semantic memory. We discuss three ways to improve the storage capacity of the network: adding uninformative neurons, removing informative neurons, and introducing popularity-modulated Hebbian learning. We show that if the strength of synapses is modulated by an exponential decay of the *popularity* of the presynaptic neuron, any distribution of patterns can be stored and retrieved with approximately an optimal storage capacity: $C_{min} \propto I_f p$ – that is, the minimum number of connections per neuron needed to sustain the retrieval of a pattern is proportional to the information content of the pattern multiplied by the number of patterns stored in the network.

1 Introduction

The Hebbian learning principle (Hebb, 1949) states that given two neurons – one pre-synaptic, the other post-synaptic – their coactivity results in the enhancement of the synaptic strength that transmits information from the first to the latter. Initially formulated in a speculative way, it has since its conception found several physiological counterparts, nowadays widely studied, such as the phenomena of long-term potentiation or spike-timing dependent plasticity. In general terms, the Hebbian rule is the only learning principle known to be massively applied throughout the cortex.

Memory is one of the most direct applications of the Hebbian principle, and one of the first to be proposed by neural network modelers. Autoassociative memory networks can store patterns of neural activity by modifying the synaptic weights that interconnect neurons following the Hebbian rule (Amit, 1989; Hopfield, 1982). Once a pattern of activity is stored, it becomes an attractor of the dynamics of the system and can be retrieved by presenting the network with partial or corrupted versions of it. Direct evidence showing attractor behavior in the hippocampus of *in vivo* animals has been reported (Wills, Lever, Cacucci, Burgess, & O'Keefe, 2005). These kinds of memory systems have been proposed to be present at all levels along the cortex of higher order brains, where Hebbian plasticity plays a major role.

Most models of autoassociative memory studied in the literature store patterns that are obtained from some random distribution. Some exceptions appeared during the 1980s when interest grew around the storage of patterns derived from hierarchical trees (Gutfreund, 1988; Parga & Virasoro, 1986), presumably underlying the structure of semantic memory. Of particular interest, Virasoro (1988) relates the behavior of networks of general architecture with *prosopagnosia*, an impairment that impedes a patient from individuating certain stimuli without affecting its capacity to categorize them. Interestingly, the results from this model indicate that prosopagnosia is not present in Hebbian-plasticity-derived networks. Some other developments have used perceptron-like or other arbitrary local rules for storing generally correlated patterns (Diederich & Opper, 1987; Gardner, Stroud, & Wallace, 1989) or patterns with spatial correlation (Monasson, 1992). More recently, Tsodyks and collaborators (Blumenfeld, Preminger, Sagi, & Tsodyks, 2006) have studied a Hopfield memory in which a sequence of morphs between two uncorrelated patterns are stored. In this work, the use of a saliency function favoring unexpected over expected patterns during learning results in the formation of a continuous one-dimensional attractor that spans the space between the two original memories. The fusion of basins of attraction can be an interesting phenomenon that we are not going to treat in this work, since we assume that the elements stored in a memory such as the semantic one are separable by construction.

In this work we are interested in the storage of generally correlated patterns of activity in an autoassociative neural network equipped with Hebbian learning mechanisms, focusing on its possible applications in the field of semantic memory. Semantic representations, and in particular of those concepts related to objects, are thought to be strongly correlated. It has been proposed that some of the consequences of this correlation are priming effects or category-specific deficits in semantic memory (Tyler, Moss, Durrant-Peatfield, & Levy, 2000). However, it was only recently that the first proposal appeared of a mechanistic way in which the cortex could store effectively correlated patterns of activity as autoassociative memories (Kropff & Treves, 2007). The cited work shows that a simple extension of the classical learning rule, which additionally uses the mean activity of neurons (popularity) as learning thresholds, solves the task, with performance levels similar to those of classical models applied to uncorrelated data. Furthermore, it presents analytical predictions of this performance in the approximation of very low connectivity (the number of synaptic connections per neuron scaling at most like the logarithm of the number of neurons in the network), when the noise influence of the network on any neuron at any moment can be considered as resulting from a random Gaussian distribution. Interestingly, Kropff and Treves conclude that because the stored memories are correlated, random damage of the network results in a selective impairment for retrieval, a phenomenon that is observed in patients with various types and degrees of cortical damage involving areas where semantic information is represented. It is our intention in this chapter to deepen this line of research by presenting analytical predictions that do not assume low levels of connectivity, in order to apply them to feature norms; the actual data comes from cognitive experimental studies.

Feature norms are a way to get an insight on how semantic information is organized in the human brain (Garrard, Ralph, Hodges, & Patterson, 2001; McRae et al., 2005; Vinson & Vigliocco, 2002). The information is collected by asking different types of questions about particular concepts to a large population of subjects. Representations of the concepts are obtained in terms of the features that appear more often in the subjects' descriptions. In this work we analyze the feature norms of McRae and colleagues (2005), which were downloaded from the Psychonomic Society Archive of Norms, Stimuli, and Data website (www.psychonomic.org/archive) with consent of the authors.

The rest of this work is organized as follows. In section 2 we define a simple binary associative network, showing how it can be modified from its classical form so as to allow the effective storage of correlated representations. In section 3 we solve the equilibrium equations for the stable attractor states of the system using a self-consistent signal-to-noise approach, generalizing the findings of Kropff and Treves (2007) for the case of arbitrary connectivity levels in the network. Finally, in section 4 we study the storage and retrieval in our network of the feature norms of McRae et al. (2005), discussing whether or not the information they provide is enough to model semantic memory phenomena and what further modifications to the model can be proposed in order to increase performance.

2 The model

We assume a network with N neurons and $C \leq N$ synaptic connections per neuron. If the network stores p patterns, the parameter $a = p / C$ is a measure of the memory load normalized by the size of the network. In classical models, the equilibrium properties of large-enough networks depends on p, C, and N only through a, which allows the definition of the thermodynamic limit ($p \to \infty$, $C \to \infty$, $N \to \infty$, a constant).

The activity of neuron i is described by the variable σ_i, with $i = 1 \ldots N$. Each of the p patterns is a particular state of activation of the network. The activity of neuron i in pattern μ is described by ξ_i^μ, with $\mu = 1 \ldots p$. The perfect retrieval of pattern μ is thus characterized by $\sigma_i = \xi_i^\mu$ for all i. We will assume binary patterns, where $\xi_i^\mu = 0$ if the neuron is silent and $\xi_i^\mu = 1$ if the neuron fires. Consistently, the activity states of neurons will be limited by

$0 \leq \sigma_i \leq 1$. We will further assume a fraction a of the neurons being activated in each pattern. This quantity receives the name of *sparseness*.

Each neuron receives C synaptic inputs. To describe the architecture of connections, we use a random matrix with elements $C_{ij} = 1$ if a synaptic connection between post-synaptic neuron i and pre-synaptic neuron j exists and $C_{ij} = 0$ otherwise, with $C_{ij} = 0$ for all i. In addition to this, synapses have associated weights J_{ij}.

The influence of the network activity on a given neuron i is represented by the field

$$h_i = \sum_{j=1}^{N} C_{ij} J_{ij} \sigma_j, \tag{1}$$

which enters a sigmoidal activation function in order to update the activity of the neuron

$$\sigma_i = \{1 + \exp \beta(U - h_i)\}^{-1}, \tag{2}$$

where β is inverse to a temperature parameter and U is a threshold favoring silence among neurons (Buhmann, Divko, & Schulten, 1989; Tsodyks & Feigel'Man, 1988).

The learning rule that defines the weights J_{ij} must reflect the Hebbian principle: every pattern in which both neurons i and j are active will contribute positively to J_{ij}. In addition to this, the rule must include, in order to be optimal, some prior information about pattern statistics. In a one-shot learning paradigm, the optimal rule uses the sparseness a as a learning threshold,

$$J_{ij} = \frac{1}{Ca} \sum_{\mu=1}^{p} \left(\xi_i^\mu - a \right) \left(\xi_j^\mu - a \right) \tag{3}$$

However, as we have shown in previous work (Kropff & Treves, 2007), in order to store correlated patterns this rule must be modified using a_j, or the *popularity* of the pre-synaptic neuron, as a learning threshold,

$$J_{ij} = \frac{1}{Ca} \sum_{\mu=1}^{p} \xi_i^\mu \left(\xi_j^\mu - a_j \right), \tag{4}$$

with

$$a_i \equiv \frac{1}{p} \sum_{\mu=1}^{p} \xi_i^\mu. \tag{5}$$

This requirement comes from splitting the field into a signal part and a noise part,

$$h_i = \frac{1}{Ca} \xi_i^1 \sum_{j=1}^{N} C_{ij} \left(\xi_j^1 - a_j \right) \sigma_j + \frac{1}{Ca} \sum_{\mu=2}^{p} \xi_i^\mu \sum_{j=1}^{N} C_{ij} \left(\xi_j^\mu - a_j \right) \sigma_j,$$

and, under the hypothesis of Gaussian noise, setting the average to zero and minimizing the variance. This last is

$$var = \frac{1}{C^2 a^2} \sum_{\mu=1}^{p} \xi_i^\mu \sum_{j=1}^{N} C_{ij} \sigma_j^2 \left(\xi_j^\mu - a_j \right)^2$$

$$+ \frac{1}{C^2 a^2} \sum_{\mu \neq \nu = 1}^{p} \xi_i^\mu \xi_i^\nu \sum_{j=1}^{N} C_{ij} \sigma_j^2 \left(\xi_j^\mu - a_j \right) \left(\xi_j^\nu - a_j \right)$$

$$+ \frac{1}{C^2 a^2} \sum_{\mu=1}^{p} \xi_i^\mu \sum_{j \neq k = 1}^{N} C_{ij} C_{ik} \sigma_j \sigma_k \left(\xi_j^\mu - a_j \right) \left(\xi_k^\mu - a_k \right)$$

$$+ \frac{1}{C^2 a^2} \sum_{\mu \neq \nu = 1}^{p} \xi_i^\mu \xi_i^\nu \sum_{j \neq k = 1}^{N} C_{ij} C_{ik} \sigma_j \sigma_k \left(\xi_j^\mu - a_j \right) \left(\xi_k^\nu - a_k \right) \tag{6}$$

If statistical independence is granted between any two neurons, only the first term in Equation 6 survives when averaging over $\{\xi\}$.

In Figure 11.1 we show that the rule in Equation 3 can effectively store uncorrelated patterns taken from the distribution

$$P(\xi_i^\mu) = a\delta \left(\xi_i^\mu - 1 \right) + (1-a) \delta \left(\xi_i^\mu \right) \tag{7}$$

but cannot handle less trivial distributions of patterns, suffering a storage collapse. The storage capacity can be brought back to normal by using the learning rule in Equation 4, which is also suitable for storing uncorrelated patterns.

Having defined the optimal model for the storage of correlated memories, we analyze in the following sections the storage properties and its consequences through mean-field equations.

3 Self-consistent analysis for the stability of retrieval

We now proceed to derive the equations for the stability of retrieval, similarly to what we have done earlier (Kropff & Treves, 2007) but in a network with an arbitrary level of random connectivity, where the approximation $C \ll N$ is no longer valid (Roudi & Treves, 2004; Shiino & Fukai,

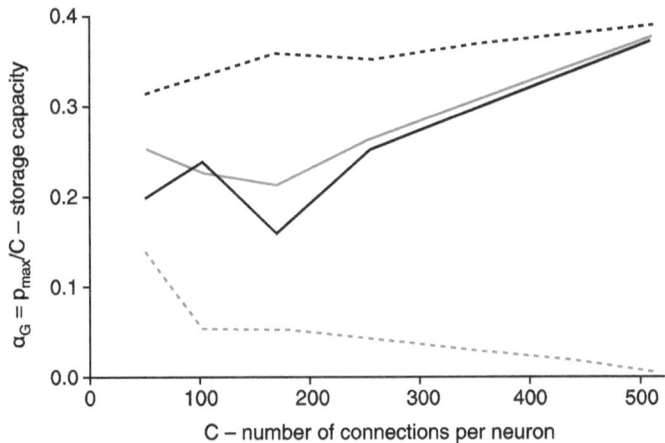

Figure 11.1 The four combinations of two learning rules and two types of dataset. *Light:* one-shot "standard" learning rule of Equation 3. *Dark:* modified rule of Equation 4. *Solid:* trivial distribution of randomly correlated patterns obtained from Equation 7. *Dashed:* nontrivially correlated patterns obtained using a hierarchical algorithm. In three cases, the storage capacity (the maximum number of retrievable patterns normalized by C) with C (the number of connections per neuron) is finite and converges to a common value as C increases. Only in the case of one-shot learning of correlated patterns is there a storage collapse.

1992, 1993). Furthermore, we introduce patterns with variable mean activation, given by

$$d_\mu \equiv \frac{1}{N} \sum_{j=1}^{N} \xi_j^\mu \tag{8}$$

for a generic pattern μ. As a result of this, the optimal weights are given by

$$J_{ij} = g_j \sum_{\mu=1}^{p} \frac{C_{ij}}{Cd_\mu} \xi_i^\mu (\xi_j^\mu - a_j), \tag{9}$$

which ensures that patterns with different overall activity will have not only a similar noise but also a similar signal. In addition, we have introduced a factor $g_j = g(a_j)$ in the weights that may depend on the popularity of the pre-synaptic neuron. We will consider $g_j = 1$ for all but the last section of this work.

If the generic pattern 1 is being retrieved, the field in Equation 1 for neuron i can be written as a signal and a noise contribution:

$$h_i = \xi_i^1 m_i^1 + \sum_{\mu \neq 1} \xi_i^\mu m_i^\mu \tag{10}$$

with

$$m_i^\mu = \frac{1}{Cd_\mu} \sum_{j=1}^{N} g_j C_{ij} (\xi_j^\mu - a_j)\sigma_j \tag{11}$$

We hypothesize that in a stable situation the second term in Equation 10, the noise, can be decomposed into two contributions

$$\sum_{\mu \neq 1} \xi_i^\mu m_i^\mu = \gamma_i \sigma_i + \rho_i z_i \tag{12}$$

The second term in Equation 12 represents a Gaussian noise with standard deviation, ρ_i, and z_i is a random variable taken from a normal distribution of unitary standard deviation. The first term is proportional to the activity of the neuron i and results from closed synaptic loops that propagate this activity through the network back to the original neuron, as shown in Roudi and Treves (2004). As is typical in the self-consistent method, we will proceed to estimate m_i^μ from the ansatz in Equation 12, inserting it into Equation 10, and validating the result with, again, Equation 12, checking the consistency of the ansatz.

Since Equation 12 is a sum of $p \rightarrow \infty$ microscopic terms, we can take a single term ν out and assume that the sum changes only to a negligible extent. In this way, the field becomes

$$h_i \simeq \xi_i^1 m_i^1 + \xi_i^\nu m_i^\nu + \gamma_i \sigma_i + \rho_i z_i \tag{13}$$

If the network has reached stability, which we assume, updating neuron i does not affect its state. This can be expressed by inserting the field into Equation 2,

$$\sigma_i = \{1 + \exp(-\beta(h_i - U))\}^{-1} \equiv G[\xi_i^1 m_i^1 + \xi_i^\nu m_i^\nu + \rho_i z_i] \tag{14}$$

In the right-hand side (RHS) of Equation 14, the contribution of $\gamma_i \sigma_i$ to the field has been reabsorbed into the definition of $G[x]$. At first order in $\xi_j^\nu m_j^\nu$, Equation 14 corresponding to neuron j can be written as

$$\sigma_j \simeq G[\xi_j^1 m_j^1 + \rho_j z_j] + G'[\xi_j^1 m_j + \rho_j z_j] \xi_j^\nu m_j^\nu \tag{15}$$

To simplify the notation we will further use $G_j \equiv G[\xi_j^1 m_j^1 + \rho_j z_j]$ and $G_{j'} \equiv G'[\xi_j^1 m_j + \rho_j z_j]$. To this order of approximation, Equation 11 becomes

$$m_i^\mu = \frac{1}{Cd_\mu} \sum_{j=1}^N Ng_j C_{ij}(\xi_j^\mu - a_j)\{G_j + G_{j'} \xi_j^\mu m_j^\mu\}$$ (16)

Other terms of the same order in the Taylor expansion could have been introduced in Equation 15, corresponding to the derivatives of G with respect to $\xi_j^\mu m_j^\mu$ for $\mu \neq v$. It is possible to show, however, that such terms give a negligible contribution to the field, since they are averaged out similarly to what happens with the noncontributing terms of Equation 6.

If we define

$$L_i^\mu = \frac{1}{Cd_\mu} \sum_{j=1}^N g_j C_{ij}(\xi_j^\mu - a_j)G_j$$

$$K_{ij}^\mu = \frac{1}{Cd_\mu} g_j C_{ij}(\xi_j^\mu - a_j)\xi_j^\mu G_{j'},$$ (17)

Equation 16 can be simply expressed as

$$m_i^\mu = L_i^\mu + \sum_{j=1}^N K_{ij}^\mu m_j^\mu$$ (18)

This equation can be applied recurrently to itself, renaming indexes,

$$m_i^\mu = L_i^\mu + \sum_{j=1}^N K_{ij}^\mu L_j^\mu + \sum_{j=1}^N \sum_{k=1}^N K_{ij}^\mu K_{jk}^\mu m_k^\mu$$ (19)

If applied recurrently infinite times, this procedure results in

$$m_i^\mu = L_i^\mu + \sum_{j=1}^N K_{ij}^\mu L_j^\mu + \sum_{j=1}^N \sum_{k=1}^N K_{ij}^\mu K_{jk}^\mu L_k^\mu + \ldots$$ (20)

which, by exchanging mute variables, can be rewritten as

$$m_i^\mu = L_i^\mu + \sum_{j=1}^N L_j^\mu \left\{ K_{ij}^\mu + \sum_{k=1}^N K_{ik}^\mu K_{kj}^\mu + \sum_{k,l=1}^N K_{ik}^\mu K_{kl}^\mu K_{lkj}^\mu + \ldots \right\}$$ (21)

Equation 21 can be decomposed into the contribution of the activity of G_i on one side and that of the rest of the neurons on the other, which will correspond to the first and the second term on the RHS of Equation 12. To re-obtain this equation we multiply by ξ_i^μ and sum over μ, using the definition of L_i^μ from Equation 17,

$$\sum_{\mu \neq 1} m_i^\mu \xi_i^\mu = G_i g_i \sum_{\mu \neq 1} \frac{\xi_i^\mu (1 - a_i)}{C d_\mu} \left(C_{ii} + \sum_{j=1}^{N} C_{ji} \left\{ K_{ij}^\mu + \sum_{k=1}^{N} K_{ik}^\mu K_{kj}^\mu + \ldots \right\} \right)$$

$$+ \sum_{l \neq i} G_l g_l \sum_{\mu \neq 1} \frac{\xi_i^\mu (\xi_l^\mu - a_l)}{C d_\mu} \left(C_{il} + \sum_{j=1}^{N} C_{jl} \left\{ K_{ij}^\mu + \sum_{k=1}^{N} K_{ik}^\mu K_{kj}^\mu + \ldots \right\} \right) \quad (22)$$

Let us first treat the first term of Equation 22, corresponding to $\gamma_i \sigma_i$ in Equation 12. Taking into account that $C_{ij} = 0$ (no self-excitation), only the contribution containing the curly brackets survives. As shown in Roudi and Treves (2004), each term inside the curly brackets, containing the product of multiple Ks, is different only to a vanishing order from the product of independent averages, each one corresponding to the sum of K_{ab} over all presynaptic neurons b. In this way,

$$G_i g_i (1 - a_i) \sum_{\mu \neq 1} \frac{\xi_i^\mu}{C d_\mu} \sum_{j, l_1 \ldots l_n = 1}^{N} C_{ji} K_{il_1}^\mu \left[\prod_{o=1}^{n-2} K_{l_o l_{o+1}}^\mu \right] K_{l_n j}^\mu$$

$$\approx a G_i g_i a_i (1 - a_i) \frac{C}{N} \left\langle \frac{1}{d_\mu^{n+1}} \right\rangle_\mu (a\Omega)^n, \quad (23)$$

where we have introduced $a \equiv p/C$, or the memory load normalized by the number of connections per neuron. The $\langle \ldots \rangle_\mu$ brackets symbolize an average over the index μ, and Ω is a variable of order 1 defined by

$$\Omega \equiv \frac{1}{aN} \sum_{j=1}^{N} a_j (1 - a_j) G_j g_j \quad (24)$$

Adding up all the terms with different powers of Ω in Equation 23 results in

$$\gamma_i \sigma_i = a a_i (1 - a_i) g_i \frac{C}{N} \left\langle \frac{\Omega}{d_\mu (d_\mu/a - \Omega)} \right\rangle_\mu G_i \quad (25)$$

Since Ω does not depend on μ, if $d_\mu = a$ for all μ the average results simply in the classical $\Omega/(1 - \Omega)$ factor.

As postulated in the ansatz, the second term in Equation 22 is a sum of many independent contributions and can thus be thought of as a Gaussian noise. Its mean is zero by virtue of the factor $(\xi_i^\mu - a_l)$, uncorrelated with both ξ_i^μ (by hypothesis) and d_μ (negligible correlation). Its variance is given by

$$\langle\langle\rho_i^2\rangle\rangle = \left\langle\left\langle \left[\sum_{l \neq i} G_i^2\, g_i^2 \sum_{\mu \neq 1} \frac{\xi_i^\mu\,(\xi_l^\mu - a_l)^2}{C^2 d_\mu^2} \left(C_{il} + \sum_{j=1}^{N} C_{jl}\left\{ K_{ij}^\mu + \right.\right.\right.\right.\right.$$

$$\left.\left.\left.\left.\left. \sum_{k=1}^{N} K_{ik}^\mu K_{kj}^\mu + \ldots \right\}\right)\right]^2 \right\rangle\right\rangle, \tag{26}$$

which corresponds to the first and only surviving term of Equation 6, the other three terms vanishing for identical reasons. Distributing the square in the big parentheses and repeating the steps of Equation 23, this results in

$$\langle\langle\rho_i^2\rangle\rangle = a a_i \left\{ \left\langle \frac{1}{d_\mu^2} \right\rangle_\mu + 2\frac{C}{N}\left\langle \frac{\Omega}{d_\mu^2\,(d_\mu/a - \Omega)} \right\rangle_\mu + \frac{C}{N}\left\langle \frac{\Omega^2}{d_\mu^2\,(d_\mu/a - \Omega)^2} \right\rangle_\mu \right\}$$

$$\times \sum_{\mu \neq 1} \frac{1}{C} \sum_{l \neq i} (\xi_l^\mu - a_l)^2\, g_l^2\, \frac{C}{N}\, G_l^2. \tag{27}$$

If we define

$$q \equiv \{\ldots\} \frac{1}{N} \sum_{l=1}^{N} G_l^2\, a_l\,(1 - a_l)\, g_l^2, \tag{28}$$

including the whole content of the curly brackets from the previous equation, then the variance of the Gaussian noise is simply $a a_i q$, and the second term of Equation 12 becomes

$$\rho_i z_i = \sqrt{a a_i q}\, z_i, \tag{29}$$

with z_i, as before, an independent normally distributed random variable with unitary variance. The initial hypothesis of Equation 12 is, thus, self-consistent.

Taking into account these two contributions, the mean field experienced by a neuron i when retrieving pattern 1 is

$$h_i = \xi_i^1 m + a a_i(1 - a_i) G_i g_i \frac{C}{N}\left\langle \frac{\Omega}{d_\mu\,(d_\mu/a - \Omega)} \right\rangle_\mu + \sqrt{a q a_i}\, z_i, \tag{30}$$

where we have used $m_i^1 \simeq m$ and

$$m \equiv \frac{1}{N d_1} \sum_{j=1}^{N} (\xi_j^1 - a_j)\, g_j\, \sigma_j \tag{31}$$

is a variable measuring the weighted overlap between the state of the network and the pattern 1, which together with q (Equation 28) and Ω (Equation 24) form the group of macroscopic variables describing the possible stable states of the system. While m is a variable related to the signal that pushes the activity toward the attractor, q and Ω are noise variables. Diluted connectivity is enough to make the contribution of Ω negligible (in which case the diluted equations (Kropff & Treves, 2007) are re-obtained), while q gives a relevant contribution as long as the memory load is significantly different from zero, $a = p / C > 0$.

To simplify the analysis we adopt the zero temperature limit ($\beta \rightarrow \infty$), which turns the sigmoidal function of Equation 2 into a step function. To obtain the mean activation value of neuron i, the field h_i defined by Equation 30 must be inserted into Equation 2 and the equation in the variable σ_i solved. This equation is

$$\sigma_i = \Theta\left[\xi_i^1 m + aa_i\,(1 - a_i)\sigma_i\,g_i\,\frac{C}{N}\left\langle\frac{\Omega}{d_\mu\,(d_\mu/a - \Omega)}\right\rangle_\mu + \sqrt{aqa_i}z_i - U\right], \tag{32}$$

where $\Theta[x]$ is the Heaviside function yielding 1 if $x > 0$ and 0 otherwise. When z_i has a large enough modulus, its sign determines one of the possible solutions, $\sigma_i = 1$ or $\sigma_i = 0$. However, for a restricted range of values, $z_- \leq z_i \leq z_+$, both solutions are possible. Using the definition of γ_i in Equation 25 to simplify notation, we can write $z_+ = (U - \xi_i^1 m) / \sqrt{aqa_i}$ and $z_- = (U - \xi_i^1 m - \gamma_i) / \sqrt{aqa_i}$. A sort of Maxwell rule must be applied to choose between the two possible solutions (Shiino & Fukai, 1993), by virtue of which the point of transition between the $\sigma_i = 0$ and the $\sigma_i = 1$ solutions is the average between the two extremes

$$y_\xi \equiv \frac{z_+ + z_-}{2} = \frac{U - \xi_i^1 m - \gamma_i / 2}{\sqrt{aqa_i}}. \tag{33}$$

Inserting Equation 32 into Equation 31 yields

$$m = \frac{1}{Nd_1}\sum_{j=1}^{N}(\xi_j^1 - a_j)\,g_j\int_{-\infty}^{\infty}Dz\,\Theta\,[z - y_\xi], \tag{34}$$

where we have introduced the average over the independent normal distribution Dz for z_j. This expression can be integrated, resulting in

$$m = \frac{1}{Nd_1}\sum_{j=1}^{N}(\xi_j^1 - a_j)\,g_j\phi[y_\xi], \tag{35}$$

where we define

$$\phi\left(y_{\xi}\right) \equiv \tfrac{1}{2}\left\{1 + \mathrm{erf}\left[\frac{y_{\xi}}{\sqrt{2}}\right]\right\}. \tag{36}$$

Following the same procedure, Equation 28 can be rewritten as

$$q = \left\{\left\langle\frac{1}{d_{\mu}^2}\right\rangle_{\mu} + 2\frac{C}{N}\left\langle\frac{\Omega}{d_{\mu}^2\left(d_{\mu}/a - \Omega\right)}\right\rangle_{\mu} + \frac{C}{N}\left\langle\frac{\Omega^2}{d_{\mu}^2\left(d_{\mu}/a - \Omega\right)^2}\right\rangle_{\mu}\right\}$$

$$\times \frac{1}{N}\sum_{j=1}^{N} a_j\left(1 - a_j\right) g_j^2 \,\phi\left(y_{\xi}\right) \tag{37}$$

Before repeating these steps for the variable Ω, we note that

$$\int Dz G_{j'} = \frac{1}{\sqrt{aqa_j}}\int Dz\,\frac{\partial\sigma_j}{\partial z} = \frac{1}{\sqrt{aqa_j}}\int Dz z\sigma_j, \tag{38}$$

where we have applied integration by parts. Equation 23 results then in

$$\Omega = \frac{1}{Na}\sum_{j=1}^{N}\frac{a_j(1 - a_j)g_j}{\sqrt{2\pi aqa_j}}\exp\left\{-\frac{y_{\xi}^2}{2}\right\}. \tag{39}$$

Equations 35, 37, and 39 define the stable states of the network. Retrieval is successful if the stable value of m is close to 1. In Figure 11.2 we show the performance of a fully connected network storing the feature norms of McRae et al. (2005) in three situations: theoretical prediction for a diluted network as in Kropff and Treves (2007), theoretical prediction for a fully connected network calculated from Equations 35–39, and the actual simulations of the network. Figure 11.2 shows that the fully connected theory better approximates the simulations, performed with random subgroups of patterns of varying size p and full connectivity for each neuron, $C = N$, equal to the total number of features involved in the representation of the subgroup of concepts.

Finally, we can rewrite Equations 35–39 in a continuous way by introducing two types of popularity distribution across neurons:

$$F\left(x\right) = P\left(a_i = x\right) \tag{40}$$

as the global distribution, and

$$f\left(x\right) = P(a_i = x \mid \xi_i^1 = 1) \tag{41}$$

as the distribution related to the pattern that is being retrieved.

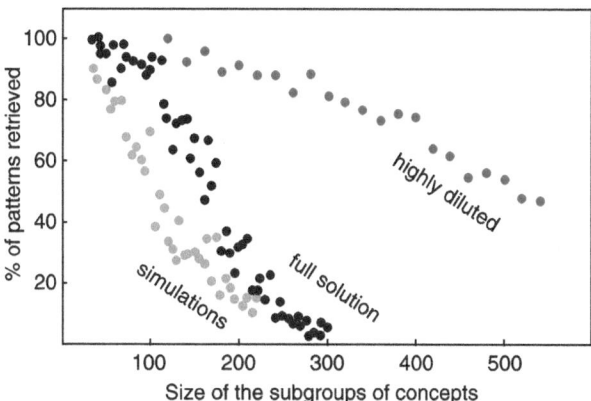

Figure 11.2 Simulations and numerical solutions of the equations of a network
storing random subgroups of patterns taken from the feature norms
of McRae et al. (2005). The performance of the network depends
strongly on the size of the subgroup. Though this is observed in the
highly diluted approximation, the decay in performance is not
enough to explain the data. It is the full solution with $g(x) = 1$ that
results in a good fit of the simulations. In each simulation, the number of neurons equals the number of features describing some of
the stored concepts, and there is full connectivity between neurons,
$C = N$.

The equations describing the stable values of the variables become

$$m = \int_0^1 f(x)\,g(x)(1-x)\phi(y_1) - \frac{1}{d_1}\int_0^1 [F(x) - d_1 f(x)]\,g(x)x\phi(y_0)$$

$$q = \left\{ \left\langle \frac{1}{d_\mu^2} \right\rangle_\mu + 2\frac{C}{N}\left\langle \frac{\Omega}{d_\mu^2\,(d_\mu/a - \Omega)} \right\rangle_\mu + \frac{C}{N}\left\langle \frac{\Omega^2}{d_\mu^2\,(d_\mu/a - \Omega)^2} \right\rangle_\mu \right\}$$

$$\times \left\{ d_1 \int_0^1 f(x)g^2(x)x(1-x)\phi(y_1) + \int_0^1 [F(x) - d_1\,f(x)]g^2(x)x(1-x)\phi(y_0) \right\}$$

$$\Omega = \frac{d_1}{a}\int_0^1 f(x)g(x)\,\frac{x(1-x)}{\sqrt{2\pi aqx}}\,\exp\left(-y_1^2/2\right)$$

$$+ \frac{1}{a}\int_0^1 [F(x) - d_1 f(x)]g(x)\,\frac{x(1-x)}{\sqrt{2\pi aqx}}\,\exp\left(-y_0^2/2\right), \qquad (42)$$

where, adapted from Equation 33,

$$y_\xi = \frac{1}{\sqrt{aqx}}\left(U - \xi m - ax(1-x)g(x)\,\frac{C}{2N}\left\langle \frac{\Omega}{d_\mu\,(d_\mu/a - \Omega)} \right\rangle_\mu \right) \qquad (43)$$

4 The storage of feature norms

In Kropff and Treves (2007) we have shown that the *robustness* of a memory in a highly diluted network is inversely related to the *information* it carries. More specifically, a stored memory needs a minimum number of connections per neuron C_{min} that is proportional to

$$I_f \equiv \int_0^1 f(x)x(1-x)dx \tag{44}$$

In this way, if connections are randomly damaged in a network, the most informative memories are selectively lost.

The distribution $F(x)$ affects the retrievability of all memories. As we have shown in Kropff and Treves (2007), it is typically a function with a maximum near $x = 0$. The relevant characteristic of $F(x)$ is its tail for large x. If $F(x)$ decays fast enough, the minimal connectivity scales like

$$C_{min} \propto pI_f \log\left[\frac{I_F}{aI_f}\right], \tag{45}$$

where I_F corresponds to the same pseudo-information function as in Equation 44, but using the distribution $F(x)$. If $F(x)$ decays exponentially ($F(x) \sim \exp(-x/a)$), the scaling of the minimal connectivity is the same, with only a different logarithmic correction,

$$C_{min} \propto pI_f \log^2\left[\frac{I_F}{aI_f}\right] \tag{46}$$

The big difference appears when $F(x)$ has a tail that decays as slowly as a power law ($F(x) \sim x^{-\gamma}$). The minimal connectivity is then much larger,

$$C_{min} \propto \frac{pI_f}{a} \log\left[\frac{a^{\gamma-2}}{I_f}\right], \tag{47}$$

since the sparseness, measuring the global activity of the network, is in cortical networks $a \ll 1$. Unfortunately, as can be seen in Figure 11.3, the distribution of popularity $F(x)$ for the feature norms of McRae et al. (2005) is of this last type. This is the reason why, as shown in Figure 11.2, the performance of the network is very poor in storing and retrieving patterns taken from this dataset. In a fully connected network as the one shown in the figure, a stored pattern can be retrieved as long as its minimal connectivity $C_{min} \leq N$, the number of connections per neuron. Along the x axis of Figure 11.2, representing the number of patterns from the norms stored in the network, the average of I_f is rather constant, p and N increase proportionally, and a decreases, eventually taking C_{min} over the full connectivity limit.

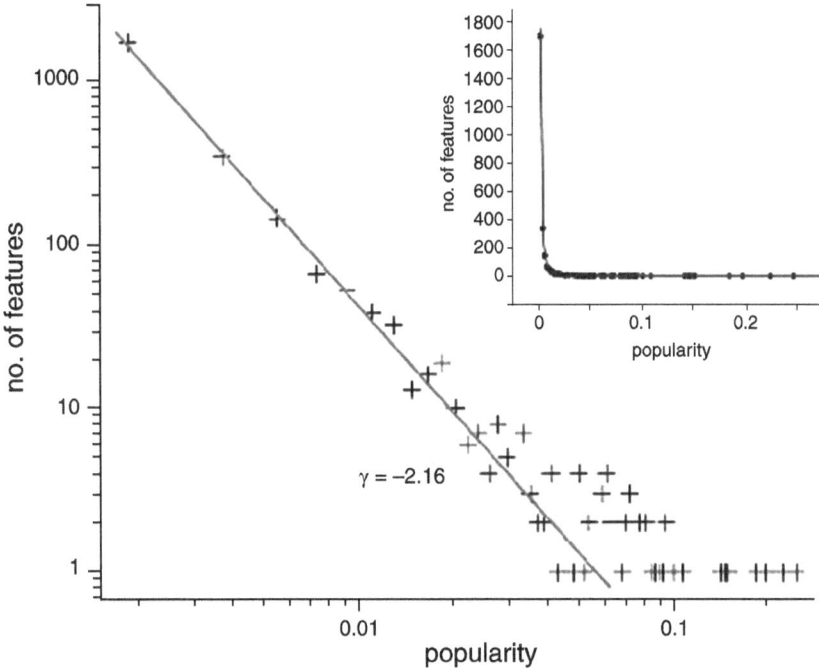

Figure 11.3 The popularity distribution $F(x)$ of the feature norms is a power law, with $\gamma \gg 2.16$. Note that both axes are logarithmic. In the inset, the same plot appears with linear axes, including the corresponding fit.

In the following subsections, we analyze different ways to increase this poor storage capacity and effectively store and retrieve the feature norms in an autoassociative memory.

4.1 Adding uninformative neurons

As discussed in Kropff and Treves (2007), a way to increase the storage capacity of the network in general terms is to push the distribution $F(x)$ toward the smaller values of x. One possibility is to add neurons with low information value (i.e., with low popularity) so as to make I_f smaller in average without affecting the sparseness a too much. In Figure 11.4a we show that the full set of patterns from the feature norms can be stored and retrieved if 5 new neurons per pattern are added, active in that particular pattern and in no other one.

4.2 Removing informative neurons

A similar effect on the distribution $F(x)$ can be obtained by eliminating selectively the most informative neurons. In Figure 11.4b we show that if the full set of patterns is stored, a retrieval performance of ~80% is achieved if the 40

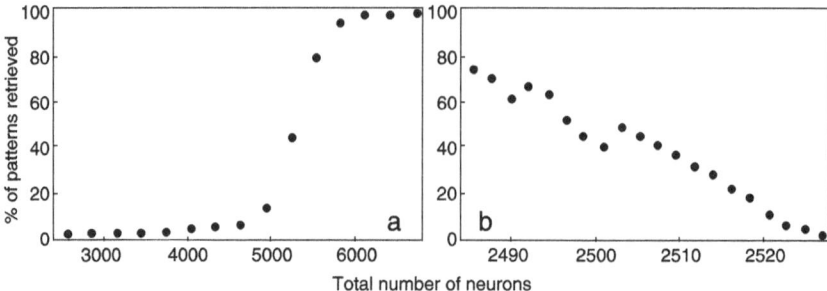

Figure 11.4 Adding or removing neurons affects the overall distribution F(x) and, thus, the performance of the network. The starting point for both situations is 2526 neurons corresponding to all the features in the norms. (a) Adding 5 neurons with minimal popularity per pattern is enough to get 100% performance. Note that the transition is sharp. (b) Removing the 40 most informative neurons also results in an improved performance, in this case of 80% of the stored patterns.

more informative features are eliminated. We estimate that 100% performance should be achieved if around 60 neurons were selectively eliminated.

It is not common in the neural literature to find a poor performance that is improved by damaging the network. This must be interpreted in the following way. The connectivity of the network is not enough to sustain the retrieval of the stored patterns, which are too informative to be stable states of the system. By throwing away information, the system can be brought back to work. However, a price is being paid: the representations are impoverished since they no longer contain the most informative features.

4.3 *Popularity-modulated weights*

A final way to push the distribution F(x) toward low values of x can be understood from Equation 42. Indeed, g(x) can be thought of as a modulator of the distributions F(x) and f(x). Inspired by Kropff and Treves (2007), if g(x) decays exponentially or faster, the storage capacity of a set of patterns with any decaying F(x) distribution should be brought back to a $C_{min} \propto pI_f$ dependence, without the $a^{-1} \gg 1$ factor.

In Figure 11.5 we analyze two possible g(x) functions that favor low over high values of x:

$$g_1(x) = \frac{a(1-a)}{x(1-x)} \tag{48}$$

$$g_2(x) = \sqrt{\frac{a(1-a)}{x(1-x)}} \tag{49}$$

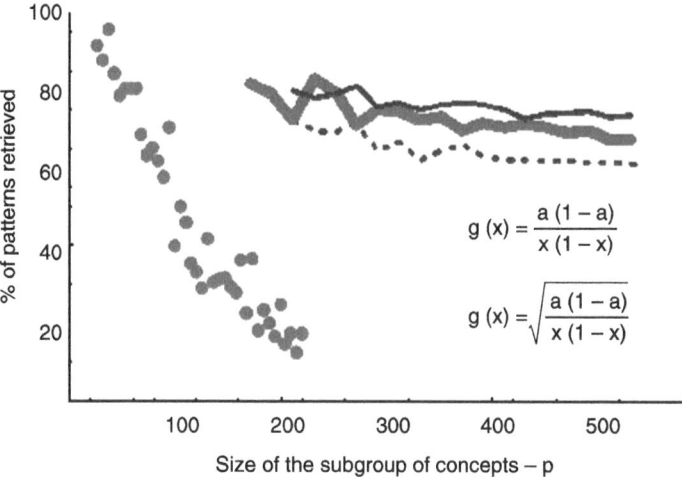

$$g(x) = \frac{a(1-a)}{x(1-x)}$$

$$g(x) = \sqrt{\frac{a(1-a)}{x(1-x)}}$$

Figure 11.5 Simulations (dashed line) and theoretical predictions (solid line) of a network storing subgroups of patterns of varying size taken from McRae et al. (2005) feature norms with a popularity-modulated Hebbian learning rule. The thin solid and dashed lines use a value of $g(x)$ inversely proportional to $x(1-x)$, normalized so as to maintain the average field of order 1. The thick solid line corresponds to a $g(x)$ inversely proportional to $x(1-x)$. Following our predictions, the exact form of $g(x)$ does not affect the general performance, which is substantially improved with respect to the simulations with $g(x) = 1$, copied from Figure 11.3 in gray dots.

The storage capacity of the network increases drastically in both cases. Furthermore, we estimate that ~60% of the lost memories in the figure suffer from a too high value of the threshold U, set, as in all simulations in this chapter, to 0.6. This value was chosen to maximize the performance in the previous simulations. However, with a much more controlled noise, the optimal threshold should be lower, generally around $m/2$ (so as to distinguish optimally fields of order m plus noise from those of order 0 plus noise). Setting the threshold at this level could possibly improve even further the performance of the network.

5 Discussion

We have presented the full nondiluted solution describing the stable states of a network that stores correlated patterns. A simple Hebbian learning rule is applicable as long as neurons can be treated as statistically independent. In order to analyze the storage of the patterns taken from the feature norms of McRae et al. (2005), we include in the learning rule the possibility that the global activity is different for each pattern. The full solution explains the poor performance of autoassociative networks storing the feature norms (Cree, McNorgan, & McRae, 2006; Cree, McRae, & McNorgan, 1999;

McRae, de Sa, & Seidemberg, 1997). We show that this data has a popularity distribution decaying as a power law – the worst of the cases analyzed in Kropff and Treves (2007).

The specific mechanisms that control learning thresholds in biological implementations of Hebbian protocols are, unfortunately, still obscure. Neurons seem to have a complex and versatile set of criteria to adapt their learning capabilities, put together under the common denomination of "metaplasticity" (Abraham & Bear, 1996). In our proposal, the threshold is set to the mean activity of the post-synaptic neuron, which is a plausible and not original idea (the same is proposed already in Bienenstock, Cooper, & Munro, 1982, with different modeling purposes), though too simplistic to aim to capture the full richness of cortical meta-learning rules.

The three proposed solutions aiming to improve the storage capacity of the network have a very different scope. In the first place, the adding of new uninformative neurons can be interpreted not only as a possible brain mechanism (e.g., the recombination of information in association areas would be a source of uninformative features) but also as a measure of the quality of currently available feature norms. Uninformative features are, on average, less likely to appear repeatedly in people's descriptions and thus are more likely to be left out of the norms. If this effect was corrected in the production of norms, it would perhaps be possible to obtain descriptions of concepts that were effectively storable as patterns of activity in a network with a structure of correlations reflecting to some extent that of semantic memories.

The elimination of very informative neurons in a damaged network could be achieved by damaging selectively the most active ones. This effect might sound counter-intuitive, since generally the damage of a network is associated with the loss of performance. The way to understand this idea is to think that the network is overloaded with information, so that by removing the most informative (or, as proposed, the more active) neurons, the level of information per neuron is brought down to normal levels. A well known practice where something similar happens is the treatment of epileptic patients, consisting in the removal of the brain tissue where the source of the seizure is located. Once "damaged" in this way, the network comes back to normal functioning.

Finally, the modulation of synaptic strength following pre-synaptic popularity can be considered to be an intermediate solution between the two extremes. Again, information is sacrificed by silencing partially the synapses related to very informative neurons, which results in an overall improvement in the network performance. Whether or not it is a strategy applied by the cortex to deal with correlated representations is a question for which we have still no experimental evidence.

6 References

Abraham, W. C., & Bear, M. F. (1996). Metaplasticity: The plasticity of synaptic plasticity. *Trends in Neurosciences, 19*(4), 126–130.

Amit, D. J. (1989). *Modelling brain function: The world of attractor neural networks.* Cambridge, UK: Cambridge University Press.

Bienenstock, E. L., Cooper, L. N., & Munro, P. W. (1982). Theory for the development of neuron selectivity: Orientation specificity and binocular interaction in visual cortex. *Journal of Neuroscience, 2*(1), 32–48.

Blumenfeld, B., Preminger, S., Sagi, D., & Tsodyks, M. (2006). Dynamics of memory representations in networks with novelty-facilitated synaptic plasticity. *Neuron, 52*(2), 383–394.

Buhmann, J., Divko, R., & Schulten, K. (1989). Associative memory with high information content. *Physical Review, A, 39*, 2689–2692.

Cree, G. S., McNorgan, C., & McRae, K. (2006). Distinctive features hold a privileged status in the computation of word meaning: Implications for theories of semantic memory. *Journal of Experimental Psychology, 32*, 643–658.

Cree, G. S., McRae, K., & McNorgan, C. (1999). An attractor model of lexical conceptual processing: Simulating semantic priming. *Cognitive Science, 23*(3), 371–414.

Diederich, S., & Opper, M. (1987). Learning of correlated patterns in spin–glass networks by local learning rules. *Physical Review Letters, 58*(9), 949–952.

Gardner, E. J., Stroud, N., & Wallace, D. J. (1989). Training with noise and the storage of correlated patterns in a neural network model. *Journal of Physics, A: Mathematical and General, 22*, 2019–2030.

Garrard, P., Ralph, M. A. L., Hodges, J. R., & Patterson, K. (2001). Prototypicality, distinctiveness, and intercorrelation: Analyses of the semantic attributes of living and nonliving concepts. *Cognitive Neuropsychology, 18*, 125–174.

Gutfreund, H. (1988). Neural networks with hierarchically correlated patterns. *Physical Review, A, 37*(2), 570–577.

Hebb, D. O. (1949). *The organization of behavior.* New York: Wiley.

Hopfield, J. J. (1982). Neural networks and physical systems with emergent collective computational habilities. *Proceedings of the National Academy of Sciences, USA, 79*, 2554–2558.

Kropff, E., & Treves, A. (2007). Uninformative memories will prevail: The storage of correlated representations and its consequences. *HFSP Journal, 1*(4), 249–262.

McRae, K., Cree, G. S., Seidenberg, M. S., & McNorgan, C. (2005). Semantic feature production norms for a large set of living and nonliving things. *Behavioral Research Methods, Instruments, and Computers, 37*, 547–559.

McRae, K., de Sa, V. R., & Seidemberg, M. S. (1997). On the nature and scope of featural representations of word meaning. *Journal of Experimental Psychology: General, 126*(2), 99–130.

Monasson, R. (1992). Properties of neural networks storing spatially correlated patterns. *Journal of Physics, A: Mathematical and General, 25*, 3701–3720.

Parga, N., & Virasoro, M. A. (1986). The ultrametric organization of memories in a neural network. *Journal de Physique, 47*(11), 1857–1864.

Roudi, Y., & Treves, A. (2004). An associative network with spatially organized connectivity. *Journal of Statistical Mechanics: Theory and Experiment* (July), P07010.

Shiino, M., & Fukai, T. (1992). Self-consistent signal-to-noise analysis and its application to analogue neural networks with asymmetric connections. *Journal of Physics, A: Mathematical and General, 25*, L375.

Shiino, M., & Fukai, T. (1993). Self-consistent signal-to-noise analysis of the statistical behavior of analog neural networks and enhancement of the storage capacity. *Physical Review, E, 48*, 867.

Tsodyks, M. V., & Feigel'Man, M. V. (1988). The enhanced storage capacity in neural networks with low activity level. *Europhysics Letters, 6*, 101–105.

Tyler, L. K., Moss, H. E., Durrant-Peatfield, M. R., & Levy, J. P. (2000). Conceptual structure and the structure of concepts: A distributed account of category-specific deficits. *Brain and Language, 75*(2), 195–231.

Vinson, D. P., & Vigliocco, G. (2002). A semantic analysis of grammatical class impairments: Semantic representations of object nouns, action nouns and action verbs. *Journal of Neurolinguistics, 15*, 317–335.

Virasoro, M. A. (1988). The effect of synapses destruction on categorization by neural networks. *Europhysics Letters, 7*(4), 293–298.

Wills, T. J., Lever, C., Cacucci, F., Burgess, N., & O'Keefe, J. (2005). Attractor dynamics in the hippocampal representation of the local environment. *Science, 308*(5723), 873–876.

12 A unified theory of exogenous and endogenous attentional control

Michael C. Mozer and
Matthew H. Wilder

Although diverse, theories of visual attention generally share the notion that attention is controlled by some combination of three distinct strategies: (1) exogenous cueing from locally contrasting primitive visual features, such as abrupt onsets or color singletons (e.g., Itti & Koch, 2000); (2) endogenous gain modulation of exogenous activations, used to guide attention to task relevant features (e.g., Navalpakkam & Itti, 2005; Wolfe, 1994); and (3) endogenous prediction of likely locations of interest, based on task and scene gist (e.g., Torralba, Oliva, Castelhano, & Henderson, 2006). Because these strategies can make conflicting suggestions, theories posit arbitration and combination rules. We propose an alternative conceptualization consisting of a single unified mechanism that is controlled along two dimensions: the degree of task focus, and the spatial scale of operation. Previously proposed strategies – and their combinations – can be viewed as instances of this mechanism. Thus, our theory serves not as a replacement for existing models, but as a means of bringing them into a coherent framework. Our theory offers a means of integrating data from a wide range of attentional phenomena. More importantly, the theory yields an unusual perspective on attention that places a fundamental emphasis on the role of experience and task-related knowledge.

1 Introduction

The human visual system can be configured to perform a remarkable variety of arbitrary tasks. For example, in a pile of coins, we can find the coin of a particular denomination, color, or shape, determine whether there are more heads than tails, locate a coin that is foreign, or find a combination of coins that yields a certain total. The flexibility of the visual system to task demands is achieved by control of visual attention.

Three distinct control strategies have been discussed in the literature, Earliest in chronological order, *exogenous* control was the focus of both experimental research (e.g., Averbach & Coriell, 1961; Posner & Cohen, 1984; Treisman, 1982) and theoretical perspectives (e.g., Itti & Koch, 2000; Julesz, 1984; Koch & Ullman, 1985; Neisser, 1967). Exogenous control refers

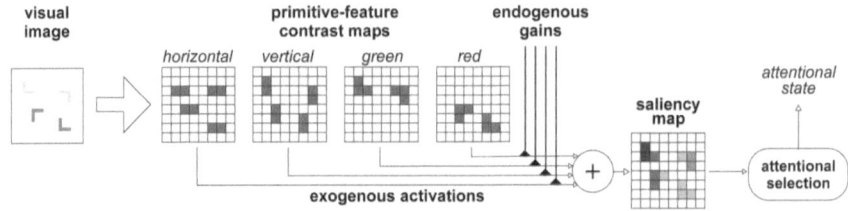

Figure 12.1 A depiction of feature-based endogenous control. [A color version of this figure can be viewed at www.cognitiveneurosciencearena.com/brain-scans]

to the guidance of attention to distinctive, locally contrasting visual features such as color, luminance, texture, and abrupt onsets. Theories of exogenous control assume a *saliency map* – a spatiotopic map in which activation in a location indicates saliency or likely relevance of that location. Activity in the saliency map is computed in the following way. Primitive features are first extracted from the visual field along dimensions such as intensity, color, and orientation. For each dimension, broadly tuned, highly overlapping detectors are assumed that yield a coarse coding of the dimension. For example, on the color dimension, one might posit spatial feature maps tuned to red, yellow, blue, and green. Next, local contrast is computed for each feature – both in space and time – yielding an activity map that specifies distinctive locations containing that feature. These *feature-contrast* maps are combined together to yield a saliency map.[1] Saliency thus corresponds to all locations that stand out from their spatial or temporal neighborhood in terms of their primitive visual features.

Subsequent research showed that attention need not be deployed in a purely exogenous manner but could be influenced by task demands (e.g., Bacon & Egeth, 1994; Folk, Remington, & Johnston, 1992; Wolfe, Cave, & Franzel, 1989). These results led to theories proposing *feature-based endogenous* control (e.g., Baldwin & Mozer, 2006; Mozer, 1991; Navalpakkam & Itti, 2005; Wolfe, 1994, 2007). In these theories, the contribution of feature-contrast maps to the saliency map is weighted by endogenous *gains* on the feature-contrast maps, as depicted in Figure 12.1. The result is that an image such as that in Figure 12.2a, which contains singletons in both color and orientation, might yield a saliency activation map like that in Figure 12.2b if task contingencies involve color or like that in Figure 12.2c if the task contingencies involve orientation.

Experimental studies support a third attentional control strategy, in which attention is guided to visual field regions likely to be of interest based on the current task and coarse properties of the scene (e.g., Biederman, 1972; Neider & Zelinsky, 2006; Torralba et al., 2006; Wolfe, 1998). This type of

---

1 The combination rule is typically summation. However, Zhaoping and Snowden (2006) have suggested that a max operation may be more appropriate.

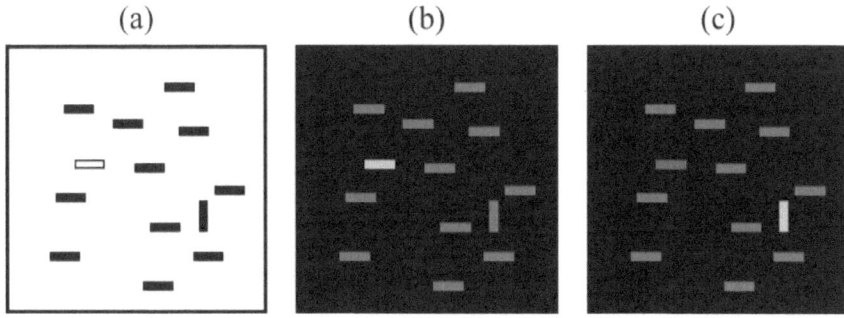

Figure 12.2 (a) A display containing two singletons, one in color and one in orientation. (b) A saliency map if color is a task-relevant feature. (c) A saliency map if orientation is a task-relevant feature.

scene-based endogenous control seems intuitive. If you are looking for your keys in the kitchen, they are likely to be on a counter. If you are waiting for a ride on a street, the car is likely to appear on the road not on a building. Even without a detailed analysis of the visual scene, its gist can be inferred – for example, whether one is in a kitchen or on the street, and the perspective from which the scene is viewed – and this gist can guide attention.

1.1 Arbitrating among control strategies

Given evidence that three distinct control strategies – exogenous, feature-based endogenous, and scene-based endogenous – can influence the allocation of attention, more than one strategy might be applied in any situation. Thus, theories must address the metacontrol issue of which strategy or combination of strategies to apply.

The saliency map can serve as a common output medium for the three strategies, allowing the strategies to operate in parallel and have their results combined in the saliency map. Alternatively, an arbitration mechanism may select which strategy to apply. For example, in guided search (Wolfe, 1994), the saliency map sums the output of distinct exogenous and feature-based endogenous control processes. And the more recent framework of Torralba et al. (2006) proposes parallel pathways for determining exogenous and scene-based endogenous control, and activity in the saliency map is the product of activity in the two pathways.

Although theories of attentional control suppose distinct control strategies operating in parallel, little experimental evidence exists to support this notion. In neuroimaging studies, large-scale neural systems for endogenous and exogenous control are indistinct (e.g., Rosen et al., 1999; Peelen, Heslenfeld, & Theeuwes, 2004). Of course, distinct mechanisms of control may operate in the same cortical area, but behavioral studies also provide evidence against multiple control strategies operating in parallel. Rather, control strategies appear to trade off. For example, increasing task difficulty via target–nontarget similarity decreases the impact of an irrelevant singleton in brightness (Proulx & Egeth, 2006; Theeuwes, 2004) – that is, as the need for feature-based endogenous control increases, exogenous control decreases.

1.2 A unifying framework

Instead of conceiving of the three control strategies as distinct and unrelated mechanisms, the principal contribution of this work is to characterize the strategies as points in a *control space*. As depicted in Figure 12.3, the control space is two dimensional, one being the task dependence of control – the degree to which control is modulated by the current goals and tasks – and the other being the spatial scale – from local to global – of featural information used for control. Exogenous control uses information at a local spatial scale and operates independently of current goals. Feature-based and scene-based endogenous control both operate with a high degree of task dependence, but utilizing information at different spatial scales. (Figure 12.3 shows two other points in the control space, which we discuss later in the chapter.)

What does laying out the three strategies – which we refer to as the *primary* strategies – in a control space buy for us? The control space offers us the means to reconceptualize attentional control. Rather than viewing attentional control as combining or arbitrating among the primary strategies, one

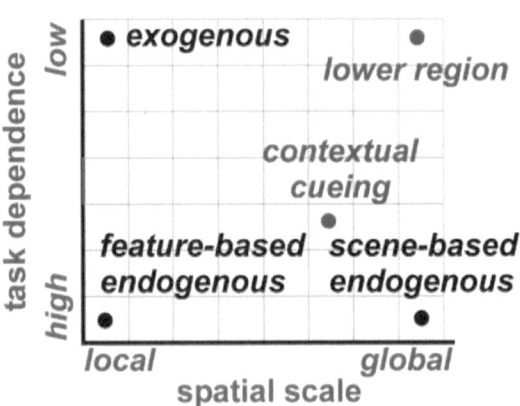

Figure 12.3 A two-dimensional control space that characterizes exogenous, feature-based endogenous, and scene-based endogenous control of visual attention. Other cases of attentional control can also be characterized in terms of this space, including contextual cueing and lower-region figure–ground. The details are discussed in the text.

might view control as choosing where to operate in this continuous two-dimensional space. By this scheme, control would always involve selecting a single strategy – a single point in the space – but the resulting strategy could appear to combine aspects of the primary strategies. Ultimately, we would like to argue for this strong claim, but at the moment we do not have evidence to support such a claim. Instead, we focus in this chapter on showing that the control space offers a general perspective that encompasses existing notions of attentional control. We make this point by presenting a model that implements the notion of the control space and can account for key experimental results in the literature. The model therefore offers a unified perspective on attentional control and into the relationships among the primary strategies, which have heretofore been conceptualized as distinct and unrelated. We call the model *spatial-scale and task-dependence control space*, or *STACS* for short.

1.3 An illustration of saliency over the control space

To give a concrete intuition about the operation of STACS, we present several examples illustrating the model's behavior. The input to STACS is an image – natural or synthetic – and the output from STACS is a saliency map. To operate at different points in the control space, STACS must be given: (1) the spatial scale at which it should process the image, (2) the degree of task dependence, and, (3) when there is some task dependence, a specification of the task. The focus is on visual search, and we therefore define a task to be search for a particular object or class of objects.

Figures 12.4–12.6 show sample images and tasks, along with the saliency map produced by STACS at different points in the control space. Figure 12.4 shows a street scene, and the task is to search for the people. Examining the grid of saliency maps, the map in the upper-left corner reflects local scale, task-independent processing, or what we referred to as exogenous control – a saliency map that indicates high contrast locations in the image independent of the task. The map in the lower-left corner reflects feature-based endogenous control – a saliency map that predicts the presence of people based on local features. And the map in the lower-right corner reflects scene-based endogenous control – a saliency map that predicts the likely locations of people based on the scene gist. The middle row of maps in Figure 12.4 correspond to an intermediate degree of task specificity, and the middle columns correspond to an intermediate spatial scale.

Figures 12.5 and 12.6 show a similar grid of saliency maps for the tasks of searching for cars and buildings, respectively. As in Figure 12.4, it is apparent that the task plays a significant role in modulating the saliency in the task-dependent regions of the control space. The point of these three examples is to show that a continuum of saliency maps can be achieved, and how this continuum encompasses the primary control strategies.

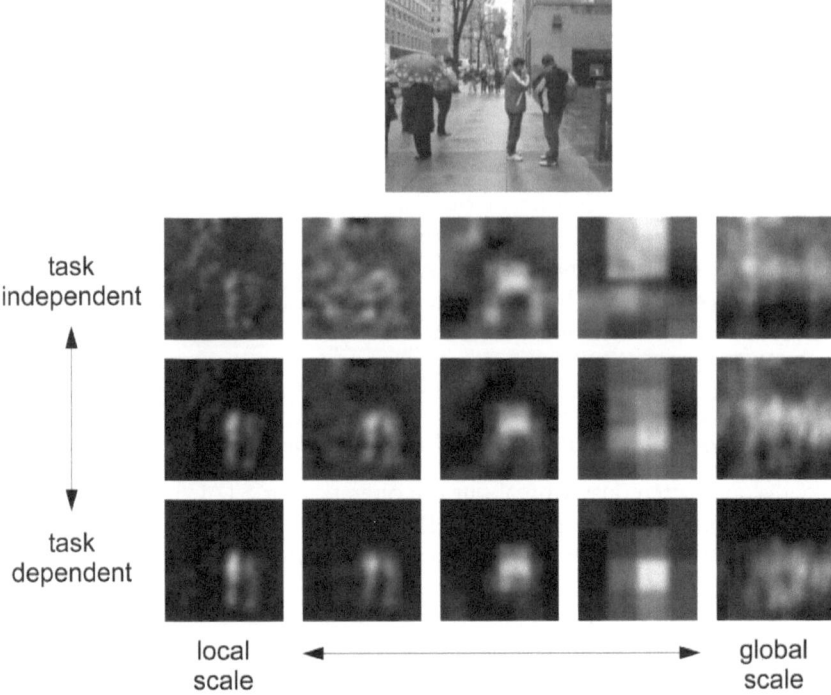

task
independent

↑

↓

task
dependent

local
scale   ◄————————————►   global
scale

Figure 12.4 A street scene, plus saliency maps produced by STACS in response to
this input for five spatial scales (*columns*) and three levels of task dependence (*rows*). The task is to find the *people* in the image. The saliency maps
are interpreted in the text.

2 An implementation of the unified framework for attentional control

As with all models of attention, STACS assumes that computation of the
saliency map can be performed in parallel across the visual field with relatively simple operations. If computing saliency was as computationally complex as recognizing objects, there would not be much use for the saliency
map, because the purpose of the saliency map is to provide rapid heuristic
guidance to the visual system.

Given the control settings of spatial scale and a specific task, STACS
is configured to perform the mapping from image to saliency map. Rather
than creating yet another model of attention, our aim in developing
STACS was to generalize existing models, such that existing models can be
viewed as instantiations of STACS with different control settings. We focus
on two classes of models in particular. One class combines exogenous and

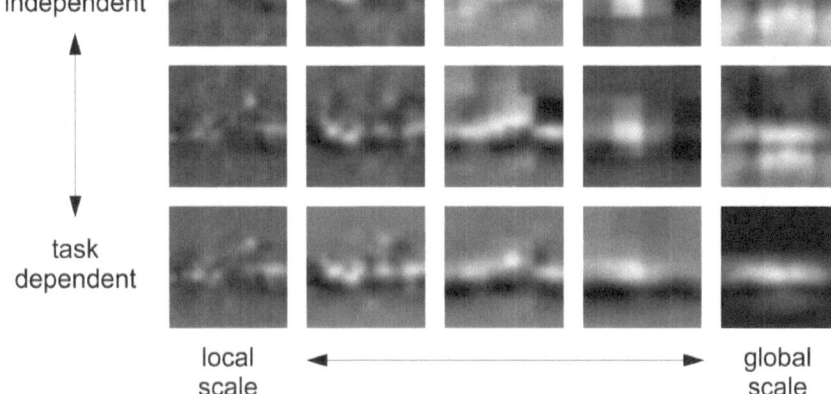

Figure 12.5 A street scene, plus saliency maps produced by STACS in response to this input for five spatial scales (*columns*) and three levels of task dependence (*rows*). The task is to find the *cars* in the image.

feature-based endogenous control, as exemplified by Navalpakkam and Itti (2005) – hereafter *NI* – and the Wolfe (1994) guided search model which is quite similar to NI in its key claims. Another class combines exogenous and scene-based endogenous control, as most recently represented by Torralba, Oliva, Castelhano, and Henderson (2006) – hereafter, *TOCH*. Table 12.1 shows that the basic processing stages of NI and TOCH are quite similar, and the union of the two models yields a more general model, STACS. We now describe these stages in more detail and explain how STACS can be configured to implement various control strategies. Figure 12.7 provides a schematic depiction of the STACS architecture.

2.1 Feature extraction

All images presented to STACS are normalized to be 256×256 pixels. Color filters (red, green, blue, and yellow) are applied to each pixel in the image, as are Gabor filters centered on each pixel at four orientations (0°, 45°, 90°, and

task
independent

↑
|
↓

task
dependent

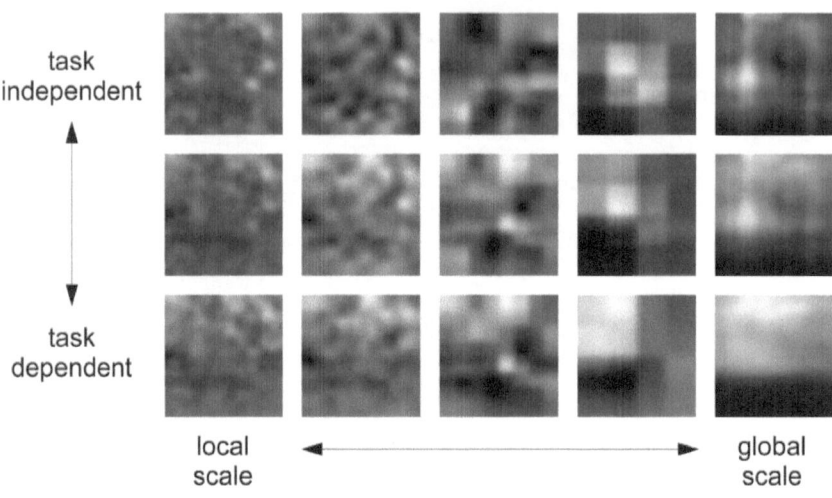

local   global
scale   scale

Figure 12.6 A street scene, plus saliency maps produced by STACS in response to this input for five spatial scales (*columns*) and three levels of task dependence (*rows*). The task is to find the *buildings* in the image.

Table 12.1 Processing stages of two existing models of attentional control

Stage	Navalpakkam & Itti (2005); Wolfe (1994)	Torralba et al. (2006)
Parallel feature extraction with detectors having broad, overlapping tuning curves	color, orientation, luminance	color, orientation at multiple spatial scales
Contrast enhancement	via center–surround differencing	via cross-dimensional normalization
Dimensionality reduction	no	yes
Associative network to compute saliency	linear	mostly linear with a Gaussian squashing function

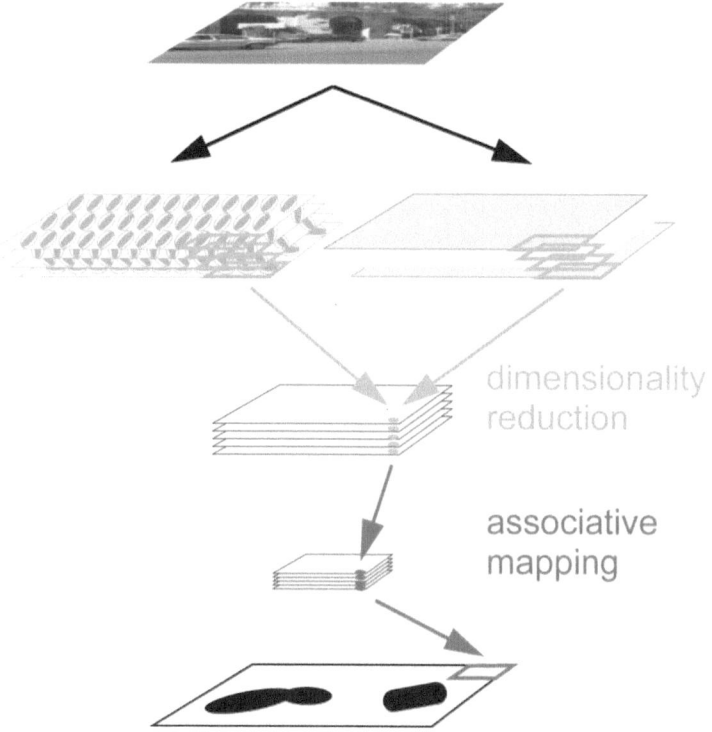

Figure 12.7 A schematic depiction of the STACS architecture, showing the patchwise analysis of an image to yield a saliency map.

135°) with a diameter of 9 pixels. Feature extraction is performed using the matlab Saliency Toolbox, which essentially implements the NI model, with a few minor simplifications.[2] Figure 12.8a shows a sample image, and Figure 12.8b shows the feature representation at the finest spatial scale.

2.2 Contrast enhancement

All subsequent analysis is based on local *patches* of the image. The size of the patch depends on the spatial-scale control setting given to the model. We implemented five scales, with patches at the scales being 16×16, 32×32, 64×64, 128×128, and 256×256. However, the smallest scale was too fine to provide much interesting information for the images we studied. The columns in Figures 12.4–12.6 correspond to the five spatial scales of our implementation. In the remainder of the chapter, we present results only for

2 The Saliency Toolbox (www.saliencytoolbox.net) was developed by Dirk Walther.

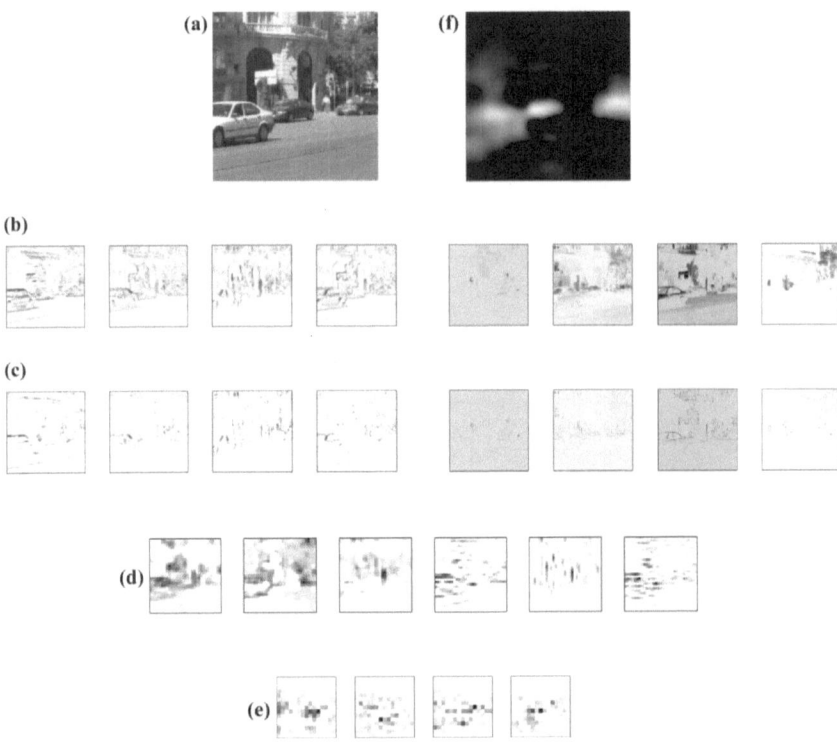

Figure 12.8 Activities in STACS at the finest spatial scale at various stages of processing: (a) an input image; (b) feature activities for four orientation maps and
four color maps; (c) activities following contrast enhancement; (d) activities from six principal components following subsampling; (e) activities
of four hidden units; and (f) the output saliency map.

the four coarsest spatial scales, because the finest scale produced no useful
output.

To enhance distinctive regions in color or orientation, STACS – like nearly
every other model of early visual processing – performs local contrast
enhancement, increasing the activities of features that are distinct from their
neighborhood. We explored several different methods, including that used
by NI (see Itti & Koch, 2000). The method we found to yield the best feature
pop-out involves multiplying the activity of each feature detector by the ratio
of the Gaussian-weighted mean activity of that detector in the local neighborhood to the Gaussian-weighted activity of that detector in a broader
neighborhood. The standard deviation of the "center" and "surround"
Gaussians are 5% and 50% of the patch size, respectively, for all spatial
scales. Figure 12.8c shows the contrast-enhanced image representation.

2.3 Dimensionality reduction

From this stage of analysis and beyond, the image is analyzed in terms of local patches that overlap one another by 50%, with the patch size dependent on the spatial scale (as specified in the previous section). The representation of an image patch has $8d^2$ elements, where d is the diameter of the patch. For large patches, the dimensionality of this representation is quite large. Consequently, models processing at the coarser spatial scales, such as TOCH, include some type of dimensionality reduction of the representation. STACS incorporates the two dimensionality reduction techniques of TOCH: subsampling and principal components analysis (PCA). For the four coarsest spatial scales of STACS, the representation was subsampled along the x- and y-axes by factors f of 4, 8, 8, and 16, and a resulting representation of size $8d^2/f^2$. The subsampled representation was fed into the PCA, and the c components with highest eigenvalues were used to represent the patch, where, for the four coarsest scales, c is 16, 32, 48, and 64. The principal components are extracted in a location-invariant manner using training on random patches of a training image corpus of real-world, naturalistic images. Figure 12.8d shows the representation of the first six principal components across the sample image in Figure 12.8a.

2.4 Associative mapping

The last stage of STACS computes the saliency of each location in the visual field based on the dimensionality-reduced, contrast-enhanced representation of the image. In the spirit of using simple, minimal computation, the saliency computation is performed via an associative linear mapping. This mapping is performed patchwise – from a patch of the image representation to the corresponding patch of the saliency representation. The patches overlap, and the saliency representation is therefore formed by summing the patchwise outputs. A supervised learning procedure obtains a set of linear coefficients for each patch (location) and each set of control settings (spatial scale and task).

Training is based on a labeled set of images: images in which each pixel is classified as to the presence or absence of a target object at that particular location. Figure 12.9 shows two sample images and labeling (via different colors) as to the locations of cars, people, and buildings. The images and target labels are from the LabelMe database (http://labelme.csail.mit.edu). We trained models for the tasks of searching for cars, people, buildings, lamps, trees, roads, windows, and signs. Each of these objects was found in at least 80 images in the LabelMe database. However, learning was hindered somewhat by a sparsely labeled database – that is, many targets in the image are not labeled as targets.

STACS is trained to perform object localization: given a task of searching for a target object in an image, STACS should identify where in the image the object appears. This goal is a quick and rough cut at object recognition.

Figure 12.9 Top: Two images from the LabelMe database. *Bottom:* Pixel-by-pixel labeling of three objects – cars, people, and buildings. [A color version of
this figure can be viewed at www.cognitiveneurosciencearena.com/brain-
scans]

STACS is quick because of the simplistic architecture, but STACS is also
accuracy-limited because of the simplistic architecture: the associative mapping is linear, and, furthermore, it is rank-limited. We achieve the rank limitation by implementing the mapping as a neural network with a linear hidden
layer (depicted as the next-to-last layer in the architecture of Figure 12.7). For
the four coarsest spatial scales, we use a bottleneck of 1, 3, 12, and 25 hidden
units. Because they are linear, the hidden units provide no additional computational power but, rather, serve as a bottleneck on the flow of information
used to compute saliency. Figure 12.8e shows the activity of hidden units
across image patches, and Figure 12.8f shows the saliency activity pattern
resulting from the associative mapping.

2.5 *Implementing task-dependent control*

For each task and each spatial scale, the necessary knowledge to perform the
associative mapping is contained in the connection strengths feeding to and
from the hidden units. One can conceptualize the model as having a pool of
hidden units for each task and spatial scale, and the appropriate pool is
enabled depending on attentional control settings.

Because the associative mapping is linear, STACS produces sensible
behavior when more than one pool of hidden units is enabled. For example,
if the bicycle and car pools are both enabled, the saliency map will be a

superimposition of the maps for bicycles and cars. Thus, linearity makes it possible to combine multiple tasks. One could imagine a hierarchy of tasks, from the very specific – corresponding to a single pool – or very general – corresponding to multiple pools (e.g., the pools for all wheeled vehicles). At an extreme, if the pools for all tasks are enabled, then STACS operates in a *task-independent* manner. Thus, the control dimension of task-specific to task-independent is achieved by the enabling of different pools of hidden units. In Figures 12.4–12.6, the saliency maps with an intermediate degree of task dependence – in the second row of the figures – are obtained by enabling the units for the specific task and weakly enabling all other task units.

3 Simulation results

Having described STACS, we now to turn to data and phenomena that lie within its scope. At this point in our development, our primary goal is to show that STACS serves as a unifying theory that can subsume existing models and the phenomena those models are designed to accommodate.

3.1 Exogenous control

Exogenous control is achieved in STACS via control settings that specify local spatial scale and task-independent processing. Because task-independent saliency is obtained by superimposing the outputs of task-specific saliency maps, exogenous control in STACS depends on the set of tasks on which it has been trained. We have a fairly small corpus of eight tasks; one should not expect STACS's exogenous control to be terribly robust in our preliminary implementation. Nonetheless, the basic approach is a novel claim of STACS: exogenous control deploys attention to locations that are of interest in *any* task. The upper-left saliency maps in Figures 12.4–12.6 are examples of exogenous control in STACS. As with all models of exogenous control, it is difficult to evaluate performance of STACS. Nonetheless, the sort of locations picked by STACS seem reasonable. Ultimately, we will validate STACS's performance via a collection of eye movement data in free viewing, such as that of Bruce and Tsotsos (2006). However, we suspect that all models of exogenous control will yield about the same performance, and it will be impossible to discriminate models on the basis of their fit to free-viewing eye movements alone. STACS has the additional virtue of explaining all varieties of attentional control.

3.2 Scene-based endogenous control

STACS can perform in a mode like the TOCH model of scene-based endogenous control. TOCH computes regions of interest based on the task and a scene gist. STACS performs in a manner qualitatively similar to TOCH when operating in a task-dependent manner at a coarse spatial scale. Indeed, the computations are nearly identical except for two properties. First, TOCH incorporates a Gaussian nonlinearity at the final stage of determining saliency. However, it is far from clear from the output of TOCH and STACS that this nonlinearity has a significant effect. Second, TOCH operates on a multi-scale representation, whereas the current implementation of STACS operates on a dimensionality-reduced fine-scale representation (which is a component of the TOCH representation). It is a minor modification to STACS to make it a more faithful recreation of TOCH when operating at the coarse spatial scale.

The lower-right saliency maps in Figures 12.4–12.6 are examples of scene-based endogenous control in STACS. STACS yields sensible behavior, qualitatively like that of TOCH: the upper region of the image is salient when searching for buildings, small bright discontinuities are salient when searching for people, and the middle region of the image along the side of the road is salient when searching for cars.

3.3 Feature-based endogenous control

STACS can perform in a mode like the NI model of feature-based endogenous control. In STACS, feature-based control requires essentially the same training procedure as we described for scene-based control, except that rather than targets being complex objects such as people and buildings, the target is a simple feature, such as the presence of the color red or a vertical line. We train a STACS hidden layer for each possible target feature. The training data consists of random synthetic images of oriented, colored lines of variable length. The target saliency output indicates the locations containing the target feature. (The learning that takes place in STACS is analogous to learning the meaning of a word such as "red" or "vertical". Assigning saliency to a specific feature is computationally trivial. The challenge is in determining which task is associated with which feature.)

We now show how STACS performs the basic tasks in the visual search literature. First consider the simple-feature search task of finding a red item among green items. The top row of Figure 12.10 shows an instance of such a display. Beside the image are saliency maps for four spatial scales, all indicating high saliency for the target in this example. Now consider the conjunction-search task of finding a red vertical (dark gray on figure) among green verticals (light gray) and red horizontals (dark gray), as shown in the bottom-row image of Figure 12.10. To perform conjunction search, STACS can either be trained on conjunction targets (red verticals) or the task can be defined by conjoining the hidden-unit pools for the two component features (red and vertical). In both implementations, STACS is unable to clearly distinguish the target from the distractors, at any spatial scale. The reason for this difficulty is the linearity and bottleneck of the associative mapping: with limited resources, STACS is unable to accurately localize complex combinations of primitive features.

To perform a more formal analysis of visual search tasks, and to model

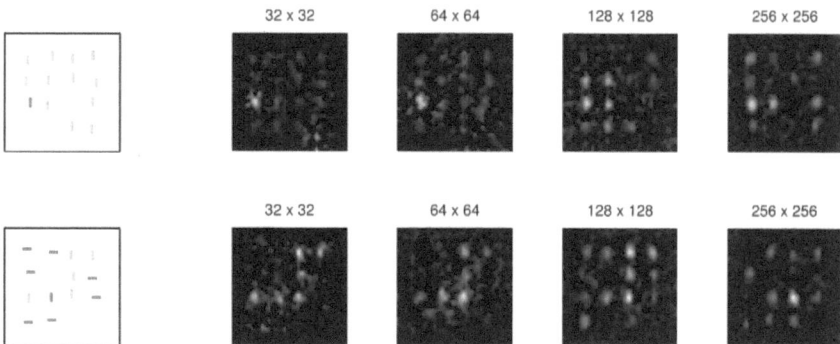

Figure 12.10 Top row: Modeling simple-feature search. On the left is an image containing a single dark gray item among light grays; on the right are saliency maps at different spatial scales for this image. *Bottom row:* Modeling conjunction search. On the left is an image containing a single dark gray vertical among light gray verticals and dark gray horizontals; on the right are saliency maps at different spatial scales for this image. [A color version of this figure can be viewed at www.cognitiveneurosciencearena. com/brain-scans]

human data, it is necessary to read response times from STACS, which in turn requires an additional assumption concerning how the saliency map is used in search. Guided search (Wolfe, 1994) makes the assumption that locations are searched in order from most salient to least salient. However, in Wolfe's model, there is a one-to-one correspondence between objects and locations. In our images, the notion of "location" is ill defined. Should a location correspond to a pixel? If so, then are neighboring pixels two distinct locations? To avoid such ambiguities, we make an alternative assumption. We assume that the probability of attending to a given pixel is proportional to its saliency. Based on this assumption, we can compute the probability of attending to a region in the neighborhood of the target (i.e., the pixels containing the target and some immediately surrounding pixels). And we can make the standard assumption that response times are related to probability via a negative log transform – that is, $RT = -log(P(n))$, where RT is the response time and n denotes the target neighborhood.

Figure 12.11a graphs feature-search performance as a function of the number of distractors in the display. There are four lines, one for each of the four coarsest spatial scales of operation. The response time (negative log probability) of the model is independent of the number of distractors, regardless of the spatial scale at which the model is operating. This simulation result is in accord with human experiments that find efficient search for features. Note that in absolute terms, the local spatial scales are most efficient for performing the search. One expects this result because the target can be identified based on local features alone (global context does not help), and because STACS has greatest spatial resolution at the local scale.

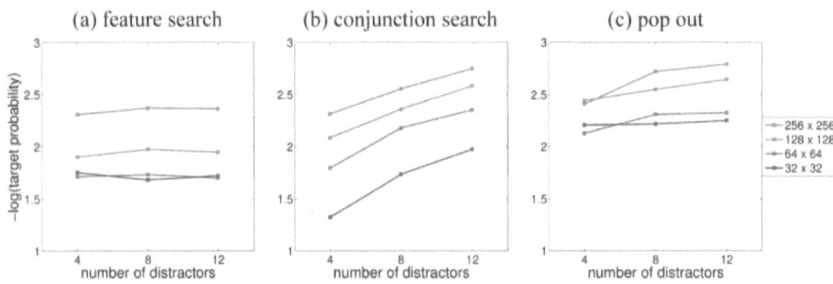

Figure 12.11 Simulation results from STACS for (a) feature, (b) conjunction, and
(c) pop-out search. Each graph contains one line for each of the four
coarsest spatial scales. The *x*-axis of each graph indicates the number of
distractor elements in the search displays. The *y*-axis represents
response-time read out from the model, assumed to be proportional to
the negative log proportion of saliency in the neighborhood of the
target. [A color version of this figure can be viewed at www.
cognitiveneurosciencearena.com/brain-scans]

Figure 12.11b graphs conjunction-search performance as a function of
the number of distractors. Consistent with human experiments, search is
inefficient in that response times increase linearly with the number of elements in the display. As with feature search, conjunction search is fastest if
STACS operates at the local spatial scale. Given that search in these artificial
displays is inefficient, one might wonder what STACS predicts for search in
naturalistic environments on tasks for which it has received significant training. Just as in artificial displays, the performance of STACS will depend on
the spatial scale. At the coarse spatial scale, we have already shown that
STACS can only roughly identify regions where objects of interest will
appear. We expect a similar result at fine scales, because objects in the image
are larger than the patch size that corresponds to the fine scale.

Figure 12.11c graphs the outcome of a pop-out or oddball-detection task.
The task is simply to find the element that differs from others in the display –
for example, the red one among green, or the vertical among horizontals.
Although this task involves detecting a single feature, it is distinct from
feature search in that the feature dimension and value are not known in
advance of a trial. Consequently, STACS must operate in a task-independent
manner (i.e., without a specific target). We trained STACS on four feature-
search tasks – for verticals, horizontals, red, and green – and treated the pop-
out task as the union of these four specific tasks. As Figure 12.11c shows,
STACS produces flat search slopes, at least for the local spatial scales. Note
that although the slopes are flat, the absolute response time is slower than for
feature search, not surprising considering that the feature-search task is more
narrowly delineated. The simulation findings are consistent with human
experiments (e.g., Lamy, Carmel, Egeth, & Leber, 2006).

3.4 Other phenomena

We have shown that STACS provides a unified account of varied phenomena that were previously explained in terms of distinct control strategies. By viewing these control strategies as points in a control space (Figure 12.3), STACS leads one to question whether other points in the control space correspond to known attentional phenomena. In this section, we mention two examples.

3.4.1 Lower region

Vecera, Vogel, and Woodman (2002) found a novel cue to figure–ground assignment. In displays like Figure 12.12a, viewers tended to perceive the lower region as the foreground figure (the gray region in this image). A rational explanation of this phenomenon is that because viewers are ordinarily upright, the lower region of their visual field contains objects of interest. We interpret the lower-region cue to figure–ground assignment as a sort of endogenous control that operates on a global spatial scale – that is, it is based on the overall scene properties – and is task-independent – that is, it operates in the absence of a specific target of search. We have indicated this point in the control space in Figure 12.3. The lower region, occupying the fourth corner of the control space, is a natural complement to the three control strategies described previously.

When shown images like that in Figure 12.12a, STACS, configured for global spatial scale and task-independent processing, yields a saliency map like that in Figure 12.12b. Because this saliency map is formed by combining task-specific maps, and we trained STACS on only eight tasks, one can justifiably be cautious in making strong claims about this result. Nonetheless, the result was an emergent consequence of training. STACS produces a similar result regardless of the colors of the two regions.

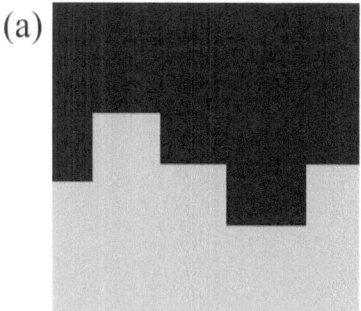 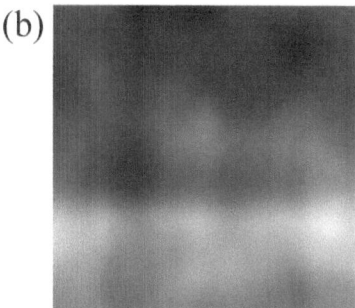

Figure 12.12 (a) A sample display from Vecera et al. (2002). (b) Sample output from STACS operating in a task-independent mode at the global spatial scale.

3.4.2 *Contextual cueing*

Chun and Jiang (1998) found that repeating configurations in a visual search task led to a speed up of response times. Participants were shown displays like that in Figure 12.13a containing a single target – the letter T – among distractors (letter L at various orientations). The participants' task was to report the orientation of the T (pointing to the left or the right). Unbeknownst to participants, some display configurations were repeated over the course of the experiment. In these *predictive* displays, the target and distractors appeared in the same locations, although their orientations and colors could change from trial to trial. After several exposures, participants are roughly 60 ms faster to predictive displays than to random or *nonpredictive* displays.

Figure 12.13b shows results from a STACS simulation of Chun and Jiang (1998). The simulation was trained for the task of detecting the specific targets (a T facing to either the left or the right). As in earlier simulations, the model's response time is assumed to be proportional to the negative log of the relative saliency (or probability) of the target. The four sets of bars are for different spatial scales. At all spatial scales, the model produces faster response times for predictive than nonpredictive displays. However, only at the most global scales is the effect large in magnitude. The small effect at the local scales is due to the fact that the model can learn to suppress locations less likely to contain the target. The effect at the global scales is due to learning about display configurations. The model obtains a contextual cueing effect not only at the scale in which the entire display is processed as a whole (256×256), but also at the scale of half of the display (128×128), suggesting that contextual cueing might occur for displays in which only the distractors in the quadrant closest to the target are predictive.

We therefore characterize contextual cueing as a phenomenon in the

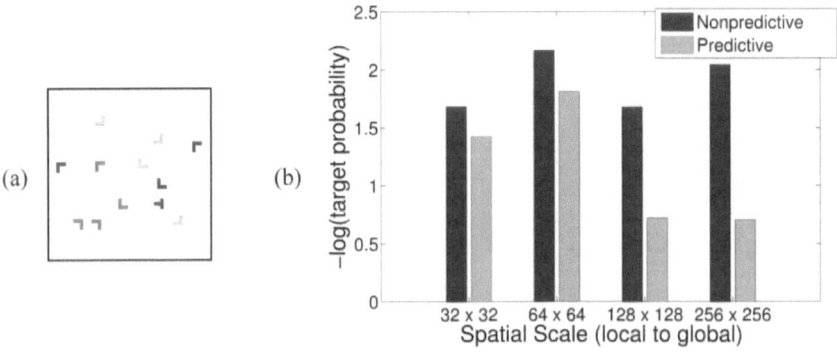

Figure 12.13 (a) A sample display from Chun and Jiang (1998). (b) Performance of STACS operating at various spatial scales for predictive and nonpredictive displays. [A color version of this figure can be viewed at www.cognitiveneurosciencearena.com/brain-scans]

control space (Figure 12.3) that occurs with an intermediate-to-global spatial scale and that has at least some degree of task dependence.

4 Neurobiological implications of the model

We introduced a perspective on attentional control, STACS, that attempts to integrate theoretical ideas from existing models and to provide a unified framework for considering a wide range of attentional phenomena. STACS can be seen as a generalization and unification of existing models of attentional control, in particular TOCH and NI.

Although designed to identify salient locations in the visual field, STACS effectively performs a crude sort of object detection. For each location in the visual field, STACS estimates the probability that a target is present at that location given the visual features in the neighborhood. What distinguishes STACS from a full-blown model of object recognition is the fact that STACS is computationally limited: dimensionality-reduction bottlenecks and the linearity of the model restrict what can be computed. The virtue of these limitations is that STACS is computationally bounded. Computational simplicity should go hand in hand with speed in the brain, and a quick response is essential if saliency is to provide useful guidance for more detailed, computationally intensive processing of regions of the visual field most likely to be relevant.

Beyond the virtue of speed, linearity in the model has two important additional benefits. First, training STACS with gradient descent works well because the linear connections have no local optima. As a result, gradient descent learning procedures can be incremental and ongoing. In contrast, nonlinear neural networks tend to get stuck in a region of weight space from which they cannot escape, even if the training data change over time. Because training can be incremental and ongoing, it is easy to model the effects of recent experience on performance, as we did with the contextual cueing simulation.

A second benefit of linearity of the associative mapping is that it allows STACS, trained on single tasks, to perform – in principle – arbitrary combinations of tasks dynamically simply by enabling pools of task units. For example, search for a red vertical can be specified by enabling the combination of red and vertical tasks, each of which has a dedicated pool of processing units. STACS can search for wheeled vehicles by enabling car, bus, bike, and train tasks. And, STACS can operate in an exogenous control mode simply by enabling *all* tasks in parallel. In practice, the ability of STACS to perform combinations of tasks is limited both by its ability to perform individual tasks and by the fact that the linear combination can produce locations with high saliency that are not the most salient in any one task.

This view of control suggests a specific role of cortical feedback. A neural implementation of STACS would require feedback connections that convey the control settings – the spatial scale at which the model is operating and the set of tasks that are enabled. These control settings modulate ongoing processing; specifically, they enable or disable different neural pools. In contrast, a model such as SAIM (Heinke & Humphries, 2003) suggests a very different view of cortical feedback. SAIM uses top-down connectivity to obtain interactive constraint-satisfaction dynamics; as a result, the feedback connections are integral to object recognition. STACS's notion of cortical feedback as modulating basically bottom-up processing seems better in accord with current conceptions of cortical dynamics. Nonetheless, the substantial differences between the models should lead to interesting experimental tests.

5 Acknowledgments

The authors thank Zhaoping Li for comments on an earlier draft of the manuscript. This research was supported by NSF BCS 0339103, NSF CSE-SMA 0509521, and NSF BCS-0720375.

6 References

Averbach, E., & Coriell, A. S. (1961). Short-term memory in vision. *Bell Systems Technical Journal, 40,* 309–328.

Bacon, W. F., & Egeth, H. E. (1994). Overriding stimulus-driven attentional capture. *Perception & Psychophysics, 55,* 485–496.

Baldwin, D., & Mozer, M. C. (2006). Controlling attention with noise: The cue-combination model of visual search. In R. Sun & N. Miyake (Eds.), *Proceedings of the twenty-eighth annual conference of the Cognitive Science Society* (pp. 42–47). Hillsdale, NJ: Lawrence Erlbaum Associates.

Biederman, I. (1972). Perceiving real-world scenes. *Science, 177,* 77–80.

Bruce, N., & Tsotsos, J. (2006). Saliency based on information maximization. In Y. Weiss, B. Schölkopf, & J. Platt (Eds.), *Advances in neural information processing systems 18* (pp. 155–162). Cambridge, MA: MIT Press.

Chun, M. M., & Jiang, Y. (1998). Contextual cueing: Implicit learning and memory of visual context guides spatial attention. *Cognitive Psychology, 36,* 28–71.

Folk, C. L., Remington, R. W., & Johnston, J. C. (1992). Involuntary covert orienting is contingent on attentional control settings. *Journal of Experimental Psychology: Human Perception & Performance, 18,* 1030–1044.

Heinke, D., & Humphreys, G. W. (2003). Attention, spatial representation and visual neglect: Simulating emergent attention and spatial memory in the selective attention for identification model (SAIM). *Psychological Review, 110,* 29–87.

Itti, L., & Koch, C. (2000). A saliency-based search mechanism for overt and covert shifts of visual attention. *Vision Research, 40,* 1489–1506.

Julesz, B. (1984). Toward an automatic theory of preattentive vision. In G. M. Edelman, W. E. Gall, & W. M. Cownan (Eds.), *Dynamic aspects of neocortical function* (pp. 595–612). New York: Neurosciences Research Foundation.

Koch, C., & Ullman, S. (1985). Shifts in selective visual attention: Towards the underlying neuronal circuitry. *Human Neurobiology, 4,* 219–227.

Lamy, D., Carmel, T., Egeth, H. E., & Leber, A. B. (2006). Effects of search mode and intertrial priming on singleton search. *Perception & Psychophysics, 68,* 919–932.

Mozer, M. C. (1991). *The perception of multiple objects: A connectionist approach.* Cambridge, MA: MIT Press.

Navalpakkam, V., & Itti, L. (2005). Modeling the influence of task on attention. *Vision Research, 45,* 205–231.

Neider, M. B., & Zelinsky, G. J. (2006). Scene context guides eye movements during visual search. *Vision Research, 46,* 614–621.

Neisser, U. (1967). *Cognitive psychology.* New York: Appleton-Century-Crofts.

Peelen, M. V., Heslenfeld, D. J., & Theeuwes, J. (2004). Endogenous and exogenous attention shifts are mediated by the same large scale neural network. *NeuroImage, 22,* 822–830.

Posner, M. I., & Cohen, Y. (1984). Components of visual orienting. In H. Bouma & D. G. Bouwhuis (Eds.), *Attention and performance X* (pp. 531–556). Hillsdale, NJ: Lawrence Erlbaum Associates.

Proulx, M. J., & Egeth, H. E. (2006). Target–nontarget similarity modulates stimulus-driven control in visual search. *Psychonomic Bulletin & Review, 2006,* 524–529.

Rosen, A. C., Rao, S. M., Cafarra, P., Scaglioni, A., Bobholz, J. A., Woodley, S. J., et al. (1999). Neural basis of endogenous and exogenous spatial orienting: A functional MRI study. *Journal of Cognitive Neuroscience, 33,* 135–152.

Theeuwes, J. (2004). Top-down search strategies cannot override attentional capture. *Psychonomic Bulletin & Review, 11,* 65–70.

Torralba, A., Oliva, A., Castelhano, M. S., & Henderson, J. M. (2006). Contextual guidance of eye movements and attention in real-world scenes: The role of global features on object search. *Psychological Review, 113,* 766–786.

Treisman, A. (1982). Perceptual grouping and attention in visual search for features and objects. *Journal of Experimental Psychology: Human Perception and Performance, 8,* 194–214.

Vecera, S. P., Vogel, E. K., & Woodman, G. F. (2002). Lower region: A new cue for figure–ground assignment. *Journal of Experimental Psychology: General, 131,* 194–205.

Wolfe, J. M. (1994). Guided search 2.0: A revised model of visual search. *Psychonomic Bulletin and Review, 1,* 202–238.

Wolfe, J. M. (1998). Visual memory: What do you know about what you saw? *Current Biology, 8,* R303–R304.

Wolfe, J. M. (2007). Guided search 4.0: Current progress with a model of visual search. In W. Gray (Ed.), *Integrated models of cognitive systems.* New York: Oxford University Press.

Wolfe, J. M., Cave, K. R., & Franzel, S. L. (1989). Guided search: An alternative to the feature integration model for visual search. *Journal of Experimental Psychology: Human Perception and Performance, 15,* 419–433.

Zhaoping, L., & Snowden, R. J. (2006). A theory of a saliency map in primary visual cortex (V1) tested by psychophysics of colour-orientation interference in texture segmentation. *Visual Cognition, 14,* 911–933.

13 Free-energy, value and neuronal systems

Karl J. Friston, Klaas E. Stephan, and Stefan Kiebel

This report summarises our recent attempts to integrate action and perception within a single optimisation framework. We start with a statistical formulation of Helmholtz's ideas about neural energy to furnish a model of perceptual inference and learning that can explain a remarkable range of neurobiological facts. Using constructs from statistical physics it can be shown that the problems of inferring what cause our sensory inputs and learning causal regularities in the sensorium can be resolved using exactly the same principles. Furthermore, inference and learning can proceed in a biologically plausible fashion. The ensuing scheme rests on empirical Bayes and hierarchical models of how sensory information is generated. The use of hierarchical models enables the brain to construct prior expectations in a dynamic and context-sensitive fashion. This scheme provides a principled way to understand many aspects of the brain's organisation and responses.

Here, we suggest that these perceptual processes are just one aspect of systems that conform to a free-energy principle. The free-energy considered here represents a bound on the surprise inherent in any exchange with the environment, under expectations encoded by its state or configuration. A system can minimise free-energy by changing its configuration to change the way it samples the environment, or to change its expectations. These changes correspond to action and perception, respectively, and lead to an adaptive exchange with the environment that is characteristic of biological systems. This treatment implies that the system's state and structure encode an implicit and probabilistic model of the environment and that its actions suppress surprising exchanges with it. Furthermore, it suggests that free-energy, surprise and (negative) value are all the same thing. We look at models entailed by the brain and how minimisation of free-energy can explain its dynamics and structure.

1 Introduction

This chapter illustrates how ideas from theoretical physics can help understand biological systems, in particular the brain. This is not a rigorous treatment, but a series of heuristics that provide an interesting perspective on how biological systems might function. The first section motivates and describes a free-energy principle that addresses the maintenance of structural order seen in living systems. The subsequent sections use this principle to understand key functional and structural aspects of neuronal systems, with a focus on perceptual learning and inference. This work pursues an agenda established by von Helmholtz in the nineteenth century, who sought a basis for neuronal energy in his work on conservation laws in physics. This ambition underlies many energy-based approaches to neural networks (Borisyuk & Hoppensteadt, 2004), including the approach described here.

Despite the latitude for disorder, the nervous system maintains an exquisite configurational and dynamic order. This order is preserved on both an evolutionary and a somatic timescale. The premise of this chapter is that this precise structural and functional organisation is maintained by causal structure in the environment. The principles behind this maintenance and the attending neuronal mechanisms are the subject of this chapter.

The analysis presented in this chapter rests on some fairly mathematical and abstract approaches to understanding the behaviour of systems. These approaches were developed primarily in statistical physics and machine learning. The payoff for adopting this mathematical treatment is that many apparently diverse aspects of the brain's structure and function can be understood in terms of one simple principle – namely, the minimisation of a quantity (free-energy) that reflects the probability of sensory input, given the current state of the brain. We will see that this principle can be applied at different timescales to explain perpetual inference, attention and learning. Furthermore, exactly the same principle can explain how we interact with, or sample, the environment; providing a principled account of adaptive behaviour. It highlights the importance of perception for action and enforces a mechanistic view of many ethological and neuronal processes. Another payoff is the disclosure of some rather counter-intuitive conclusions about our brains – for example, it suggests that everything we do serves to minimise surprising exchanges with the environment (and other people); it suggests that perception plays a secondary role in optimising action; it suggests that the salience, attention and encoding of uncertainty in the brain are all aspects of the same underlying process; it suggests that the hierarchical structure of our brains transcribes causal hierarchies from the environment. Finally, it furnishes clear links among other important formulations of adaptive systems – for example, we will see that value, in microeconomics and reinforcement learning, is synonymous with (negative) free-energy and surprise. Similarly, adaptive fitness can be formulated in terms of free-energy, which allows one to link evolutionary and somatic timescales in terms of hierarchical co-evolution.

Many people now regard the brain as an inference machine that conforms to the same principles that govern the interrogation of scientific data (Ballard, Hinton, & Sejnowski, 1983; Dayan, Hinton, & Neal, 1995; Friston, 2003, 2005; Kawato, Hayakawa, & Inui, 1993; Kersten, Mamassian, & Yuille, 2004; Körding & Wolpert, 2004; MacKay, 1956; Mumford, 1992; Neisser, 1967;

Rao & Ballard, 1998). In everyday life, these rules are applied to information obtained by sampling the world with our senses. Over the past years, we have pursued this perspective in a Bayesian framework to suggest that the brain employs hierarchical or empirical Bayes to infer the causes of its sensations. This model of brain function can explain a wide range of anatomical and physiological facts – for example, the hierarchical deployment of cortical areas, recurrent architectures using forward and backward connections, and functional asymmetries in these connections (Angelucci, Levitt, Walton, Hupe, Bullier, & Lund, 2002a; Angelucci, Levitt, & Lund, 2002b; Friston, 2003). In terms of synaptic physiology, it predicts associative plasticity and, for dynamic models, spike-timing dependent plasticity. In terms of electro-physiology it accounts for classical and extra-classical receptive field effects and long-latency or endogenous components of evoked cortical responses (Friston, 2005; Rao & Ballard, 1998). It predicts the attenuation of responses encoding prediction error, with perceptual learning, and explains many phenomena like repetition suppression, mismatch negativity and the P300 in electroencephalography. In psychophysical terms, it accounts for the behavioural correlates of these physiological phenomena – for example, priming, and global precedence (for an overview, see Friston, 2005).

It is fairly easy to show that both perceptual inference and learning rest on a minimisation of free-energy (Friston, 2003) or suppression of prediction error (Rao & Ballard, 1998). The notion of free-energy derives from stat-istical physics and is used widely in machine learning to convert difficult integration problems, inherent in inference, into easier optimisation prob-lems. This optimisation or free-energy minimisation can, in principle, be implemented using relatively simple neuronal infrastructures. The purpose of this chapter is to suggest that perception is just one emergent aspect of free-energy minimisation and that a free-energy principle for the brain can explain the intimate relationship between perception and action. Further-more, the processes entailed by the free-energy principle cover not just infer-ence about the current state of the world, but a dynamic encoding of context that bears the hallmarks of attention and perceptual salience.

The free-energy principle states that systems change to decrease their free-energy. The concept of free-energy arises in many contexts, especially phys-ics and statistics. In thermodynamics, free-energy is a measure of the amount of work that can be extracted from a system, and it is useful in engineering applications (for discussion of free-energy theorems, see Streater, 1993). It is the difference between the energy and the entropy of a system. Free-energy also plays a central role in statistics, where, borrowing from statistical thermodynamics, approximate inference by variational free-energy mini-misation (also known as variational Bayes, or ensemble learning) has max-imum likelihood and maximum *a posteriori* methods as special cases. It should be noted that the only link between these two uses of the term "free-energy" is mathematical – that is, both appeal to the same probabilistic fundaments. It is the second sort of free-energy, which is a measure of statistical probability distributions, that we use here. Systems that minimise their statistical free-energy make implicit inferences about their surroundings. Previous treatments of free-energy in inference (e.g., predictive coding) have been framed as explanations or descriptions of the brain at work. In this chapter, we try to go a step further by suggesting that free-energy minimisation is mandatory in biological systems and has a more fundamental status. We try to do this by presenting a series of heuristics that draw from theoretical biology and statistical thermodynamics.

1.1 Overview

This chapter has two sections. In the first, we lay out the theory behind the free-energy principle, starting from a selectionist standpoint and ending with the implications of the free-energy principle for neurobiology. The second section addresses the implementation of free-energy minimisation in hierarchical neuronal architectures and considers some of the key behaviours of brain-like systems that self-organise in accord with the free-energy principle.

2 Theory

In this section we develop a series of heuristics that lead to a variational free-energy principle for biological systems and, in particular, for the brain. We start with evolutionary or selectionist considerations that transform difficult questions about how biological systems operate into simpler questions about constraints on their behaviour. These constraints lead to the important notion of an ensemble density that is encoded by the state of the system. This density is used to construct a free-energy for any system that is in exchange with its environment. We then consider the implications of minimising this free-energy with regard to quantities that determine the system's (i.e., brain's) state and, critically, its action upon the environment. We will see that this minimisation leads naturally to perceptual inference about the world, encoding of perceptual uncertainty (i.e., attention or salience), perceptual learning about the causal structure of the environment and, finally, a principled exchange with, or sampling of, that environment.

In what follows, free-energy becomes a Lyapunov function for the brain. A Lyapunov function is a scalar function of a system's state that decreases with time; it is also referred to colloquially as a Harmony function in the neural network literature (Prince & Smolensky, 1997). Another important Lyapunov function is the *value* function in reinforcement learning and optimal control theory. Value functions are generally optimised to approximate the payoff, reward or loss expected under some policy for changing the system's states. Usually, one tries to infer the Lyapunov function given a system's behaviour or construct a Lyapunov (i.e., value) function to optimise behaviour. However, we address the converse problem: given the Lyapunov function, what would systems that minimise free-energy look like?

2.1 Thermodynamics and biological systems

We start with an apparent anomaly: biological systems and especially neuronal systems appear to contravene the second law of thermodynamics. The second law states that the entropy of closed systems increases with time. Entropy is a measure of disorder or, more simply, the number of ways the elements of a system can be rearranged. The fact that the second law applies only to "closed" systems is quite important because biological systems are open, which means they have the opportunity to resist the second law – but how?

The second law applies to macroscopic or ensemble behaviour. It posits time-irreversible behaviour of a system, despite the fact that its microscopic dynamics can be time-reversible. This apparent paradox is resolved with the fluctuation theorem (see Evans & Searles, 2002). The fluctuation theorem shows that the entropy of small systems can decrease, but as the system's size or the observation time gets longer, the probability of this happening decreases exponentially. The fluctuation theorem is important for non-equilibrium statistical mechanics and includes the second law as a special case. Critically, the fluctuation theorem holds for dissipative, non-equilibrium systems. A dissipative system is an open system that operates far from equilibrium by exchanging energy or entropy with the environment. Recently, the fluctuation theorem has been applied to non-equilibrium transitions between equilibrium states to show how free-energy differences can be computed from thermodynamic path integrals (Crooks, 1999). Equivalent derivations for deterministic systems highlight the close relationship between non-equilibrium free-energy theorems and the fluctuation theorem (Evans, 2003). These non-equilibrium free-energy theorems are of particular interest because they apply to dissipative systems like biological systems.

2.2 The nature of biological systems

If the fluctuation theorem is so fundamental, why do we see order emerging all around us? Specifically, why are living systems apparently exempt from these thermodynamic laws? How do they preserve their order (i.e., configurational entropy),[1] immersed in an environment that is becoming irrevocably more disordered? The premise here is that the environment unfolds in a thermodynamically structured and lawful way, and biological systems embed these laws into their anatomy. The existence of environmental order is assured, at the level of probability distributions, through thermodynamics. For example, although disorder always increases, the second law *per se* is invariant. This invariance is itself a source of order. In short, organisms

---

1 Configurational entropy measures randomness in the distribution of matter in the same way that thermal entropy measures the distribution of energy.

could maintain configurational order if they transcribed physical laws governing their environment into their structure. One might ask how this transcription occurs. However, a more basic question is not how biological systems arise, but what are they?

What is the difference between a plant and a stone? The obvious answer is that the plant is an open non-equilibrium system, exchanging matter and energy with the environment, whereas the stone is an open system that is largely at equilibrium. Morowitz (1968) computed the thermal bonding energy required to assemble a single *Escherichia coli* bacterium. He concluded that "if equilibrium process alone were at work, the largest possible fluctuation in the history of the universe is likely to have been no longer than a small peptide" (p. 68). In short, biological systems must operate far from equilibrium: the flow of matter and energy in open systems allows them to exchange entropy with the environment and self-organise. Self-organisation (Ashby, 1947; Haken, 1983) refers to the spontaneous increase in the internal organisation of open systems. Typically, self-organising systems also exhibit emergent properties. Self-organisation only occurs when the system is far from equilibrium (Nicolis & Prigogine, 1977). The concept of self-organisation is central to the description of biological systems and also plays a key role in chemistry, where is it often taken to be synonymous with self-assembly.[2]

2.2.1 Beyond self-organisation

Biological systems are thermodynamically open, in the sense that they exchange energy and entropy with the environment. Furthermore, they operate far from equilibrium, showing self-organising behaviour (Ashby, 1947; Haken, 1983; Kauffman, 1993; Nicolis & Prigogine, 1977). However, biological systems are more than simply dissipative self-organising systems. They can negotiate a changing or non-stationary environment in a way that allows them to endure over substantial periods of time. This means that they avoid phase-transitions that would otherwise change their physical structure. A key aspect of biological systems is that they act upon the environment to change their position within it, or relation to it, in a way that precludes extremes of temperature, pressure or other external fields. By sampling or navigating the environment selectively, they keep their exchange within bounds and preserve their physical integrity.

---

2 The theory of dissipative structures was developed to understand structure formation in far-from-equilibrium systems. Examples include turbulence and convection in fluid dynamics (e.g., Bénard cells), percolation and reaction-diffusion systems such as the Belousov–Zhabotinsky reaction. Self-assembly is another important example from chemistry that has biological implications (e.g., for pre-biotic formation of proteins). Self-organisation depends on a (reasonably) stationary environment that couples to the system to allow an appropriate exchange of entropy and energy.

It seems as though the special thing about biological systems is that they restrict themselves to a bounded domain of their state-space (i.e., an environmental niche) which admits their existence. In other words, a particular class of agent is its own existence proof and induces a small set of states in which it will be found with high probability and a large number of states in which it is unlikely to be found. For example, a snowflake can only occupy states within a specific temperature range; on falling from the sky, it encounters a phase-boundary and melts. Although the snowflake is self-organising, it is not adaptive, in the sense that it will not navigate its environment to maintain a temperature that is consistent with being a snow-flake (see Figure 13.1). Conversely, snowflakes that maintain their altitude and regulate their temperature may survive indefinitely, with a qualitatively recognisable form. The key difference between the normal and adaptive snowflake is the ability to change their relationship with the environment and attain some equilibrium in their exchange with the world.

Similar mechanisms can be envisaged in an evolutionary setting, wherein systems that avoid phase-transitions will be selected above those that cannot (cf. the selection of chemotaxis in single-cell organisms). By considering the

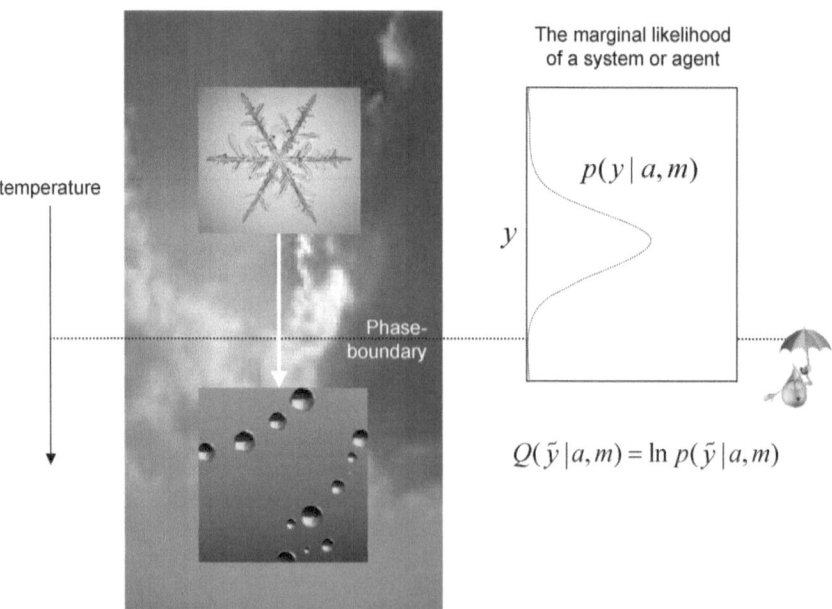

Figure 13.1 Schematic highlighting the idea that dissipative, self-organising systems (like normal snowflakes) are defined by a probabilistic relationship to the environment. By occupying a particular environmental niche, systems are confined to a domain that is far from phase-boundaries. The phase-boundary depicted here is a temperature phase-boundary that would cause the snowflake to melt (i.e., induce a phase-transition).

nature of biological systems in terms of selective pressure, one can replace difficult questions about how biological systems emerge with questions about what behaviours they must exhibit to exist. In other words, selection explains how biological systems arise; the only outstanding issue is what characteristics they must possess. The snowflake example suggests that biological systems act on the environment to preclude phase-transitions. It is therefore sufficient to define a principle that ensures this sort of exchange. We will see that free-energy minimisation is one such principle.

2.3 A population density formulation

To develop these arguments formally, we need to define some quantities that describe an agent, phenotype or system, m, and its exchange with the environment. This exchange rests on quantities that describe the system, the effect of the environment on the system, and the effect of the system on the environment. We will denote these as λ, $\tilde{y}$ and a, respectively. $\tilde{y}$ can be thought of as system states that are caused by environmental forces – for example, the state of sensory receptors. This means that $\tilde{y}$ can be regarded as sensory input. The quantities a represent forces exerted by effectors that act on the environment to change sensory samples. We will represent this dependency by conditioning the sensory samples $p(\tilde{y}) \rightarrow p(\tilde{y} \mid a)$ on action. Sometimes, this dependency can be quite simple – for example, the activity of stretch receptors in muscle spindles is affected directly by muscular forces causing that spindle to contract. In other cases, the dependency can be more complicated – for example, the oculomotor system, controlling eye position, can influence the activity of every photoreceptor in the retina.

The tilde means that $\tilde{y} = y, \dot{y}, \ddot{y}, \ldots$ covers generalised motion in terms of high-order temporal derivatives. This allows a to change the motion or trajectory of sensory input through its higher derivatives by interacting with forces that cause $\tilde{y}$. We will call these environmental causes δ. This formulation means that sensory input is a generalised convolution of the action and unknown or hidden causes. We will unpack these quantities later. At the moment, we will simply note that they can be high-dimensional and time-varying.

Figure 13.2 shows these quantities in terms of a Bayesian graph or network. Here, directed connections indicate conditional dependence. The key thing to note is that the quantities describing the state and structure of the system, λ, are insulated from environmental causes by the sensory and action variables; technically speaking, these are the system's Markov blanket.[3] The Markov blanket of any node is sufficient to predict that node's behaviour. This means that changes in the state or structure of an agent

---

3 The Markov blanket for a node, in a Bayesian network, is the set of nodes comprising its parents, its children, and its children's parents.

Exchange with the environment

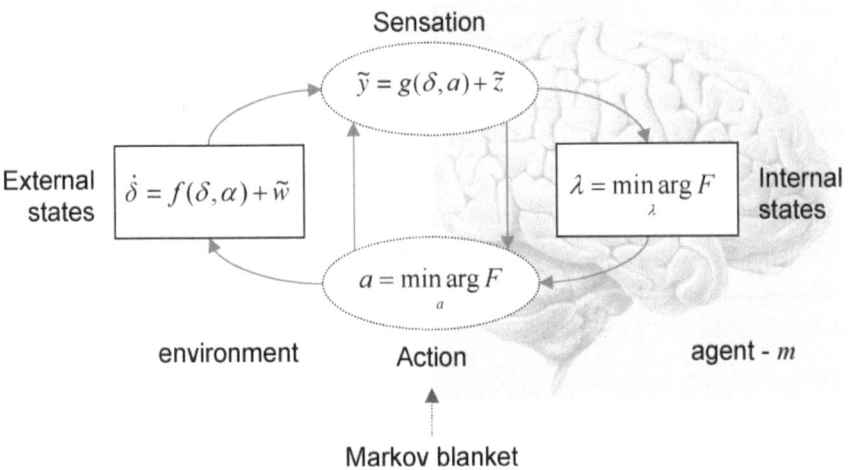

Sensation

$$\tilde{y} = g(\delta, a) + \tilde{z}$$

External states
$$\dot{\delta} = f(\delta, \alpha) + \tilde{w}$$

$$\lambda = \min_{\lambda} \arg F$$ Internal states

$$a = \min_{a} \arg F$$

environment   Action   agent - *m*

Markov blanket

Figure 13.2 Schematic detailing the quantities that define the free-energy. These quantities refer to the internal configuration of the brain and quantities that determine how a system is influenced by the environment. This influence is encoded by the variables ỹ that could correspond to sensory input or any other changes in the systems state due to external environmental forces or fields. The parameters *a* correspond to physical states of the system that change the way the external forces act upon it or, more simply, change the way the environment is sampled. A simple example of these would be the state of ocular motor systems controlling the direction of eye gaze. $p(\tilde{y} \mid \delta, a)$ is the conditional probability of sensory input given its causes, δ, and the state of effectors (i.e., action).

can, in principle, be expressed as a function of $p(\tilde{y} \mid a)$. The free-energy is one such function.

2.3.1 Surprise and value

The basic premise we started with is that biological systems must keep ỹ within bounds (i.e., phase-boundaries) through adaptive changes in *a*. Put simply, adaptive systems or agents should minimise unlikely or surprising exchanges with the environment. We can express this more formally by requiring adaptive systems to minimise surprise, or maximise value:

$$Q(\tilde{y} \mid a) = \ln p(\tilde{y} \mid a, m) \tag{1}$$

In fact, it is fairly simple to show, using the Fokker–Planck equation, that any member of a population, whose population density $p(\tilde{y} \mid m)$ is at equilibrium, must increase $Q(\tilde{y} \mid a)$, on average. Heuristically; if an ensemble or population of agents is in equilibrium (i.e., the probability of finding a particular sort of agent in the particular state does not change with time), each agent must oppose the dispersive effects of random fluctuations in their states. This implies that each agent must move towards the most densely populated regions of state-space to offset any dispersion. This entails increasing $Q(\tilde{y} \mid a)$ so that the population density shrinks at the same rate it is dispersed.

The surprise, $- \ln p(\tilde{y} \mid a, m)$, measures the conditional improbability of exchange given a particular agent and its action. Each point in the space of exchange, $\tilde{y}, a \in \mathcal{R}^{|\tilde{y}| \times |a|}$, will have a measure of this sort, which will be high if the exchange is compatible with m and low if not (i.e., high in domains populated by m). More intuitively, we would be surprised to find particular agents in some environments (e.g., a snowflake in a sauna). In a selectionist setting, the quantity $Q(\tilde{y} \mid a)$ could be regarded as the *adaptive value* of a particular exchange. From a statistical perspective, $Q(\tilde{y} \mid a)$ is also known as the *log-evidence* or marginal likelihood (marginal because it obtains by integrating out dependencies on the causes, δ). These two perspectives are useful because they link evolutionary selection in theoretical biology to Bayesian model selection in machine learning; we will exploit this link below by treating the system or agent as a model of its sensory input.

Finally, $Q(\tilde{y} \mid a)$ also plays the role of the value in microeconomics, value-learning and optimal control theory. Value-learning is a branch of computational neuroscience that deals with the reinforcement of actions and optimisation of policies; it borrows from optimal control theory to explain optimal behaviour. In this context, the value function $Q(\tilde{y} \mid a)$ encodes the expected reward or payoff, $\ell(\tilde{y} \mid a)$, under a policy prescribing the trajectory though state-space (i.e., a path integral, which may or may not be discounted the further one integrates into the future). The basic idea is to find an optimum value function that can be used to guide action from any state. In the current context, this value function is given and the underlying payoff or loss-function $\ell(\tilde{y} \mid a)$ is implicit; the fact that a trajectory has a high log-likelihood (see Equation 1) means it visits rewarding states that are necessary for the agent's survival. There are a number of interesting connections between the free-energy principle and optimal control theory. For example, both surprise and value can be formulated as energies (see Todorov, 2006). Furthermore, the free-energy principle speaks to an estimation-control duality: the best-known example of this is when the Kalman filter and the linear-quadratic regulator become identical, under certain transformations (Kalman, 1960). The outstanding issue here is, how can a system minimise its surprise?

2.3.2 A free-energy bound

To optimise surprise or value, we require action to maximise[4]

$$Q(\tilde{y} \mid a) = \ln \int p(\tilde{y}, \delta \mid a) d\delta \qquad (2)$$

where $p(\tilde{y}, \delta \mid a)$ is the joint density of environmental effects and their unknown causes, conditioned on action. However, this maximisation must be accomplished by changes in action, which can only be a function of $\tilde{y}$ and the internal states, λ, of the agent, because these are the only variables to which it has access.

Clearly, the system cannot perform the integration in Equation 2, because it does not know the causes. However, it can optimise a bound on the integral using a relatively simple gradient descent. One such bound is the free-energy, which is a scalar function of sensory and internal states[5]

$$
\begin{aligned}
F(\tilde{y}, \lambda \mid a) &= -\langle \ln p(\tilde{y}, \delta \mid a) \rangle_q + \langle \ln q(\delta; \lambda) \rangle_q \\
&= -\int q(\delta; \lambda) \ln \frac{p(\tilde{y}, \delta \mid a)}{q(\delta; \lambda)} d\delta \qquad (3) \\
&\geq -\ln \int q(\delta; \lambda) \frac{p(\tilde{y}, \delta \mid a)}{q(\delta; \lambda)} d\delta = -Q(\tilde{y} \mid a)
\end{aligned}
$$

The inequality is an example of something called Jensen's inequality, which follows simply from the concavity of the log function. To make this bound, $F(\tilde{y}, \lambda \mid a)$, a function of internal states λ, we have introduced $q(\delta; \lambda)$, which is an arbitrary density function on the causes that is encoded by the system's internal states. Usually, $q(\delta; \lambda)$ is called an *ensemble density*[6] and can be regarded as the probability density that the causes δ would be selected from an ensemble of environments. For example, λ could be the mean and variance of a Gaussian distribution on temperature, δ.

The free-energy (i.e., the bound) above comprises two terms. The first is the energy expected under the ensemble density. This energy is simply the surprise or information about the joint occurrence of the sensory input and its causes. The second term is the negative entropy of the ensemble density. Notice that action can be considered causes of sensory input that are not covered by the ensemble density. In what follows, we look at the ensemble density and its role in adaptive behaviour.

---

4 Dropping the dependency on *m* for clarity.
5 $\langle . \rangle_q$ means the expectation under the density *q*.
6 In statistical physics, an "ensemble" denotes a fictitious collection of replicas of the system in question, each of which represents a possible state that the system might be in.

2.3.3 The ensemble and generative densities

The free-energy formulation in Equation (3) has a fundamental implication: systems that minimise the surprise of their interactions with the environment by adaptive sampling can only do so by optimising a bound, which is a function of the system's states. Formulating that bound in terms of Jensen's inequality requires that function to be a probability density, which links the system's states to the hidden causes of its sensory input. In other words, the system is compelled to represent the causes of its sensorium. This means that adaptive systems, at some level, represent the state and causal architecture of the environment in which they are immersed. Conversely, this means that causal regularities in the environment are transcribed into the system's configuration.

   Note that the free-energy is defined by two densities; the ensemble density, $q(\delta;\lambda)$, and the generative density, $p(\tilde{y},\delta \mid a)$, from which one could *generate* sensory samples and their causes. The generative density factorises into a likelihood and prior density, $p(\tilde{y} \mid \delta,a)p(\delta)$, which specify a generative model. This means the free-energy formulation induces a generative model for any system and an ensemble density over the causes or parameters of that model. The functional form of these densities is needed to evaluate the free-energy. We will consider in the next section functional forms that may be employed by the brain. For the moment, we will just note that these forms enable the free-energy to be evaluated as a function, $F(\tilde{y},\lambda \mid a)$, of the system's sensory input and internal state. Figure 13.2 shows a schematic of the quantities introduced so far; and how they relate to each other.

2.4 The free-energy principle

The free-energy principle states that all the quantities that can change – that is, that are part of the system – will change to minimise free-energy. These quantities are the internal parameters λ and the action parameters a. This principle, as we will see below, is sufficient to account for adaptive exchange with the environment by ensuring that a bound on adaptive value is optimised. We now consider the implications of minimising the free-energy with respect to λ and a, respectively.

2.4.1 Perception: optimising λ

Clearly, if action is to minimise surprise, the free-energy bound should be reasonably tight. A tight bound is assured when the free-energy is minimised with respect to internal parameters. In this case, it is fairly easy to show that the ensemble density approximates the conditional density of the environmental causes, given the sensory samples. This can be seen by rearranging Equation 3 to show the dependence of the free-energy on λ.

$$F = -\ln p(\tilde{y} \mid a) + D(q(\vartheta;\lambda) \parallel p(\vartheta \mid \tilde{y},a)) \tag{4}$$

Only the second term is a function of λ; this is a Kullback–Leibler cross-entropy or divergence that measures the difference between the ensemble density and the conditional density of the causes. Because this measure is always positive, minimising the free-energy corresponds to making the ensemble density the same as the conditional density, at which point the free-energy becomes the surprise: $F = -Q = -\ln p(\tilde{y} \mid a)$. This is quite a fundamental result that underlies free-energy optimisation schemes in statistical physics and machine learning and rests on the fact that the divergence cannot be less than zero (in the sense that a distance cannot be negative). This means that if one has minimised the free-energy, one has implicitly minimised surprise, because the second term in Equation 4 will be zero.

Put simply, when the free-energy is minimised, the ensemble density encoded by the system's parameters becomes an approximation to the posterior probability of the causes of its sensory input. This means the system implicitly infers the causes of its sensory samples. Clearly, this approximation depends on the physical structure of the system and the implicit form of the ensemble density and on how closely this matches the causal structure of the environment. Those systems that can match their internal structure to the causal structure of the environment will attain a tighter bound (see below).

2.4.2 Action: optimising α

Changing the system to move or re-sample the environment by minimising the free-energy with respect to action enforces a sampling of the environment that is consistent with the ensemble density. This can be seen with a second rearrangement of Equation 3 that shows how the free-energy depends upon a:

$$F = -\langle \ln p(\tilde{y} \mid \vartheta,a) \rangle_q + D(q(\vartheta) \parallel p(\vartheta)) \tag{5}$$

In this instance, only the first term is a function of action. Minimising this term corresponds to maximising the log-probability of sensory input, expected under the ensemble density. In other words, the system will sample sensory inputs that are the most likely under the ensemble density. However, as we have just seen, the ensemble density approximates the conditional distribution of the causes, given sensory inputs. This inherent circularity obliges the system to fulfil its own expectations. In other words, the system will expose itself selectively to causes in the environment that it expects to encounter. However, these expectations are limited to the repertoire of physical states the system can occupy, which specify the ensemble density. Therefore, systems with a low free-energy can only sample parts of the environment they can encode with their repertoire of physical states.

Because the free-energy is low, the inferred causes approximate the real causes. This means the system's physical state must be sustainable under these causes, because each system is its own existence proof (where a system can be any unit of selection – that is, a phenotype or a species). In short, low free-energy systems will look like they are responding adaptively to changes in the external or internal milieu, to maintain a stationary or homeostatic exchange with the environment.

This chapter is concerned largely with perceptual inference and learning in neural systems. However, there are many intriguing issues that arise if the free-energy principle is served by sampling from the environment selectively to maximise the predictability of sensory input. This sort of behaviour is found in many biological systems, ranging from the chemotactic movement of single-cell organisms to the phototropic behaviour of plants. In nervous systems there are numerous examples of sensory homeostasis, ranging from simple reflexes that reverse proprioceptive perturbations, to smooth-pursuit eye movements responsible for stabilisation of the retinal image. Heuristically, these mechanisms can be viewed as suppressing free-energy by re-sampling the environment to minimise the prediction error incurred by a mismatch between what is sampled and the prediction afforded by perceptual inference. This suggests that motor and sensory systems in the brain should be in intimate anatomic relation. This is the case at spinal, subcortical and cortical levels. For example, the primary motor and sensory cortex are juxtaposed along the central sulcus and are strongly interconnected (Huffmann & Krubitzer, 2001). Similarly, at a subcortical level, the superior collicullus represents a point of convergence for sensory information (through direct projections from the retina) and visual predictions (from visual, parietal and frontal cortex to the intermediate and deep layers). Neuronal discharges in the deep layers, which initiate saccades, define motor fields that coincide with visual receptive fields in the superficial layers (Andersen, 1989).

In summary, the free-energy principle can be motivated, quite simply, by noting that systems that minimise their free-energy respond to environmental changes adaptively. It follows that minimisation of free-energy may be a necessary, if not sufficient, characteristic of evolutionary successful systems. The attributes that ensure that biological systems minimise their free-energy can be ascribed to selective pressure, operating at somatic (i.e., the lifetime of the organism) or evolutionary timescales (Edelman, 1993). These attributes include the functional form of the densities entailed by the system's architecture. Systems that fail to minimise free-energy will have sub-optimal representations or ineffective mechanisms for action and perception. These systems will not restrict themselves to specific domains of their milieu and may ultimately experience a phase-transition (e.g., death). Note that in this formulation, adaptive action depends on perception; perception *per se* is only necessary to ensure a tight bound on the value-function minimised by action. Before returning to selective mechanisms, we will unpack the quantities describing the system and relate their dynamics to processes in the brain.

2.4.3 The mean-field approximation

Clearly, the quantities describing hidden causes in the environment could be enormous in number and variety. A key difference among them is the time-scales over which they change. We will use this separability of timescales to partition causes into three sets $\delta = \delta_u$, δ_y, δ_θ that change on a timescale of milliseconds, seconds and minutes, and we will factorise the ensemble density in terms of marginal densities:

$$q(\delta) = \prod_i q(\delta_i; \lambda_i)$$

$$= q(\delta_u; \lambda_u)q(\delta_y; \lambda_y)q(\delta_\theta; \lambda_\theta) \tag{6}$$

This induces a partitioning of the system's parameters into $\lambda = \lambda_u, \lambda_y, \lambda_\theta$, which encode time-varying marginals of the ensemble density (i.e., represent the causes). The first, λ_u, are system quantities that change rapidly. These could correspond to neuronal activity or electromagnetic states of the brain that change with a timescale of milliseconds. The causes δ_u they encode could correspond to evolving environmental states – for example, changes in the environment caused by structural instabilities or other agents. The second partition, λ_y, changes more slowly, over seconds. These could correspond to the kinetics of molecular signalling in neurons – for example, calcium-dependent mechanisms underlying short-term changes in synaptic efficacy and classical neuromodulatory effects. The equivalent partition of causes in the environment may be contextual in nature, such as the level of radiant illumination or slowly varying fields that set the context for more rapid fluctuations in its state. Finally, λ_θ represent system quantities that change slowly – for example, long-term changes in synaptic connections during experience-dependent plasticity, or the deployment of axons that change on a neurodevelopmental timescale. The corresponding environmental quantities are (relatively) invariant aspects of its causal architecture. These could correspond to physical laws and other structural regularities that shape our interactions with the world.

In statistical physics, the factorization in Equation 6 is known as a mean-field approximation.[7] Clearly, our approximation with these marginal densities is a little arbitrary, but it helps organise the functional correlates of their respective optimisation in the nervous system. More precisely, we are assuming that the brain uses the same mean-field approximation used above because it has evolved to exploit the ensuing computational efficiency; the

---

7 The basic idea of a mean-field approximation is to approximate a very high dimensional probability distribution with the product of a number of simpler (marginal) densities. This is often used to cope with problems that are otherwise computationally or analytically intractable.

mean-field approximation greatly finesses the minimisation of free-energy when considering particular mechanisms. These schemes usually employ variational techniques.[8]

2.5 Optimising variational modes

We now revisit in more detail optimisation of system parameters that underlie perception, using the mean-field approximation. Because variational techniques predominate in this approximation, the free-energy in Equation 3 is also known as the variational free-energy, and λ_i are called variational parameters. The mean-field factorisation means that the approximation cannot cover the effect of random fluctuations in one partition on the fluctuations in another. However, this is not a severe limitation, because these effects are modelled through mean-field effects (i.e., through the means of random fluctuations). This approximation is particularly easy to motivate in the present framework, because random fluctuations at fast timescales are unlikely to have a direct effect at slower timescales and their influence can be sensibly approximated with their average.

Using variational calculus, it is simple to show (see Friston, Mattout, Trujillo-Barreto, Ashburner, & Penny, 2007) that, under the mean-field approximation above, the marginal ensemble densities have the following form

$$q(\delta_i) \propto \exp(V(\delta_i))$$

$$V(\delta_i) = \langle \ln p(\tilde{y}, \delta) \rangle_{q_{\setminus i}} \tag{7}$$

where $V(\delta_i)$ is simply the log-probability of the input and its causes δ_i, expected under $q_{\setminus i}$, the ensemble density of the other partitions (or, more precisely, its Markov blanket). We will call this the variational energy. From Equation 7, it is evident that the mode (highest point) of the ensemble density maximises the variational energy. The mode is an important variational parameter. For example, if we assume $q(\delta_i)$ is Gaussian, then it is parameterised by two variational parameters, $\lambda_i = \mu_i, \Sigma_i$, encoding the mode or expectation and covariance, respectively. This is known as the Laplace approximation and will be used later. In what follows, we will focus on minimising the free-energy by optimizing μ_i, noting that there may be other variational parameters describing higher moments. Fortunately, under the Laplace approximation,

8 Variational techniques were introduced by Feynman (1972), in the context of quantum mechanics, using the path integral formulation. They have been adopted widely by the machine-learning community (e.g., Hinton & von Camp, 1993; MacKay, 1995). Established statistical methods like expectation maximisation and restricted maximum likelihood (Dempster, Laird, & Rubin, 1977; Harville, 1977) can be formulated in terms of free-energy (Friston et al., 2006; Neal & Hinton, 1998).

the only other variational parameter required is the covariance. This has a simple form, which is an analytic function of the mode and does not need to be represented explicitly (see Friston et al., 2007). We now look at the optimisation of the variational modes μ_i, which represent the most likely causes of sensory input, and at the neurobiological and cognitive processes this optimisation entails:

2.5.1 Perceptual inference: optimising μ_u

Minimising the free-energy with respect to neuronal states μ_u means maximising $V(\delta_u)$:

$$\mu_u = \max V(\delta_u)$$

$$V(\delta_u) = \langle \ln p(\tilde{y}\,|\,\delta, a) + \ln p(\delta) \rangle_{q_\gamma q_\theta} \tag{8}$$

The free-energy principle is served when the variational mode of the states (i.e., neuronal activity) changes to maximise the posterior probability of the causes. Equation 8 shows that this can be achieved, without knowing the true posterior, by maximising the expected log-likelihood and prior that specify a probabilistic generative model (second line). As mentioned above, this optimisation requires the functional form of the generative model. In the next section, we will look at hierarchical forms that are commensurate with the structure of the brain. For now, it is sufficient to note that the free-energy principle means that brain states will come to encode the most likely causes in the environment generating sensory input.

2.5.2 Generalised coordinates

Because states are time-varying quantities, it is important to think about what their ensemble density covers. This includes not just the states at one moment in time, but their high-order motion. In other words, a particular state of the environment and its probabilistic encoding can embody dynamics by representing the paths or trajectories of states in generalised coordinates. Generalised coordinates are a common device in physics and normally cover position and momentum. In the present context, a generalised state includes the current state and its generalised motion $\delta_u = u, u', u'', \ldots$ (i.e., the state and its first, second, etc. derivatives with time), with corresponding variational modes $\mu_u, \mu_u', \mu_u'', \ldots$. It is fairly simple to show (Friston, Trujillo-Barreto, & Daunizeau, 2008) that the optimisation in Equation 8 can be achieved with a rapid gradient descent, while coupling high- to low-order motion via mean-field terms

$$\dot{\mu}_u = \kappa\, \partial V(\delta_u)/\partial u + \mu_u'$$

$$\dot{\mu}_u' = \kappa\, \partial V(\delta_u)/\partial u' + \mu_u''$$

$$\ddot{\mu}''_u = \kappa \, \partial V(\delta_u)/\partial u'' + \mu'''_u$$

$$\dot{\mu}'''_u = \ldots \tag{9}$$

Here, $\dot{\mu}_u$ means the rate of change of μ_u and κ is some suitable rate constant. The simulations in the next section use this descent scheme, which can be implemented using relatively simple neural networks. Note that when the conditional mode has found the maximum of $V(\delta_u)$, its gradient is zero and the motion of the mode becomes the mode of the motion – that is, $\dot{\mu}_u = \mu'_u$. However, it is perfectly possible, in generalised coordinates, for these quantities to differ. At the level of perception, psychophysical phenomena suggest that we use generalised coordinates, at least perceptually – for example, on stopping, after looking at scenery from a moving train, the world is perceived as moving but does not change its position. The impression that visual objects change their position in accord with their motion is something that we have learned about the world. It is also something that can be unlearned, temporarily (e.g., perceptual after-effects). We now turn to how these causal regularities are learned.

2.5.3 Perceptual context and attention: optimising μ_γ

If we call the causes that change on an intermediate timescale, δ_γ contextual, then optimizing μ_γ corresponds to encoding the contingencies under which the fast dynamics of states evolve. We will assume these quantities pertain not to the states of the world *per se*, but to the amplitude of their random fluctuations. This is critical because at least one set of the agent's internal states must be devoted to representing uncertainly about sensory input and the magnitude of random effects on the causes themselves. Under Gaussian assumptions, it is sufficient to parameterise the expectation and variance (or inverse precision) of the likelihood model of sensory data. The expectations are encoded by $\mu_u, \mu'_u, \mu''_u, \ldots$ above, so we will treat μ_γ as precision parameters.

The optimisation of these precisions can proceed as above; however, we can assume that the context changes sufficiently slowly that we can make the approximation $\mu'_\gamma = 0$. Because, these variational parameters change more slowly than the neuronal states, the free-energy may change substantially over time. This means the variational parameters optimise the sum of free-energy over time.[9] This gives the simple gradient ascent:

$$\dot{\mu}_\gamma = \kappa \int \partial V(\delta_\gamma)/\partial \delta_\gamma \, dt$$

$$V(\delta_\gamma) = \langle \ln p(\tilde{y}, \delta) \rangle_{q_u q_\theta} \tag{10}$$

9 In the simulations below, we use peristimulus time. The integral of energy over time is known as action, which means that, strictly speaking, it is variational action that is optimised (see below).

We will see later that the conditional mode μ_γ encoding precision might correspond to the strength of lateral interactions among neurons in the brain. These lateral interactions control the relative effects of top-down and bottom-up influences on perceptual inference. This suggests that attention could be thought of in terms of optimising contextual parameters of this sort. In the present context, the influence of context on perceptual inference can be cast in terms of encoding uncertainty. We will look at neuronal implementations of this in the next section.

2.5.4 Perceptual learning: optimising μ_θ

Optimising the variational mode encoding δ_θ corresponds to inferring and learning structural regularities in the environment's causal architecture. As above, this learning can be implemented as a gradient ascent on the time integral of $V(\delta_\theta)$, which represents an expectation under the ensemble density encoding the generalised states and context:

$$\dot{\mu}_\theta = \kappa \int \partial V(\delta_\theta)/\partial \delta_\theta dt$$
$$V(\delta_\theta) = \langle \ln p(\tilde{y}, \delta) \rangle_{q_u q_\gamma}. \tag{11}$$

In the brain, this descent can be formulated as changes in connections that are a function of pre-synaptic prediction and post-synaptic prediction error (see Friston, 2003, 2005; see also the next section). The ensuing learning rule conforms to simple associative plasticity or, in dynamic models, spike-timing dependent plasticity. In the sense that optimising the variational parameters that correspond to connection strengths in the brain encodes causal structure in the environment, this instance of free-energy minimisation corresponds to learning. The implicit change in the brain's connectivity endows it with a memory of past interactions with the environment that affects the free-energy dynamics underlying perception and attention. This is through the mean-field effects in Equation 8 and Equation 10. Put simply, sustained exposure to environmental inputs causes the internal structure of the brain to recapitulate the causal structure of those inputs. In turn, this enables efficient perceptual inference. This formulation provides a transparent account of perceptual learning and categorisation, which enables the system to remember associations and contingencies among causal states and context.

2.5.5 Implications of the mean-field approximation

The mean-field approximation allows for inferences about perceptual states, context and causal regularities, without representing the joint distribution over states, context and structure explicitly (see Rao, 2005). However, the optimisation of one set of variational parameters is a function of the ensemble densities on the others (see Equations 9, 10 and 11). This has a fundamental implication: it means that irrespective of the mean-field partition, changes in one set depend on the remaining sets. This is almost obvious for two sets. For example, concepts like "activity-dependent plasticity" and "functional segregation" speak to the reciprocal influences between changes in states and connections (see Figure 13.3), in that changes in connections depend upon activity and changes in activity depend upon connections. The situation gets more interesting when we consider three sets: internal states encoding precision must be *affected by and affect both activity and plasticity.* This places strong constraints on the neurobiological candidates for these precision parameters. Happily, the ascending neuromodulatory neuro-transmitter systems, such as dopaminergic and cholinergic projections, have exactly the right characteristics. They are driven by activity in pre-synaptic connections and, crucially, can affect activity through classical neuromodulatory effects at the post-synaptic membrane (e.g., Tseng & O'Donnell, 2004), while also enabling short- and long-term potentiation of connection strengths

Mean-field interactions

$$q(\delta;\lambda) = q(u;\lambda_u)q(\theta;\lambda_\theta)q(\gamma;\lambda_\gamma)$$

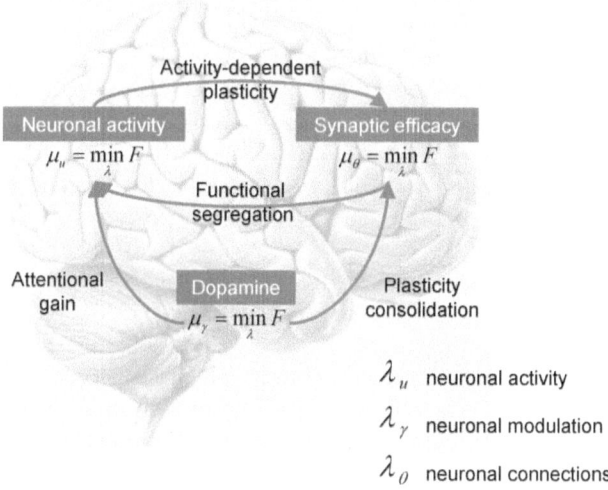

Figure 13.3 The ensemble density and its mean-field partition: $q(\delta;\lambda)$ is called an ensemble density and is encoded by the system's internal parameters, λ. These parameters (e.g., mean or expectation) change to minimise free-energy, F, and, in so doing, make the ensemble density an approximate conditional density on the causes of sensory input. The mean-field partition corresponds to a factorisation over the sets comprising the partition. Here we have used three sets (neural activity, modulation and connectivity). Critically, the optimisation of the parameters of any one set depends on the parameters of the other sets. In this figure, we have focused on means or expectations, μ_i, of the marginal densities.

(Brocher, Artola, & Singer, 1992; Gu, 2002). Interestingly, the dynamics of these systems have time constants that are intermediate between fast neuronal dynamics and slow plasticity. Furthermore, it is exactly these systems that have been implicated in value-learning (Friston, Tononi, Reeke, Sporns, & Edelman, 1994; Montague, Dayan, Person, & Sejnowski, 1995), attention and the encoding of uncertainty (Niv, Duff, & Dayan, 2005; Schultz, 2007; Yu & Dayan, 2005). Before turning to optimisation in the brain, we will consider the optimisation not of internal states, but of the agents themselves.

2.6 Model optimisation

Hitherto, we have considered only the quantitative optimisation of variational parameters given a particular agent and its implicit generative model. Exactly the same free-energy principle can be applied to optimise the agent itself. Different agents or models can come from populations of models or from qualitative changes in one model over time. A model here corresponds to a particular architecture that can be enumerated with the same set of variational parameters. Removing a part of the model or adding, for example, a synaptic connection changes the model and the variational parameters in a qualitative or categorical fashion.

Model optimisation involves maximising the marginal likelihood of the model itself. In statistics and machine learning, this is equivalent to Bayesian model selection, where the free-energy is used to approximate the log-evidence or marginal likelihood, $Q \approx \ln p(\tilde{y} \mid m_i)$, for a particular agent or model, m_i. This approximation can be motivated easily using Equation 4: if the system has minimised its free-energy and the divergence term is nearly zero, then the free-energy approaches the negative log-evidence; therefore, agents that maintain a low free-energy are likely to have a high marginal likelihood.

An evolutionary perspective might consider $Q \approx \ln p(\tilde{y} \mid m_i)$ in terms of adaptive fitness, which is defined for any agent's exchange with the environment and is independent of its internal state, λ. An adaptive agent will keep this exchange within bounds that ensure its physical integrity. Agents that fail to suppress free-energy will encounter surprising interactions with the environment that may remove them from the population. Notice that the ensuing hierarchical selection rests upon interplay between optimising the parameters of each model and optimising an ensemble of models. Optimisation at both levels is prescribed by the free-energy principle. In the theory of genetic algorithms, similar schemes are referred to as hierarchical co-evolution (e.g., Maniadakis & Trahanias, 2006). A similar relationship is found in Bayesian inference, where model selection is based on the free-energy bound on model evidence that is furnished by optimising the parameters of each model. In short, free-energy may be a useful surrogate for adaptive fitness in an evolutionary setting and for the log-evidence in model selection.

In short, within an agent's lifetime its parameters minimise free-energy, given the model implicit in its phenotype. At a supraordinate level, the agents themselves may be selected, enabling the population to explore model space and find optimal models. This exploration depends on the heritability of key model components, which could be viewed as priors about environmental niches they can model.

2.6.1 Summary

The above arguments suggest that biological systems sample their environment to fulfil expectations that are generated by the model implicit in their structure. The free-energy principle explains adaptive behaviour without invoking notions of reinforcement or operant conditioning: From the point of view of the agent, it is simply sampling the environment so that its sensory input conforms to its expectations. From its perspective, the environment is an accommodating place; fluctuations or displacements caused by environmental forces are quickly explained away by adaptive re-sampling. Because action is not encoded by the ensemble density, these adaptive responses may not be perceived. However, for someone observing this system, it will appear to respond adaptively to environmental changes and avoid adverse conditions. In other words, it will seem as if certain stimulus–response links are selectively reinforced to ensure the homeostasis of its internal milieu, where this reinforcement emerges spontaneously in the larger context of action and perception under the free-energy principle.

The assertion that adaptive systems should minimise unlikely or surprising exchanges with the environment may seem implausible at first glance. The key thing to note here is that surprise is conditioned on the agent's model of the world: it is the surprise given the expectations embodied in its phenotype or current state. Clearly, if a phenotype expects to die and it conforms to a free-energy principle, it will die. The argument is that when natural selection operates on a population, such phenotypes will disappear, leaving those that expect to live (there may be exceptions to this, if death entails progeny; other interesting exceptions are phase-transitions in developmental trajectories, e.g., in metamorphic insects).

It might be thought that the relationship between value and surprise is ambiguous, in the sense that some valuable events are *surprising* (e.g., winning a lottery), whereas value is the converse of surprise. Again, this is resolved by noting that surprise is conditional on the agent (who may have entered the lottery); it is not simply something that is unlikely to happen (e.g., winning lottery you had not entered). In this sense, it is conceptually related to "Bayesian surprise", invoked to explain visual search and the deployment of attention (Itti & Baldi, 2006). The definition of Bayesian surprise rests on the divergence between the prior and conditional densities elaborated during perceptual inference. This again emphasises the role of prior expectations in shaping surprise or value. The distinction between surprise and unlikely events suggests that, *a priori*, we expect to be rich and are chronically surprised that we are not (but this surprise could be alleviated by wining a lottery).

A further counter-intuitive aspect of minimising surprise is that it seems to preclude exploratory behaviour, novelty-seeking and risk-taking. However, optimisation of free-energy may engage different mechanisms at different timescales. Below, we will focus on dynamics and gradient descent that may be used in the brain. However, at an ethological or executive level, different schemes may operate – for example, stochastic explorations of the free-energy function (cf. genetic algorithms). This would entail sampling the environment in a stochastic fashion to find samples with the least surprise. From an observer's point of view, this would appear like random or exploratory behaviour. From the agent's point of view, everything is surprising, so it might as well sample randomly until something familiar is encountered. The trade-off between exploration and exploitation is a central theme in evolutionary theory, learning theory, microeconomics and optimisation theory (e.g., March, 1991) and can be applied easily to free-energy functions.

In this section, we have developed a free-energy principle for the evolution of an agent's state and structure and have touched upon minimisation of free-energy at the population level, through hierarchical selection. Minimising free-energy corresponds to optimising an agent's configuration, which parameterises an ensemble density on the causes of sensory input, and optimising the agent itself in somatic or evolutionary time. Factorisation of the ensemble density to cover quantities that change on different timescales provides an ontology that maps nicely onto perceptual inference, attention and learning. In the next section, we consider how the brain might instantiate the free-energy principle, with a special focus on the likelihood models implied by its structure.

3 Generative models in the brain

In this section, we will look at how the rather abstract principles of the previous section might be applied to the brain. We have already introduced the idea that a biological structure encodes a model of its environment. We now look at the form of these models implied by the structure of the brain and try to understand how evoked responses and associative plasticity emerge naturally with minimisation of free-energy. In the current formulation, attributes or quantities describing the brain parameterise an ensemble density of environmental causes. To evaluate the free-energy of this density we need to specify the functional form of the ensemble and generative densities. We will assume a Gaussian form for the ensemble densities (i.e., the Laplace approximation), which is parameterised by its mode or expectation and covariance. The generative density is specified by its likelihood and priors. Together these constitute a generative model. If this model is specified properly, we should be able to predict, using the free-energy principle, how the brain behaves in different contexts. In a series of previous papers (e.g., Friston, 2003, 2005; Friston & Price, 2001) we have described the form of hierarchical generative models that might be employed by the brain. In this section, we will cover briefly the main points again.

3.1 Perception and sensation

This section is about trying to understand cortical responses in terms of perceptual inference and learning. The specific model considered here rests on empirical Bayes, using generative models that are embodied in cortical hierarchies. This model can be regarded as a mathematical formulation of the long-standing notion (Locke, 1690/1976) that "our minds should often change the idea of its sensation into that of its judgement, and make one serve only to excite the other". In a similar vein, Helmholtz (1860/1962) distinguished between perception and sensation: "It may often be rather hard to say how much from perceptions as derived from the sense of sight is due directly to sensation, and how much of them, on the other hand, is due to experience and training" (see Pollen, 1999). In short, there is a distinction between percepts, which are the products of recognising the causes of sensory input, and sensation *per se*. Recognition – that is, inferring causes from sensation – is the inverse of generating sensory data from their causes. It follows that recognition rests on models, learned through experience, of how sensations are caused. In this section, we will consider hierarchical generative models and how cortical responses can be understood as part of the recognition process. The particular recognition scheme we will focus on is empirical Bayes, where prior expectations are abstracted from the sensory input, using a hierarchical model of how those data were caused.

Conceptually, empirical Bayes and generative models are related to "analysis-by-synthesis" (Neisser, 1967). This approach to perception, from cognitive psychology, involves adapting an internal model of the world to match sensory input and was suggested by Mumford (1992) as a way of understanding hierarchical neuronal processing. The idea is reminiscent of MacKay's epistemological automata (MacKay, 1956), which perceive by comparing expected and actual sensory input (Rao, 2005). These models emphasise the role of backward connections in mediating predictions of lower level input, based on the activity of higher cortical levels. Recognition is simply the process of solving an inverse problem, by jointly minimising prediction error (i.e., free-energy) at all levels of the cortical hierarchy. This perspective explains many physiological and behavioural phenomena – for example, extra-classical receptive field effects and repetition suppression in unit recordings, the mismatch negativity (MMN) and P300 in event-related potentials (ERPs), priming and global precedence effects in psychophysics. Critically, many of these emerge from the same basic principles governing inference with hierarchical generative models.

To finesse the inverse problem, posed by non-invertible generative models, constraints or priors are required. These resolve the ill-posed problems that confound recognition based on purely forward architectures. It has long been assumed that sensory units adapt to the statistical properties of the signals to which they are exposed (for review, see Simoncelli & Olshausen, 2001). The Bayesian framework for perceptual inference has its origins in Helmholtz's notion of perception as unconscious inference. Helmholtz realised that retinal images are ambiguous and that prior knowledge was required to account for perception (Kersten et al., 2004). Kersten et al. (2004) provide an excellent review of object perception as Bayesian inference and ask a fundamental question: "Where do the priors come from? Without direct input, how does image-independent knowledge of the world get put into the visual system?" We now answer this question and show how empirical Bayes allows most of the necessary priors to be learned and induced online, during inference.

3.2 Hierarchical dynamic models in the brain

A key architectural principle of the brain is its hierarchical organisation (Felleman & Van Essen, 1991; Hochstein & Ahissar, 2002; Mesulam, 1998; Zeki & Shipp, 1988). This organisation has been studied most thoroughly in the visual system, where cortical areas can be regarded as forming a hierarchy, with lower areas being closer to primary sensory input and higher areas adopting a multimodal or associational role. The notion of a hierarchy rests on the distinction between forward and backward connections (Angelucci et al., 2002a; Felleman & Van Essen, 1991; Murphy & Sillito, 1987; Rockland & Pandya, 1979; Sherman & Guillery, 1998). The distinction between forward and backward connections is based on the specificity of cortical layers that are the predominant sources and origins of extrinsic connections in the brain. Forward connections arise largely in superficial pyramidal cells, in supra-granular layers, and terminate in spiny stellate cells of Layer 4 or the granular layer of a higher cortical area (DeFelipe, Alonso-Nanclares, & Arellano, 2002; Felleman & Van Essen, 1991). Conversely, backward connections arise largely from deep pyramidal cells in infra-granular layers and target cells in the infra- and supra-granular layers of lower cortical areas. Intrinsic connections are both intra- and inter-laminar and mediate lateral interactions between neurons that are a few millimetres away. Due to convergence and divergence of extrinsic forward and backward connections, receptive fields in higher areas are generally larger than in lower areas (Zeki & Shipp, 1988). There is a key functional distinction between forward and backward connections that renders backward connections more modulatory or non-linear in their effects on neuronal responses (e.g., Sherman & Guillery, 1998). This is consistent with the deployment of voltage-sensitive and non-linear NMDA receptors in the supra-granular layers (Rosier, Arckens, Orban, & Vandesande, 1993) that are targeted by backward connections. Typically, the synaptic dynamics of backward connections have slower time constants. This has led to the notion that forward connections are driving and elicit an obligatory response in higher levels, whereas backward connections have both driving and modulatory effects and operate over greater spatial and temporal scales.

The hierarchical structure of the brain speaks to hierarchical models of sensory input. For example

$$y = g\left(x^{(1)}, v^{(1)}\right) + z^{(1)}$$
$$\dot{x}^{(1)} = f\left(x^{(1)}, v^{(1)}\right) + w^{(1)}$$
$$\vdots$$
$$v^{(i-1)} = g\left(x^{(i)}, v^{(i)}\right) + z^{(i)}$$
$$\dot{x}^{(i)} = f\left(x^{(i)}, v^{(i)}\right) + w^{(i)}$$
$$\vdots \tag{14}$$

In this model, sensory states y are caused by a non-linear function of internal states, $g(x^{(1)}, v^{(1)})$, plus a random effect, $z^{(1)}$. The dynamic states $x^{(1)}$ have memory and evolve according to equations of motion prescribed by the non-linear function $f(x^{(1)}, v^{(1)})$. These dynamics are subject to random fluctuations, $w^{(1)}$, and perturbations from higher levels that are generated in exactly the same way. In other words, the input to any level is the output of the level above. This means that causal states $v^{(i)}$ link hierarchical levels and that dynamic states $x^{(i)}$ are intrinsic to each level, linking states over time. The random fluctuations can be assumed to be Gaussian, with a covariance encoded by some hyper-parameters, $\delta_\gamma^{(i)}$, and independent across levels. The functions at each level are parameterised by $\delta_\theta^{(i)}$. This form of hierarchical dynamical model is very generic and subsumes most models found in statistics and machine learning as special cases.

This model specifies the functional form of the generative density in generalised coordinates of motion (see Friston et al., 2007, 2008) and induces an ensemble density on the generalised states $\delta_u^{(i)} = \tilde{x}^{(i)}, \tilde{v}^{(i)}$. If we assume that neuronal activity is the variational mode, $\tilde{\mu}_u^{(i)} = \tilde{\mu}_v^{(i)}, \tilde{\mu}_x^{(i)}$, of these states and that the variational mode of the model parameters $\delta_\gamma^{(i)}$ and $\delta_\theta^{(i)}$ corresponds to synaptic efficacy or connection strengths, we can write down the variational energy as a function of these modes using Equation 8, with $y = \mu_u^{(0)}$:

$$I(\tilde{\mu}_u) = -\tfrac{1}{2}\sum_i \tilde{\varepsilon}^{(i)T}\,\Pi^{(i)}\tilde{\varepsilon}^{(i)}$$

$$\tilde{\varepsilon}^{(i)} = \begin{bmatrix} \tilde{\varepsilon}_v^{(i)} \\ \tilde{\varepsilon}_x^{(i)} \end{bmatrix} = \begin{bmatrix} \tilde{\mu}_v^{(i-1)} - \tilde{g}(\tilde{\mu}_u^{(i)}, & \mu_\theta^{(i)}) \\ \tilde{\mu}_x^{(i)} - \tilde{f}(\tilde{\mu}_u^{(i)} & \mu_\theta^{(i)}) \end{bmatrix} \tag{15}$$

$$\Pi(\mu_\gamma^{(i)}) = \begin{bmatrix} \Pi_z^{(i)} & 0 \\ 0 & \Pi_w^{(i)} \end{bmatrix}.$$

Here, $\tilde{\varepsilon}^{(i)}$ is a generalised prediction error for the states at the ith level. The generalised predictions of the causal states and motion of the dynamic states are $\tilde{g}^{(i)}$ and $\tilde{f}^{(i)}$, respectively (see Friston et al., 2007, 2008). Here, $\tilde{\mu}_x'^{(i)} = \mu_x'^{(i)}$, $\mu_x''^{(i)}, \mu_x'''^{(i)}, \ldots$ represents the generalised velocity of $\tilde{\mu}_x^{(i)}$. $\Pi(\mu_y^{(i)})$ are the precisions of the random fluctuations that control their amplitude and smoothness. For simplicity, we have omitted terms that depend on the conditional covariance of the parameters; this is the same approximation used by expectation maximisation (Dempster, Laird, & Rubin, 1977).

3.2.1 The dynamics and architecture of perceptual inference

As mentioned above, we will focus on the optimisation of the ensemble density covering the states, implicit in perception or perceptual inference. From Equation 8 we obtain an expression that describes the dynamics of neuronal activity under the free-energy principle.

$$\dot{\tilde{\mu}}_u^{(i)} = h(\tilde{\varepsilon}^{(i)}, \tilde{\varepsilon}^{(i+1)})$$

$$= \tilde{\mu}_u'^{(i)} - \kappa \frac{\partial \tilde{\varepsilon}^{(i)T}}{\partial \tilde{\mu}_u^{(i)}} \Pi^{(i)} \tilde{\varepsilon}^{(i)} - \kappa \frac{\partial \tilde{\varepsilon}^{(i+1)T}}{\partial \tilde{\mu}_u^{(i)}} \Pi^{(i+1)} \tilde{\varepsilon}^{(i+1)} \tag{16}$$

These dynamics describe how neuronal states self-organise when the brain is exposed to sensory input. The form of Equation 16 is quite revealing; it is principally a function of prediction error – namely, the mismatch between the expected state of the world, at any level, and that predicted on the basis of the expected state in the level above. Critically, inference only requires the prediction error from the lower level, $\tilde{\varepsilon}^{(i)}$, and the higher level, $\tilde{\varepsilon}^{(i+1)}$. This drives conditional expectations $\tilde{\mu}_u^{(i)}$ to provide a better prediction, conveyed by backward connections, to explain the prediction error away. This is the essence of the recurrent dynamics that self-organise to suppress free-energy or prediction error – that is, recognition dynamics: $\dot{\tilde{\mu}}_u^{(i)} = h(\tilde{\varepsilon}^{(i)}, \tilde{\varepsilon}^{(i+1)})$.

Critically, the motion of the expected states is a linear function of the bottom-up prediction error. This is exactly what is observed physiologically, in the sense that bottom-up driving inputs elicit obligatory responses in higher levels that do not depend on other bottom-up inputs. In fact, the forward connections in Equation 16 have a simple form:[10]

$$\frac{\partial \tilde{\varepsilon}^{(i)T}}{\partial \tilde{\mu}_u^{(i)}} \Pi^{(i)} = \begin{bmatrix} -I \otimes g_v^{(i)} & -I \otimes g_x^{(i)} \\ -I \otimes f_v^{(i)} & D - (I \otimes f_x^{(i)}) \end{bmatrix} \Pi^{(i)} \tag{17}$$

This comprises block diagonal repeats of the derivatives $g_x = \partial g / \partial x$ (similarly

---

10 $\otimes$ is the Kronecker tensor product, and I denotes the identity matrix.

for the other derivatives). D is a block matrix with identity matrices in its first diagonal that ensure the internal consistency of generalised motion. The connections are modulated by the precisions encoded by $\mu_\gamma^{(i)}$. The lateral interactions within each level have an even simpler form,

$$\frac{\partial \tilde{\varepsilon}^{(i+1)T}}{\partial \tilde{\mu}_u^{(i)}} \Pi^{(i+1)} = \begin{bmatrix} \Pi_v^{(i+1)} & 0 \\ 0 & 0 \end{bmatrix}, \tag{18}$$

and reduce to the precisions of the causes at that level. We will look at the biological substrate of these interactions below.

The form of Equation 16 allows us to ascribe the source of prediction error to superficial pyramidal cells, which means we can posit these as encoding prediction error. This is because the only quantity that is passed forward from one level in the hierarchy to the next is prediction error, and superficial pyramidal cells are the major source of forward influences in the brain (Felleman & Van Essen, 1991; Mumford, 1992). Attributing this role to superficial pyramidal cells is useful because these cells are primarily responsible for the genesis of electroencephalographic (EEG) signals, which can be measured non-invasively. The prediction error itself is formed by predictions conveyed by backward connections and dynamics intrinsic to the level in question. These influences embody the non-linearities implicit in $\tilde{g}^{(i)}$ and $\tilde{f}^{(i)}$ (see Equation 17). Again, this is entirely consistent with the non-linear or modulatory role of backward connections that, in this context, model interactions among inferred states to predict lower level inferences. See Figure 13.4 for a schematic of the implicit neuronal architecture.

In short, the dynamics of the conditional modes are driven by three things. The first links generalised coordinates to ensure that the motion of the mode approximates the mode of the motion. This ensures that the representation of causal dynamics is internally consistent. The second is a bottom-up effect that depends upon prediction error from the level below. This can be thought of as a likelihood term. The third term, corresponding to an empirical prior, is mediated by prediction error at the current level. This is constructed using top-down predictions. An important aspect of hierarchical models is that they can construct their own empirical priors. In the statistics literature, these models are known as parametric empirical Bayes models (Efron & Morris, 1973), and they rely on the conditional independence of random fluctuation at each level (Kass & Steffey, 1989).

In summary, the dynamics of perceptual inference at any level in the brain are moderated by top-down priors from the level above. This is recapitulated at all levels, enabling self-organisation through recurrent interactions to minimise free-energy by suppressing prediction error throughout the hierarchy. In this way, higher levels provide guidance to lower levels and ensure an internal consistency of the inferred causes of sensory input at multiple levels of description.

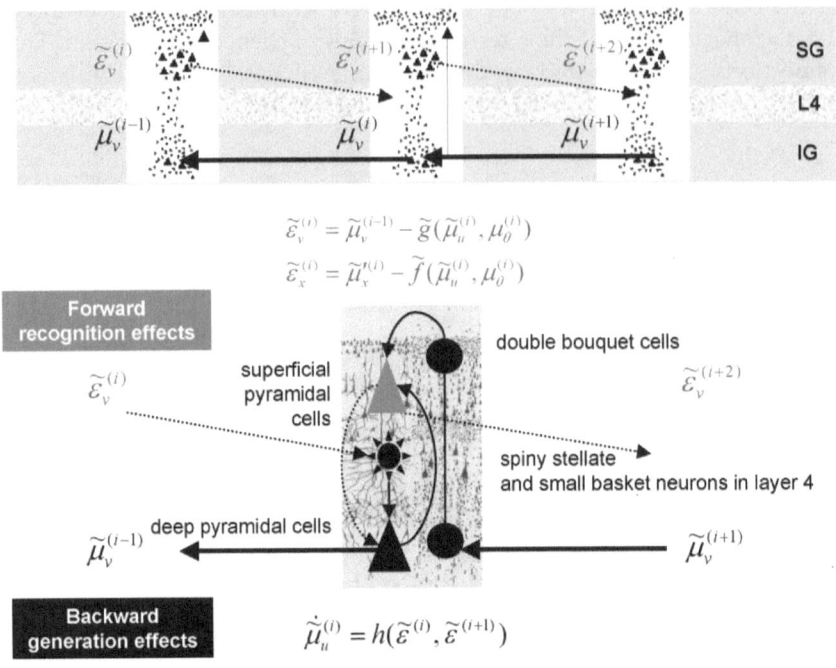

$$\widetilde{\varepsilon}_v^{(i)} = \widetilde{\mu}_v^{(i-1)} - \widetilde{g}(\widetilde{\mu}_u^{(i)}, \mu_\theta^{(i)})$$

$$\widetilde{\varepsilon}_x^{(i)} = \widetilde{\mu}_x^{'(i)} - \widetilde{f}(\widetilde{\mu}_u^{(i)}, \mu_\theta^{(i)})$$

Forward recognition effects

$\widetilde{\varepsilon}_v^{(i)}$

superficial pyramidal cells

double bouquet cells

$\widetilde{\varepsilon}_v^{(i+2)}$

spiny stellate and small basket neurons in layer 4

deep pyramidal cells

$\widetilde{\mu}_v^{(i-1)}$

$\widetilde{\mu}_v^{(i+1)}$

Backward generation effects

$$\dot{\widetilde{\mu}}_u^{(i)} = h(\widetilde{\varepsilon}^{(i)}, \widetilde{\varepsilon}^{(i+1)})$$

Figure 13.4 Schematic detailing the neuronal architectures that encode an ensemble density on the states and parameters of hierarchical models. *Upper panel:* Deployment of neurons within three cortical areas (or macro-columns). Within each area, the cells are shown in relation to the laminar structure of the cortex that includes supra-granular (SG), granular (L4) and infra-granular (IG) layers. *Lower panel:* Enlargement of a particular area and the speculative cells of origin of forward-driving connections that convey prediction error from a lower area to a higher area and the backward connections that carry predictions. These predictions try to explain away input from lower areas by suppressing the mismatch or prediction error. In this scheme, the source of forward connections is the superficial pyramidal cell population, and the source of backward connections is the deep pyramidal cell population. The differential equations relate to the free-energy minimisation scheme detailed in the main text.

3.3 Perceptual attention and learning

The dynamics above describe the optimisation of conditional or variational modes describing the most likely cause of sensory inputs. This is perceptual inference and corresponds to Bayesian inversion of the hierarchical generative model described in Equation 14. In this simplified scheme, in which conditional covariances have been ignored, minimising the free-energy is equivalent to suppressing hierarchical prediction error. Exactly the same treatment can be applied to changes in extrinsic and intrinsic connectivity encoding the conditional modes μ_γ and μ_θ. As above, the changes in these modes or synaptic efficacies are relatively simple functions of prediction error and lead to forms that are recognisable as associative plasticity. Examples of these derivations, for static systems, are provided in Friston (2005).

3.3.1 Attention uncertainty and precision

The precision parameters are interesting because of their role in moderating perceptual inference. Equation 16 shows that the influence of prediction error from the level below and the current level is scaled by the precisions $\Pi(\mu_\gamma^{(i)})$ and $\Pi(\mu_\gamma^{(i+1)})$ that are functions of their parameters, μ_γ. This means that the relative influence of the bottom-up likelihood term and top-down prior is controlled by modulatory influences encoded by μ_γ. This selective modulation of afferents is exactly the same as gain-control mechanisms that have been invoked for attention (e.g., Martinez-Trujillo & Treue, 2004; Moran & Desimone, 1985; Treue & Maunsell, 1996). It is fairly simple to formulate neuronal architectures in which this gain is controlled by lateral interactions that are intrinsic to each cortical level (see Figure 13.3).

As noted in the previous section, changes in μ_γ are supposed to occur at a timescale that is intermediate between the fast dynamics of the states and slow associative changes in extrinsic connections mediating the likelihood model. One could think of μ_γ as describing the short-term changes in synaptic efficacy, in lateral or intrinsic connections that depend upon classical neuromodulatory inputs or other slow synaptic dynamics (e.g., after-hyperpolarisation potentials, slow changes in synchronised oscillations and molecular signalling). The physiological aspects of these intermediate dynamics provide an interesting substrate for attentional mechanisms in the brain (for review, see Schroeder, Mehta, & Foxe, 2001) and are not unrelated to the ideas in Yu and Dayan (2005). These authors posit a role for acetylcholine (an ascending modulatory neurotransmitter) mediating expected uncertainty. Neural modulatory neurotransmitters have, characteristically, much slower time constants, in terms of their synaptic effects, than glutamatergic neurotransmission, which is employed by forward and backward extrinsic connections.

3.4 The Bayesian brain

The similarity between the form or structure of the brain and statistical models means that perceptual inference and learning lends itself nicely to a hierarchical treatment, which considers the brain as an empirical Bayesian device. The dynamics of neurons or populations are driven to minimise error at all levels of the cortical hierarchy and implicitly render themselves posterior or conditional modes (i.e., most likely values) of the causes, given sensory inputs. In contradistinction to supervised learning, hierarchical prediction does not require any desired output. Unlike many information theoretic approaches, it does not assume independent causes. In contrast to regularised inverse solutions (e.g., in machine vision), it does not depend on *a priori* constraints. These emerge spontaneously as empirical priors from higher levels.

The scheme implicit in Equation 16 sits comfortably with the hypothesis:

> on the role of the reciprocal, topographic pathways between two cortical areas, one often a "higher" area dealing with more abstract information about the world, the other "lower", dealing with more concrete data. The higher area attempts to fit its abstractions to the data it receives from lower areas by sending back to them from its deep pyramidal cells a template reconstruction best fitting the lower level view. The lower area attempts to reconcile the reconstruction of its view that it receives from higher areas with what it knows, sending back from its superficial pyramidal cells the features in its data which are not predicted by the higher area. The whole calculation is done with all areas working simultaneously, but with order imposed by synchronous activity in the various top-down, bottom-up loops.
>
> (Mumford, 1992, p. 241)

We have tried to show that this sort of hierarchical prediction can be implemented in brain-like architectures using mechanisms that are biologically plausible. Furthermore, this sort of scheme arises from some basic principles concerning adaptive systems and free-energy.

3.4.1 Empirical Bayes and biased competition

A compelling account of visual attention called the "biased-competition hypothesis" explains associated computational processes in terms of biasing competitive interactions among multiple stimuli in the visual field towards one stimulus, so that behaviourally relevant stimuli are processed and irrelevant stimuli are filtered out (Chelazzi, Miller, Duncan, & Desimone, 1993; Desimone & Duncan, 1995). This attentional bias is produced by top-down signals that bias competition, such that when multiple stimuli appear in the visual field, cells representing the attended stimulus win, thereby suppressing the firing of cells representing distractors (Desimone & Duncan, 1995). Neurophysiological experiments are consistent with this hypothesis; showing that attention modulates suppressive interactions among stimuli within the same receptive field (Miller, Gochin, & Gross, 1993). Free-energy minimisation furnishes a principled explanation for biased competition and suggests plausible mechanisms: baseline shifts in neuronal activity prior to the stimulus may reflect top-down predictions from higher cortical levels, whereas the modulatory (i.e., suppressive) effects during the evoked responses may reflect the multiplicative effects of precision parameters. Here, these parameters balance the relative sensitivity of neurons to top-down and bottom-up afferents. It should be noted that under the free-energy principle, attention is not an endogenous attribute of the agent; it is an emergent property of hierarchical optimisation that is neither top-down nor bottom-up.

3.4.2 *Backward or feedback connections?*

There is something slightly counter-intuitive about generative models in the brain. In this view, cortical hierarchies are trying to generate sensory predictions from high-level causes. This means the causal structure of the world is embodied in the backward connections. Forward connections simply provide feedback by conveying prediction error to higher levels. In short, forward connections are the *feedback* connections. This is why we have been careful not to ascribe a functional label like "feedback" to backward connections. Perceptual inference emerges from mutually informed top-down and bottom-up processes that enable sensation to constrain perception. This self-organising process is distributed throughout the hierarchy. Similar perspectives have emerged in cognitive neuroscience on the basis of psychophysical findings. For example, *reverse hierarchy theory* distinguishes between early explicit perception and implicit low-level vision, where "our initial conscious percept – vision at a glance – matches a high-level, generalised, categorical scene interpretation, identifying forest before trees" (Hochstein & Ahissar, 2002, p. 791).

Schemes based on generative models can be regarded as arising from the distinction between forward and inverse models adopted in machine vision (Ballard, Hinton, & Sejnowski, 1983; Kawato et al., 1993). Forward models generate inputs from causes (cf. generative models), whereas inverse models approximate the reverse transformation of inputs to causes (cf. recognition models). This distinction embraces the non-invertability of generating processes and the ill-posed nature of inverse problems. As with all underdetermined inverse problems, the role of constraints is central. In the inverse literature, *a priori* constraints usually enter in terms of regularised solutions. For example: "Descriptions of physical properties of visible surfaces, such as their distance and the presence of edges, must be recovered from the primary image data. Computational vision aims to understand how such descriptions can be obtained from inherently ambiguous and noisy data" (Poggio, Torre, & Koch, 1985, p. 314). The architectures that emerge from these schemes suggest that "Feedforward connections from the lower visual cortical area to the higher visual cortical area provide an approximated inverse model of the imaging process (optics)"; conversely, "the back-projection from the higher area to the lower area provides a forward model of the optics" (Kawato et al., 1993, p. 415) (see also Harth, Unnikrishnan, & Pandya, 1987). This perspective highlights the importance of backward connections and the role of empirical priors during Bayesian inversion of generative models.

3.4.3 Summary

In conclusion, we have seen how a fairly generic hierarchical and dynamical model of environmental inputs can be transcribed onto neuronal quantities to specify the free-energy and its minimisation. This minimisation corresponds, under some simplifying assumptions, to a suppression of prediction error at all levels in a cortical hierarchy. This suppression rests upon a balance between bottom-up (likelihood) influences and top-down (prior) influences that are balanced by representations of uncertainty (precisions). In turn, these representations may be mediated by classical neuromodulatory effects or slow post-synaptic cellular processes that are driven by overall levels of prediction error. Overall, this enables Bayesian inversion of a hierarchical model of sensory input that is context-sensitive and conforms to the free-energy principle.

4 Conclusion

In this chapter, we have considered the characteristics of biological systems in relation to non-adaptive self-organizing and dissipative systems. Biological systems act on the environment and sample it selectively to avoid phase-transitions that would irreversibly alter their structure. This adaptive exchange can be formalised in terms of free-energy minimisation, in which both the behaviour of the organism and its internal configuration minimise its free-energy. This free-energy is a function of the ensemble density encoded by the organism's configuration and the sensory data to which it is exposed. Minimisation of free-energy occurs through action-dependent changes in sensory input and the ensemble density implied by internal changes. Systems that fail to maintain a low free-energy will encounter surprising environmental conditions, in which the probability of finding them (i.e., surviving) is low. It may therefore be necessary, if not sufficient, for biological systems to minimise their free-energy.

The variational free-energy is not a thermodynamic free-energy but a free-energy formulated in terms of information theoretic quantities. The free-energy principle discussed here is not a consequence of thermodynamics but arises from population dynamics and selection. Put simply, systems with a low free-energy will be selected over systems with a higher free-energy. The free-energy rests on a specification of a generative model, entailed by the organism's structure. Identifying this model enables one to predict how a system will change if it conforms to the free-energy principle. For the brain, a plausible model is a hierarchical dynamic system in which neural activity encodes the conditional modes of environmental states and its connectivity encodes the causal context in which these states evolve. Bayesian inversion of this model, to infer the causes of sensory input, is a natural consequence of minimising free-energy or, under simplifying assumptions, the suppression of prediction error.

The ideas presented in this chapter have a long history, starting with the notions of neuronal energy described by Helmholtz (1860/1962) and covering ideas like analysis by synthesis (Neisser, 1967), efficient coding (Barlow, 1961) and more recent formulations like Bayesian inversion and predictive coding (e.g., Ballard et al., 1983; Dayan et al., 1995; Lee & Mumford, 2003; Mumford, 1992; Rao & Ballard, 1998). The specific contribution of this chapter is to provide a general formulation of the free-energy principle to cover both action and perception. Furthermore, this formulation can be used to connect constructs from machine learning and statistical physics with ideas from evolutionary theory, theoretical neurobiology, biology and microeconomics.

5 Acknowledgements

The Wellcome Trust funded this work. We would like to express our thanks to Marcia Bennett for preparing this manuscript.

6 References

Andersen, R. A. (1989). Visual and eye movement factions of the posterior parietal cortex. *Annual Review of Neuroscience, 12*, 377–405.

Angelucci, A., Levitt, J. B., & Lund, J. S. (2002b). Anatomical origins of the classical receptive field and modulatory surround field of single neurons in macaque visual cortical area V1. *Progress in Brain Research, 136*, 373–388.

Angelucci, A., Levitt, J. B., Walton, E. J., Hupe, J. M., Bullier, J., & Lund, J. S. (2002a). Circuits for local and global signal integration in primary visual cortex. *Journal of Neuroscience, 22*, 8633–8646.

Ashby, W. R. (1947). Principles of the self-organising dynamic system. *Journal of General Psychology, 37*, 125–128.

Ballard, D. H., Hinton, G. E., & Sejnowski, T. J. (1983). Parallel visual computation. *Nature, 306*, 21–26.

Barlow, H. B. (1961). Possible principles underlying the transformation of sensory messages. In W. A. Rosenblith (Ed.), *Sensory communication*. Cambridge, MA: MIT Press.

Borisyuk, R., & Hoppensteadt, F. (2004). A theory of epineuronal memory. *Neural Networks, 17*, 1427–1436.

Brocher, S., Artola, A., & Singer, W. (1992). Agonists of cholinergic and noradrenergic receptors facilitate synergistically the induction of long-term potentiation in slices of rat visual cortex. *Brain Research, 573*, 27–36.

Chelazzi, L., Miller, E., Duncan, J., & Desimone, R. (1993). A neural basis for visual search in inferior temporal cortex. *Nature, 363*, 345–347.

Crooks, G. E. (1999). Entropy production fluctuation theorem and the non-equilibrium work relation for free-energy differences. *Physical Review E, 60*, 2721–2726.

Dayan, P., Hinton, G. E., & Neal, R. M. (1995). The Helmholtz machine. *Neural Computation, 7*, 889–904.

DeFelipe, J., Alonso-Nanclares, L., & Arellano, J. I. (2002). Microstructure of the neocortex: Comparative aspects. *Journal of Neurocytology, 31*, 299–316.

Dempster, A. P., Laird, N. M., & Rubin, D. B. (1977). Maximum likelihood from incomplete data via the EM algorithm. *Journal of the Royal Statistical Society, Series B, 39,* 1–38.

Desimone, R., & Duncan, J. (1995). Neural mechanisms of selective visual attention. *Annual Review of Neuroscience, 18,* 193–222.

Edelman, G. M. (1993). Neural Darwinism: Selection and reentrant signaling in higher brain function. *Neuron, 10,* 115–125.

Efron, B., & Morris, C. (1973). Stein's estimation rule and its competitors – an empirical Bayes approach. *Journal of the American Statistical Association, 68,* 117–130.

Evans, D. J. (2003). A non-equilibrium free-energy theorem for deterministic systems. *Molecular Physics, 101,* 1551–1554.

Evans, D. J., & Searles, D. J. (2002). The fluctuation theorem. *Advances in Physics, 51,* 1529–1585.

Felleman, D. J., & Van Essen, D. C. (1991). Distributed hierarchical processing in the primate cerebral cortex. *Cerebral Cortex, 1,* 1–47.

Feynman, R. P. (1972). *Statistical mechanics.* Reading, MA: Benjamin.

Friston, K. J. (2003). Learning and inference in the brain. *Neural Networks, 16,* 1325–1352.

Friston, K. J. (2005). A theory of cortical responses. *Philosophical Transactions of the Royal Society of London, B, Biological Sciences, 360,* 815–836.

Friston, K. J., Mattout, J., Trujillo-Barreto, N., Ashburner, J., & Penny, W. (2007). Variational free energy and the Laplace approximation. *NeuroImage, 34,* 220–234.

Friston, K. J., & Price, C. J. (2001). Dynamic representations and generative models of brain function. *Brain Research Bulletin, 54,* 275–285.

Friston, K. J., Tononi, G., Reeke, G. N. Jr, Sporns, O., & Edelman, G. M. (1994). Value-dependent selection in the brain: Simulation in a synthetic neural model. *Neuroscience, 59*(2), 229–243.

Friston, K. J., Trujillo-Barreto, N., & Daunizeau, J. (2008). DEM: A variational treatment of dynamic systems. *NeuroImage, 41*(3), 849–885.

Gu, Q. (2002). Neuromodulatory transmitter systems in the cortex and their role in cortical plasticity. *Neuroscience, 111,* 815–835.

Haken, H. (1983). *Synergistics: An introduction. Non-equilibrium phase transition and self-organisation in physics, chemistry and biology* (3rd ed.). New York: Springer-Verlag.

Harth, E., Unnikrishnan, K. P., & Pandya, A. S. (1987). The inversion of sensory processing by feedback pathways: A model of visual cognitive functions. *Science, 237,* 184–187.

Harville, D. A. (1977). Maximum likelihood approaches to variance component estimation and to related problems. *Journal of the American Statistical Association, 72,* 320–338.

Helmholtz, H. (1962). *Handbuch der physiologischen optik* (Vol. 3), English trans., Ed. J. P. C. Southall. New York: Dover. (Original work published 1860)

Hinton, G. E., & von Camp, D. (1993). Keeping neural networks simple by minimising the description length of weights. In: *Proceedings of the sixth annual conference on computational learning theory (COLT-93), Santa Cruz, CA* (pp. 5–13). New York: ACM.

Hochstein, S., & Ahissar, M. (2002). View from the top: Hierarchies and reverse hierarchies in the visual system. *Neuron, 36,* 791–804.

Huffman, K. J., & Krubitzer, L. (2001). Area 3a: Topographic organization and cortical connections in marmoset monkeys. *Cerebral Cortex, 11*, 849–867.

Itti, L., & Baldi, P. (2006). Bayesian surprise attracts human attention. In: *Advances in neural information processing systems* (Vol. 19, pp. 1–8). Cambridge, MA: MIT Press.

Kalman, R. (1960). A new approach to linear filtering and prediction problems, *ASME Transactions – Journal of Basic Engineering, 82*(1), 35–45.

Kass, R. E., & Steffey, D. (1989). Approximate Bayesian inference in conditionally independent hierarchical models (parametric empirical Bayes models). *Journal of the American Statistical Association, 407*, 717–726.

Kauffman S. (1993). *Self-organisation on selection in evolution.* Oxford, UK: Oxford University Press.

Kawato, M., Hayakawa, H., & Inui, T. (1993). A forward-inverse optics model of reciprocal connections between visual cortical areas. *Network, 4*, 415–422.

Kersten, D., Mamassian, P., & Yuille, A. (2004). Object perception as Bayesian inference. *Annual Review of Psychology, 55*, 271–304.

Körding, K. P., & Wolpert, D. M. (2004). Bayesian integration in sensorimotor learning. *Nature, 427*, 244–247.

Lee, T. S., & Mumford, D. (2003). Hierarchical Bayesian inference in the visual cortex. *Journal of the Optical Society of America. A, Optics, Image Science, and Vision, 20*, 1434–1448.

Locke, J. (1976). *An essay concerning human understanding.* London: Dent. (Original work published 1690)

MacKay, D. J. C. (1995). Free-energy minimisation algorithm for decoding and cryptoanalysis. *Electronics Letters, 31*, 445–447.

MacKay, D. M. (1956). The epistemological problem for automata. In C. E. Shannon & J. McCarthy (Eds.), *Automata studies* (pp. 235–251). Princeton, NJ: Princeton University Press.

Maniadakis, M., & Trahanias, P. E., (2006). Modelling brain emergent behaviours through coevolution of neural agents. *Neural Networks, 19*(5), 705–720.

March, J. G. (1991). Exploration and exploitation in organizational learning. *Organization Science, 10*(1), 299–316.

Martinez-Trujillo, J. C., & Treue, S. (2004). Feature-based attention increases the selectivity of population responses in primate visual cortex. *Current Biology, 14*, 744–751.

Mesulam, M. M. (1998). From sensation to cognition. *Brain, 121*, 1013–1052.

Miller, E., Gochin, P., & Gross, C. (1993). Suppression of visual responses of neurons in inferior temporal cortex of the awake macaque by addition of a second stimulus. *Brain Research, 616*, 25–29.

Montague, P. R., Dayan, P., Person, C., & Sejnowski, T. J. (1995). Bee foraging in uncertain environments using predictive Hebbian learning. *Nature, 377*(6551), 725–728.

Moran, J., & Desimone, R. (1985). Selective attention gates visual processing in the extrastriate cortex. *Science, 229*, 782–784.

Morowitz, H. J. (1968). *Energy flow in biology.* New York: Academic Press.

Mumford, D. (1992). On the computational architecture of the neocortex. II. The role of cortico-cortical loops. *Biological Cybernetics, 66*, 241–251.

Murphy, P. C., & Sillito, A. M. (1987). Corticofugal feedback influences the generation of length tuning in the visual pathway. *Nature, 329*, 727–729.

Neisser, U. (1967). *Cognitive psychology.* New York: Appleton-Century-Crofts.

Nicolis, G., & Prigogine, I. (1977). *Self-organisation in non-equilibrium systems*. New York: Wiley.

Niv, Y., Duff, M. O., & Dayan, P. (2005). Dopamine, uncertainty and TD learning. *Behavioral Brain Function, 4*, 1–6.

Poggio, T., Torre, V., & Koch, C. (1985). Computational vision and regularisation theory. *Nature, 317*, 314–319.

Pollen, D. A. (1999). On the neural correlates of visual perception. *Cerebral Cortex, 9*, 4–19.

Prince, A., & Smolensky, P. (1997). Optimality: From neural networks to universal grammar. *Science, 275*, 1604–1610.

Rao, R. P. (2005). Bayesian inference and attentional modulation in the visual cortex. *NeuroReport, 16*, 1843–1848.

Rao, R. P., & Ballard, D. H. (1998). Predictive coding in the visual cortex: A functional interpretation of some extra-classical receptive field effects. *Nature Neuroscience, 2*, 79–87.

Rockland, K. S., & Pandya, D. N. (1979). Laminar origins and terminations of cortical connections of the occipital lobe in the rhesus monkey. *Brain Research, 179*, 3–20.

Rosier, A. M., Arckens, L., Orban, G. A., & Vandesande, F. (1993). Laminar distribution of NMDA receptors in cat and monkey visual cortex visualized by [3H]-MK-801 binding. *Journal of Comparative Neurology, 335*, 369–380.

Schroeder, C. E., Mehta, A. D., & Foxe, J. J. (2001). Determinants and mechanisms of attentional modulation of neural processing. *Frontiers in Bioscience, 6*, D672–684.

Schultz, W. (2007). Multiple dopamine functions at different time courses. *Annual Review of Neuroscience, 30*, 259–288.

Sherman, S. M., & Guillery, R. W. (1998). On the actions that one nerve cell can have on another: Distinguishing "drivers" from "modulators". *Proceedings of the National Academy of Sciences, USA, 95*, 7121–7126.

Simoncelli, E. P., & Olshausen, B. A. (2001). Natural image statistics and neural representation. *Annual Review of Neuroscience, 24*, 1193–1216.

Streater, R. F. (1993). The free-energy theorem. In H. Araki, K. R. Ito, A. Kishimoto, & I. Ojima (Eds.), *Quantum and non-commutative analysis* (pp. 137–147). Amsterdam: Kluwer Press.

Todorov, E. (2006). *Linearly-solvable Markov decision problems*. Paper presented at 20th annual conference on advances in neural information processing systems, Vancouver, Canada.

Treue, S., & Maunsell, H. R. (1996). Attentional modulation of visual motion processing in cortical areas MT and MST. *Nature, 382*, 539–541.

Tseng, K. Y., & O'Donnell, P. (2004): Dopamine-glutamate interactions controlling prefrontal cortical pyramidal cell excitability involve multiple signaling mechanisms. *Journal of Neuroscience, 24*, 5131–5139.

Yu, A. J., & Dayan, P. (2005). Uncertainty, neuromodulation and attention. *Neuron, 46*, 681–692.

Zeki, S., & Shipp, S. (1988). The functional logic of cortical connections. *Nature, 335*, 311–317.

14 Architectural and representational requirements for seeing processes and affordances

Aaron Sloman

This chapter, combining the standpoints of philosophy and Artificial Intelligence with theoretical psychology, summarises several decades of investigation of the variety of functions of vision in humans and other animals, pointing out that biological evolution has solved many more problems than are normally noticed. Many of the phenomena discovered by psychologists and neuroscientists require sophisticated controlled laboratory settings and specialised measuring equipment, whereas the functions of vision reported here mostly require only careful attention to a wide range of everyday competences that easily go unnoticed. Currently available computer models and neural theories are very far from explaining those functions, so progress in explaining how vision works is more in need of new proposals for explanatory mechanisms than new laboratory data. Systematically formulating the requirements for such mechanisms is not easy. If we start by analysing familiar competences, that can suggest new experiments to clarify precise forms of these competences, how they develop within individuals, which other species have them, and how performance varies according to conditions. This will help to constrain requirements for models purporting to explain how the competences work. The chapter ends with speculations regarding the need for new kinds of information-processing machinery to account for the phenomena.

1 Introduction: from Kant to Gibson and beyond

1.1 The role of vision in mathematical competences

The aim of the workshop was to discuss a computational approach to "Closing the Gap between Neurophysiology and Behaviour". My approach to this topic is to focus almost entirely on what needs to be explained rather than to present any neurophysiological model, though conjectures regarding some of the design features required in such a model are offered in section 6.

I originally began studying vision to understand the role of visual processing in mathematical discovery and reasoning, for instance in proving theorems in elementary Euclidean geometry, but also in more abstract reasoning, for example about infinite structures, which can be visualised but cannot occur in the environment. Sloman (1962) was an attempt to defend the view of mathematical knowledge as both non-empirical and synthetic, proposed by Kant (1781), but rejected by many contemporary mathematicians and philosophers. This led me into topics linking many disciplines, including mathematics, psychology, neuroscience, philosophy, linguistics, education and Artificial Intelligence. A full understanding requires parallel investigations of many different kinds of vision: in insects and other invertebrates, in birds, in primates, and in different sorts of future robots.

This chapter attempts to show how the role of vision in mathematical reasoning is connected with the ability in humans, some other animals and future human-like robots, to perceive and reason about structures and processes in the environment, including *possible* processes that are not actually occurring. This requires us to focus attention on aspects of vision that are ignored by most other researchers including: those who study vision as concerned with image structure (e.g. Kaneff, 1970); those who study vision as a source of geometrical and physical facts about the environment (e.g. Marr, 1982); those who regard vision as primarily a means of controlling behaviour (e.g. Berthoz, 2000 and many researchers in AI/Robotics and psychology who regard cognition as closely tied to embodiment); and those who regard vision as acquisition of information about affordances (e.g. J. J. Gibson, 1979; E. J. Gibson & Pick, 2000).

1.2 Why complete architectures matter

The processes of seeing involve actual or potential interactions between sensory mechanisms, action-control mechanisms and more central systems. Those subsystems arise from different stages in our evolutionary history and grow during different stages in individual development. So the functions of vision differ both from one species to another and can change over time within an individual as the information-processing architecture grows. Some of those developments are culture-specific, such as what language the individual learns to read, which gestures are understood and which building designs are encountered.

A full understanding of vision requires investigation of different multifunctional architectures in which visual systems with different collections of competences can exist. An architecture with more sophisticated "central" mechanisms makes possible more sophisticated visual functions. For instance, a central mechanism able to use an ontology of causal and functional roles is required for a system that can see something causing, preventing, or enabling something else to occur. The ability to make use of an ontology including mental states (a meta-semantic ontology) is required if a visual system is to be able to perceive facial expressions, such as happiness, sadness, surprise, etc. and make use of the information. The requirements are discussed further in section 4.

Psychologists and ethologists, instead of merely asking which animals can do X, or at what age or under what conditions young children can do X, should adopt the design-based approach, and constantly ask "how could that work?" We can use that to generate collections of requirements for information-processing architectures and then investigate mechanisms that could support the observed variety of visual functions in robots. Very often, it is not obvious whether a particular theory suffices, so using a theory as a basis for designing, implementing and testing *working* artificial systems is a crucial part of the process of explaining how natural systems work. This chapter uses that approach, basing some unobvious functions of vision on detailed analysis of requirements for human-like visual systems. It ends with some speculations about sorts of mechanisms (using a large collection of multi-stable, interlinked dynamical systems) that appear to be required for those functions, which do not yet exist in any known computational models, and which may be hard to identify in neural mechanisms.

Understanding how a complex system works includes knowing what would happen if various aspects of the design were different, or missing. So understanding how humans work requires us to relate the normal human case to other products of biological evolution, to cases of brain damage or genetic brain abnormality, and to possible future engineering products. We can summarise this as follows (building on Sloman, 1982, 1984, 1994, 2000): We need to study the space of possible sets of requirements or niches, *niche space*, and we need to study the space of possible designs for working systems that can meet different sets of requirements, *design space*. Finally, we need to understand the various relationships between regions of design space and regions of niche space and the complex ecosystem loops in which changes in one design or niche produce changes in other designs or niches.

By analysing the different sets of functions supported by different information-processing architectures we can come up with a theory-based survey of possibilities for a mind or a visual system. For each system design there is a specific "logical topography", a set of possible states, processes and causal interactions that can occur within the architecture, some involving also interactions with the environment. The set of possibilities generated by each architecture can be subdivided and categorised in different ways for different purposes – producing different "logical geographies" (Ryle, 1949). E.g. the purposes of common-sense classification are different from the purposes of scientific explanation.

1.3 *Varieties of representation: Generalised Languages (GLs)*

A particularly important feature of any information-processing system is how it encodes the information it acquires and uses. Researchers designing computational models often have commitments to particular forms of representation, since those are the ones for which they have tools, e.g. tools for modelling neural nets or tools for symbolic computation using trees and graphs. Those commitments can severely restrict the kinds of questions they ask, and the answers they consider.

Pre-verbal children and many non-human animals can perceive and react to processes as they occur. That requires mechanisms providing the ability to represent changes while they happen. Perhaps the same mechanisms, or closely related mechanisms, can be used to reason about processes that are not happening. If some other primates and very young children use internal "generalised languages" (GLs) as proposed in Sloman (1979) and elaborated in Sloman and Chappell (2007), that suggests that GLs supporting structural variability and compositional semantics evolved before external human languages used for communication, and that GLs also precede the learning of communicative language in individual humans.

We shall later give several examples of human visual competences, including geometric reasoning competences, that seem to require use of GLs, for example in sections *3.1, 4.1, 4.4, 5, 5.8,* and *6.* The suggestion that GLs are used for all these purposes, including the representation of processes at different levels of abstraction, poses deep questions for brain science, as we'll see later.

2 Wholes and parts: beyond "scaling up"

The rest of this chapter addresses a subset of the requirements for visual systems. Designs meeting those requirements must be able to work with designs for other components: they must "scale out".

2.1 Putting the pieces together: "scaling out"

To study vision we need to consider a larger set of problems, including understanding information-processing requirements for a human-like (or chimp-like, or crow-like) organism or robot to perceive, act and learn in the environment. Instances of designs for mechanisms providing particular competences (e.g. visual competences) must be capable of interacting with other components in a larger design that satisfies requirements for a *complete* animal or robot.

This requirement to "scale out" contrasts with the frequently mentioned need to "scale up", namely coping successfully with larger and more complex inputs. Many human competences do not scale up, including parsing, planning, and problem-solving competences. It is possible to produce a highly efficient implementation of some competence that scales up very well but cannot be integrated with other human competences. Most current implementations of linguistic processing, vision, planning, or problem-solving do not scale out, because they are designed to work on their own in test situations, not to work with one another.

Being biologically inspired is not enough. Using observed human performance in a laboratory to specify a computing system that produces the same error rates, or learning rates or reaction times, need not lead to models that "scale out", i.e. can be extended to form part of a larger model meeting a much wider range of requirements. Not all mechanisms that perform like part of a system are useful parts of something that performs like the whole system. For example, we have machines that meet one requirement for a computer model of a good human chess player, namely that the machine is able to beat most human players. But if you add other requirements, such as that the player should be able to teach a weaker player, by playing in such a way as to help the weaker player learn both from mistakes and from successes, then the obvious designs that do well as competent chess-players are not easily extendable to meet the further requirements. Similarly, a computer system can be comparable to humans at recognising certain classes of objects in images, without being extendable so that it can intelligently use the fact that same object looks different from different viewpoints. However, the ability of a design with functionality F1 to be expanded to include functionality F2 is partly a matter of degree: it depends on how much extra mechanism is needed to provide F2.

2.2 *Biological* mechanisms *vs biological* wholes

The relevance of biological *mechanisms* for AI (e.g. neural nets and evolutionary computations) has been acknowledged for several decades. What is relatively new in the computational modelling community is moving beyond studying what *biological mechanisms* can do, to finding out what *whole animals* can do and how they do it, which is one way of studying *requirements* to be met by new designs. Brooks (1991) also recommended building complete systems, though his rejection of representations went too far, as we'll see.

Finding out how information flows around brain circuits connected to the eyes has received a lot of attention prompted by new non-invasive brain-imaging machinery, but does not tell us what the information is, how it is represented, or what the information is used for. That can include things as diverse as fine-grained motor control, triggering saccades, aesthetic enjoyment, recognition of terrain, finding out how something works, or control of intermediate processes that perform abstract internal tasks.

It is possible to make progress in the case of simple organisms where neurophysiology corresponds closely to information-processing functions. Measuring brain processes may be informative in connection with evolutionarily older, simpler, neural circuits (e.g. some reflexes), but many newer functions are extremely abstract, and probably only very indirectly related to specific neural events. In those cases, results of brain imaging may show some correlations without telling us much about what the virtual machines implemented in brains are doing. For example, consider speakers of two very different languages with different phonetic structures, grammars, and vocabularies who hear and understand reports with the same semantic content. If similar parts of their brains show increased activity: that will not answer any of the deep questions about how understanding works. E.g. we shall be no nearer knowing how to produce a working model with the same functionality.

The journey towards full understanding has far to go, and may even be endless, since human minds, other animal minds and robot minds can vary indefinitely. Evaluating intermediate results on this journey is difficult. A Popperian emphasis on the need for falsifiable hypotheses can slow down scientific creativity. Instead, as proposed by Lakatos (1980), we need to allow that distinguishing degenerative and progressive research programmes can take years, or decades, and we need to understand explanations of possibilities, as well as laws (Sloman, 1978, ch. 2).

2.3 Vision and mathematical reasoning

In his *Critique of Pure Reason*, Kant had claimed, in opposition to Hume, that there are ways of discovering new truths that extend our knowledge (i.e. they are "synthetic", not analytic) and which are not empirical. When trying to prove a theorem, mathematicians frequently use the ability to *see* both structural relationships, e.g. relations between overlapping figures, and also the possibility of changing such relationships, e.g. drawing a new construction line. Even without making the change it is often possible to visualise the consequences of such a change. If you look at a triangle with vertices labelled A, B and C, or simply imagine looking at one, you can *see* that it is possible to draw a straight line from any vertex, e.g. A, to the midpoint of the opposite side. You may also be able to work out that the two resulting triangles *must* have the same area.

Not all mathematical discoveries are based on visual reasoning. For example, very different discoveries, some of them documented in Sloman (1978, ch. 8), occur as a child learns to count, and then discovers different uses for the counting process and different features of the counting process, such as the fact that the result of counting a collection of objects is not altered by rearranging the objects but can be altered by breaking one of the objects into two objects. Such mathematical discoveries depend on perceiving invariant structures and relationships in procedures that can be followed. Simultaneous perception of spatial and temporal relationships can lead to the discovery that any one-to-one mapping between elements of two finite sets can be converted into any other such mapping by successively swapping ends of mappings, This sort of discovery requires quite abstract visual capabilities. A self-monitoring architecture (Sloman, 2008) seems to be needed, allowing one process to observe that another process has consequences that do not depend on the particularities of the example, and which are therefore *necessary* consequences of the procedure.

2.4 Development of visual competences

Many people can see and think about geometrical possibilities and relationships. Very young children cannot see all of them, let alone think about them. Why not? And what has to change in their minds and brains to enable them to see such things? There are developments that would not normally be described as mathematical, yet are closely related to mathematical competences.

For example, a very young child who can easily insert one plastic cup into another may be able to lift a number of cut-out pictures of objects from recesses, and know which recess each picture belongs to, but be *unable* to get pictures back into their recesses: the picture is placed in roughly the right location and pressed hard, but that is not enough. The child apparently has not yet extended his or her ontology to include boundaries of objects and alignment of boundaries. Such learning may include at least three related aspects: (a) developing new forms of representation; (b) extending the learner's ontology to allow new kinds of things that exist; and (c) developing new ways of manipulating representations, in perception and reasoning. What the child playing with puzzle pieces has to learn, namely facts about boundaries and how they constrain possible movements, is a precursor to being able to think mathematically about bounded regions of a plane. Later mathematical education will build on general abilities to see structures and processes and to see how structures can constrain or facilitate processes, as illustrated in Sauvy and Suavy (1974). We shall see that this is related to perception of proto-affordances.

3 Affordance-related visual competences: seeing processes and possibilities

3.1 Perceiving and reasoning about changes and proto-affordances

Visual, geometrical, reasoning capabilities depend on (a) the ability to attend to parts and relationships of a complex object, including "abstract" parts like the midpoint of a line and collinearity relationships, (b) the ability to discern the possibility of changing what is in the scene, e.g. adding a new line, moving something to a new location, (c) the ability to work out the consequences of making those changes, e.g. working out which new structures, relationships and further possibilities for change will come into existence.

Both the ability to see and make use of affordances and the ability to contemplate and reason about geometric constructions depend on a more primitive and general competence, namely the ability to see not only structures but also *processes*, and the closely related ability to see *the possibility* of processes that are not actually occurring, and also *constraints* that limit those possibilities. I call such possibilities and constraints "proto-affordances".

Gibson's affordances (J. J. Gibson, 1979) were concerned only with opportunities for *actions* that *the perceiver* could perform, whereas normal humans can see what is common between processes that they produce, processes that others produce and processes that are not parts of any intentional actions, e.g. two surfaces coming together. They can perceive and think about the *possibility* of processes occurring without regard to what they can do or what can affect them, e.g. noticing that a rock could start rolling down a hillside. This involves seeing proto-affordances, i.e. seeing that certain processes can and others cannot occur in a certain situation.

That ability underlies both the ability to perceive "vicarious" affordances, namely affordances for others, and also the ability to think about possible occurrences such as rain falling or the wind blowing tomorrow, without specifying a viewpoint. This uses an "exosomatic ontology", making it possible to refer to entities and processes that can exist outside the body, independently of any sensory or motor signals. A pre-verbal child or non-verbal animal that can see and reason about such proto-affordances and vicarious affordances is probably using a spatial GL to represent the possibilities, as suggested in section 1.3.

Perceiving vicarious affordances for predators or for immature offspring can be biologically very important. It is an open research question which animals can reason about vicarious affordances, though both pre-verbal human children and some chimpanzees can perceive and react to affordances for others, as shown by Warneken and Tomasello (2006), illustrated in videos available at http://email.eva.mpg.de/~warneken/video.htm

3.2 Evolutionary significance of independently mobile graspers

There are commonalities between affordances related to doing things with left hand, with right hand, with both hands, with teeth and with tools such as tongs or pliers. In principle it is possible that all the means of grasping are represented in terms of features of the sensorimotor signals involved as in Lungarella and Sporns (2006), but the variety of such patterns is astronomical. If, however, grasping is represented more abstractly, in terms of 3-D relations between surfaces in space, using an amodal form of representation and an exosomatic ontology referring to things outside the body, the variety of cases can be considerably reduced: for instance very many types of grasping involve two surfaces moving together with an object between. Such abstract representation can be used for high level planning of actions involving grasping, including the common requirement for the two grasping surfaces to be further apart during the approach than the diameter of the thing to be grasped. More detailed information about specific cases can then be used either when planning details, or during servo-controlled action execution. I suspect that biological evolution long ago "discovered" the enormous advantages of amodal, exosomatic, representations and ontologies as compared with representations of patterns in sensorimotor signals.

Although 2-D image projections are often helpful for controlling the fine details of an action during visual servoing (as noted in Sloman, 1982), using an exosomatic ontology including 3-D spatial structures and processes, rather than sensorimotor signal patterns, makes it possible to learn about an affordance in one situation and transfer that learning to another where sensor inputs and motor signals are quite different: e.g. discovering the consequences of grasping an object with one hand then transferring what has been learned to two-hand grasping, or biting. This assumes that the individual can acquire generic, re-usable, mappings between 3-D processes in the environment and sensor and motor signal patterns.

4 Towards a more general visual ontology

4.1 *Proto-affordances and generative process representations*

Affordances for oneself and for others depend on the more fundamental "proto-affordances". A particular proto-affordance, such as the potential of one object to impede the motion of another, can be the basis for many action affordances. E.g. it could produce a negative affordance for an agent trying to push the moving object towards some remote location, or a positive affordance for an individual wishing to terminate the motion of a moving object.

An animal's or machine's ability to discover and represent proto-affordances, to combine and manipulate their representations, allows a given set of proto-affordances to generate a huge variety of affordances, including many never previously encountered, permitting creative problem-solving and planning in new situations. The ability to combine proto-affordances to form new complex affordances apparently enabled the New Caledonian crow Betty to invent several ways of transforming a straight piece of wire into a hook in order to lift a bucket of food from a glass tube (Weir, Chappell, & Kacelnik, 2002).

Most AI vision researchers, and many psychologists, assume that the sole function of visual perception is acquiring information about objects and processes that exist in the environment, whereas a major function of vision (in humans and several other species) seems to include acquiring information from the environment about *what does not exist but could exist*. Mechanisms for seeing processes may also be involved in reasoning about consequences of possible processes. Such predictive and manipulative abilities are not innate, but develop over time. Examples in human infants and children are presented in E. J. Gibson and Pick (2000), though no mechanisms are specified.

4.2 *Complex affordances: combining process possibilities*

Some animals learn, by playing in the environment, that affordances can be *combined* to form more complex affordances, because *processes can be combined to form more complex processes*. Reasoning about such complex processes and their consequences depends on the ability to combine simpler proto-affordances to form more complex ones.

Because processes occur in space and time, and can have spatially and temporally related parts, they can be combined in at least the following ways:

- processes occurring in sequence can form a more complex process;
- two processes can occur at the same time (e.g. two hands moving in opposite directions);
- processes can overlap in time, e.g. the second starting before the first has completed;
- processes can overlap in space, for example a chisel moving forwards into a rotating piece of wood;
- one process can modify another, e.g. squeezing a rotating wheel can slow down its rotation;
- one process can launch another, e.g. a foot kicking a ball.

The ability to represent a sequence of processes is part of the ability to form plans prior to executing them. It is also part of the ability to predict future events, and to explain past events, but this is just a special case of a more general ability to combine proto-affordances. How do animal brains represent a wide variety of structures and processes?

There have been various attempts to produce systematic ways of generating and representing spatial structures. However, the demands on a system for representing spatial *processes* (e.g. translating, rotating, stretching, compressing, shearing twisting, etc.) are greater than demands on generative specifications of spatial *structures*. The extra complexity is not expressible just by adding an extra dimension to a vector. The set of possible perceivable processes can be constrained by a context, but the remaining set can be very large, e.g. in a child's playroom, a kitchen, a group of people at a dinner table, a garden, a motorway, various situations in which birds build nests, and many more. A theory of how biological vision works must explain what kinds of information about spatial processes particular animals are capable of acquiring, how the information is represented, how it is used, and how the ability to acquire and use more kinds of information develops. It seems that in order to accommodate the variety of processes humans and other animals can perceive and understand, they will need forms of representation that have the properties we ascribed to spatial GLs in section 1.3, with the benefits described in Sloman (1971).

4.3 Varieties of learning about processes

In the first few years of life, a typical human child must learn to perceive and to produce many hundreds of different sorts of spatial process, some involving its own body, some involving movements of other humans and pets, and some involving motion of inanimate objects, with various causes. Examples include topological processes where contact relationships, containment relationships, alignment relationships go into and out of existence and metrical processes where things change continuously. The very same physical process can include both metrical changes, as something is lowered into or lifted out of a container and discrete topological changes as contact, containment, or overlap relationships between objects and spatial regions or volumes change.

In each context there are different sets of "primitive" processes and different ways in which processes can be combined to form more complex processes. Some simple examples in a child's environment might include an object simultaneously moving and rotating, where the rotation may be *closely coupled* to the translation, e.g. a ball rolling on a surface, or *independent of* the translation, e.g. a frisbee spinning as it flies. Other examples of closely coupled adjacent processes are: a pair of meshed gear wheels rotating; a string unwinding as an axle turns; a thread being pulled through cloth as a needle is lifted; a pair of laces being tied; a bolt simultaneously turning and moving further into a nut or threaded hole; a sleeve sliding on an arm as the arm is stretched; and sauce in a pan moving as a spoon moves round in it.

Many compound processes arise when a person or animal interacts with a physical object. Compound 3-D processes are the basis of an enormous variety of affordances. For example, an object may afford grasping, and lifting, and as a result of that it may afford the possibility of being moved to a new location. The combination of grasping, lifting and moving allows a goal to be achieved by performing a compound action using the three affordances in sequence. The grasping itself can be a complex process made of various successive sub-processes, and some concurrent processes. However, other things in the environment can sometimes obstruct the process: as in the chair example in section 4.5 (Figure 14.2).

In addition to producing physical changes in the environment, more abstract consequences of actions typically include the existence of new positive and negative affordances. The handle on a pan lid may afford lifting the lid, but once the lid is lifted not only is there a new physical situation, there are also new action affordances and epistemic affordances, e.g. because new *information* is available with the lid off. The action of moving closer to an open door also alters epistemic affordances (Figure 14.1).

4.4 Reasoning about interacting spatial processes

Processes occurring close together in space and time can interact in a wide variety of ways, depending on the precise spatial and temporal relationships. It is possible to learn *empirically* about the consequences of such interactions by observing them happen, and collecting statistics to support future predictions, or formulating and testing universal generalisations. However, humans and some other animals sometimes need to be able to *work out* consequences of novel combinations, for example approaching a door that is shut, while carrying something in both hands, for the first time. It does not take a genius

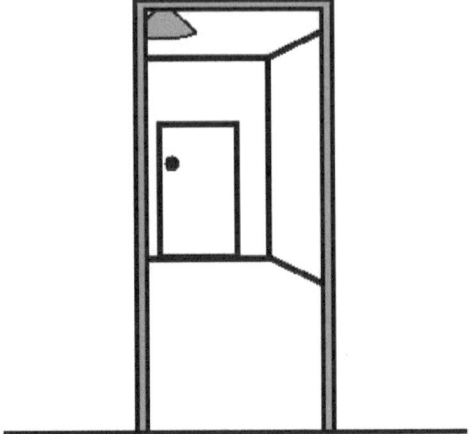

Figure 14.1 As you move nearer the door you will have access to more information about the contents of the room, and as you move further away
you will have less. Moving left or right will change the information
available in a different way.

to work out that an elbow can be used to depress the handle while pushing to
open the door.

E. J. Gibson and Pick (2000) state on page 180 that the affordance of a tool
can be discovered in only two ways, by *exploratory activities* and by *imitation*.
They apparently failed to notice a third way of discovering affordances,
namely *working out* what processes are possible when objects are manipulated, and what their consequences will be.

A requirement for human-like visual mechanisms is that they should
produce representations that can be used for *reasoning* about novel spatial
configurations and novel combinations of processes, i.e. the kind of reasoning that led to the study of Euclidean geometry. (An example of the need to
"scale out".) Likewise, young children can reason about spatial processes
and their implications long before they can do logic and algebra and to
some extent even before they can talk. As remarked in Sloman and Chappell
(2007), this has implications for the evolution and development of language.

By "exploratory activities" Gibson and Pick (2000) referred to physical
exploration and play. The missing alternative, working things out, can also
involve exploratory activities, but the explorations can be done with *representations* of the objects and processes instead of using the actual physical
objects. The representations used can either be entirely mental, e.g. visualising
what happens when some geometrical configuration is transformed, or physical, for instance 2-D pictures representing 3-D structures, with processes
represented by marks on the pictures (Sloman, 1971; Sauvy & Suavy, 1974).

Although reasoning with representations in place of the objects is
sometimes fallacious, nevertheless, when done rigorously, it is mathematical

inference rather than empirical inference. As Lakatos (1976) showed, the methods of mathematics are far from infallible. But that does not make them empirical in the same way as the methods of the physical sciences are. The issues are subtle and complex: see Sloman (2008).

The ability to think about and reason about novel combinations of familiar types of process is often required for solving new problems, for example realising for the first time that instead of going from A to B and then to C it is possible to take a short cut from A to C, or realising that a rigid circular disc can serve as well as something long and thin (like a screwdriver) to lever something up.

4.5 Creative reasoning about processes and affordances

Gibson and Pick describe various kinds of "prospectivity" that develop in children but they focus only on empirically learnt kinds of predictive rules, and ignore the child's growing ability to design and represent novel complex multi-stage processes and work out that they can achieve some goal.

The ability to work things out is facilitated and enhanced by the ability to form verbal descriptions, as in inventing stories, but the ability to use a public human language is not a prerequisite for the ability, as can be seen in the creative problem-solving of pre-verbal children and some other animals. Both seem to be able to work out the consequences of some actions using intuitive geometric and topological reasoning.

Figure 14.2 illustrates affordances that interact in complex ways when combined, because of the changing spatial relationships of objects when processes occur. A large chair may afford lifting and carrying from one place to another, and a doorway may afford passage from one room to another. But the attempt to combine the two affordances may fail when the plan is tried, e.g. if it is found during execution that the chair is too wide to fit through the doorway. After failing, a learner may discover that a combination of small rotations about different axes combined with small translations can form a compound process that results in the chair getting through the doorway. An older child may be able to see the possibility of the whole sequence of actions by visualising the whole process in advance, working out by visual reasoning how to move the chair into the next room, and then doing it. It is not at all clear what sort of brain mechanism or computer mechanism can perform that reasoning function or achieve that learning.

A child can also learn that complex affordances with positive and negative aspects can be disassembled so as to retain only the positive aspects. Children and adults often have to perform such "process de-bugging". A colleague's child learnt that open drawers could be closed by pushing them shut. The easiest way was to curl his fingers over the upper edge of the projecting drawer and push, as with other objects. The resulting pain led him to discover that the pushing could be done, slightly less conveniently, by flattening his hand when pushing, achieving the goal without the pain. Being

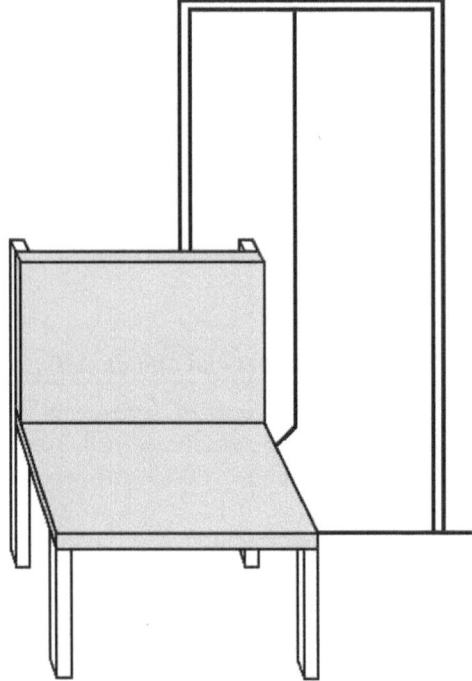

Figure 14.2 A person trying to move a chair that is too wide to fit through a door can work out how to move it through the door by combining a collection of translations and 3-D rotations about different axes, some done in parallel, some in sequence. Traditional AI planners cannot construct plans involving continuous interacting actions.

able to do geometrical reasoning enables a child who is old enough to work out *why* pushing with a flat hand prevents fingers being caught between the two surfaces.

4.6 The need for explanatory theories

Many humans can perform such reasoning by visualising processes in advance of producing them, but it is not at all clear what representations are used to manipulate information about the shapes and affordances of the objects involved.

There has been much work on giving machines with video cameras or laser scanners the ability to construct representations of 3-D structures and processes that can be projected onto a screen to show pictures or videos from different viewpoints, but, except for simple cases, robots still lack versatile spatial representation capabilities that they can use for multiple purposes such as manipulating objects, planning manipulations, and reasoning about them.

AI planning systems developed in the 1960s, such as the STRIPS planner (Fikes & Nilsson, 1971) and more complex recent planners (surveyed in Ghallab, Nau, & Traverso, 2004), all make use of the fact that knowledge about affordances can be abstracted into reusable information about the preconditions and effects of actions. Doing that provides a new kind of *cognitive* affordance: concerned with acting on information structures, demonstrating the possibility of combining knowledge about simple actions to provide information about complex actions composed of simple actions. However that work assumed that the information about actions and affordances can be expressed in terms of implications between propositions expressed in a logical formalism. Planning processes search for a sequence (or partially ordered network) of discrete actions that will transform the initial problem state into the desired goal state. But we need a richer mechanism to handle actions that involve interactions between continuously changing structural relations, like the changes that occur while an armchair is being rotated and translated simultaneously, or a sink is being wiped clean with a cloth.

Sloman (1971) challenged the then AI orthodoxy by arguing that intelligent machines would need to be able to reason geometrically as well as logically, and that some reasoning with diagrams should be regarded as being valid and rigorous, and in some cases more efficient than reasoning using logic, because logical representations are topic-neutral and sometimes lose some of the domain structure that can be used in searching for proofs. But it became clear that, although many people had independently concluded that AI techniques needed to be extended using spatial reasoning techniques, neither I nor anyone else knew how to design machines with the right kinds of abilities, even though there were many people working on giving machines the ability to recognise, analyse and manipulate images, or parts of images, often represented as 2-D rectangular arrays, though sometimes in other forms, e.g. using log-polar coordinates, e.g. Funt (1977). Other examples were presented in Glasgow, Narayanan, and Chandrasekaran (1995). More recent examples are Jamnik, Bundy, and Green (1999), Winterstein (2005).

4.7 An objection: blind mathematicians

It could be argued that the description of mathematical reasoning as "visual" must be wrong because people who have been blind from birth can reason about shapes and do logic and mathematics even though they cannot see (Jackson, 2002). That argument ignores the fact that some of the visual apparatus produced by evolution to support seeing and reasoning about structures and processes in the environment is in brain mechanisms that perform some of their functions without optical input: like the normal ability to see what *can* change in a situation when those changes are not occurring, or the ability to visualise a future sequence of actions that are not now being performed, and therefore cannot produce retinal input.

People who have been blind from birth may still be using the bulk of the visual system that evolved in their ancestors, just as sighted people may be using it when they dream about seeing things, and when they visualise diagrams with their eyes shut. The fact that individuals with different disabilities acquire a common humanity via different routes is an indication of how much of human mentality is independent of our specific form of embodiment. (Though not independent of the forms of our ancestors!)

5   Studying mechanisms vs studying requirements

5.1   *The importance of requirements*

Researchers often launch into seeking designs, on the assumption that the requirements are clear, e.g. because they think everyone knows what visual systems do, or because they merely try to model behaviours observed in particular experiments, or give too much importance to particular benchmark tests. This focus can lead researchers (and their students) to ignore the question of what else needs to be explained. As already remarked, successful models or explanations of a limited set of behaviours may not scale out. A deeper problem is that there is as yet not even a generally agreed ontology for discussing requirements and designs: we do not have an agreed set of concepts for describing cognitive functions with sufficient detail to be used in specifying requirements for testable working systems.

5.2   *Mistaken requirements*

Sloman (1989) lists nine common mistaken assumptions about vision systems. This chapter extends that list. For example it is tempting to suppose that a requirement for 3-D vision mechanisms is that they must construct a 3-D model of the perceived scene, with the components arranged within the model isomorphically with the relationships in the scene. Such a model can be used to generate graphical displays showing the appearance of the environment from different viewpoints. However, impossible figures like Escher's drawings and the Penrose triangle show that we are able to see a complex structure without building such a model, for there cannot be a model of an impossible scene. Perhaps, instead, a visual system constructs a large collection of fragments of information of various sorts about surfaces, objects, relationships, possible changes and constraints on changes in the scene, with most of the information fragments represented in registration with the optic array, though in an amodal form. Unlike an integrated model, this could support many uses of the information.

This idea generalises the notion of an "aspect graph", in which distinct 2-D views of a 3-D object are linked to form a graph whose edges represent actions that a viewer can perform, such as moving left, or right or up or down, to alter the view. This idea can be generalised so that more actions are included, such as touching or pushing, or grasping an object and more changes are produced such as two objects coming together or moving apart, or an object rotating, or sliding or tilting, or becoming unstable, etc.

Much current research in vision and robotics focuses on mechanisms that manipulate only representations of sensorimotor phenomena, e.g. statistical patterns relating multi-modal sensor and motor signals, and making no use of amodal exosomatic ontologies and forms of representation. Exceptions include SLAM (self localisation and mapping) mechanisms that create an exosomatic representation of building or terrain layout, and use that to plan routes. The great advantage of exosomatic representations is sometimes that a single representation of a process in the environment, such as two fingers grasping a berry, need not specify variations in sensor and motor signals that depend on precisely how the grasping is done and the viewpoint from which the process is observed, and which other objects may partially occlude relevant surfaces.

If a visual system's representation of a 3-D scene is made up of many piecemeal representations of fragments of the scene and the possible effects of processes involving those fragments, then in principle those fragments could form an inconsistent totality. So it would seem that an intelligent robot or animal must constantly check whether it has consistent percepts. However, since no portion of the 3-D environment is capable of containing impossible objects, there is normally no need for such a visual system to check that all the derived information is consistent, except in order to eliminate ambiguities, which can often be done more easily by a change of viewpoint. This is just as well since in general consistency checking is an intractable process.

5.3 Obvious and unobvious requirements: ontological blindness

Many people, e.g. Neisser (1967), Fidler and Leonardis (2007), have noticed the need for hierarchical decomposition of complex perceived objects and the usefulness of a mixture of top-down, bottom-up and middle-out processing in perception of such objects. In addition to such *part-whole* hierarchies there are also *ontological* layers, as illustrated in the Popeye program described in chapter 9 of Sloman (1978). The program was presented with images made of dots in a rectangular grid, such as Figure 14.3, which it analysed and interpreted in terms of:

- a layer of dot configurations (which could, for example, contain collections of collinear adjacent dots);
- a layer of line configurations, where lines are interpretations of "noisy" sets of collinear dots, and can form configurations such parallel pairs, and junctions of various sorts;
- a layer of 2-D overlapping, opaque "plates" with straight sides and rectangular corners, which in Popeye were restricted to the shapes of cut-out

Figure 14.3 This illustrates configurations of dots presented to the Popeye program in chapter 9 of Sloman (1978), which attempted to find a known word by concurrently looking for structures in several ontological layers, with a mixture of top-down, bottom-up and middle-out influences. If the noise and clutter were not too bad, the program, like humans, could detect the word before identifying all the letters and their parts. It also degraded gracefully.

capital letters, such as "A", "E", "F", "H" etc. represented in a noisy fashion by collections of straight line segments;

- a layer of sequences of capital letters represented by the plates, also in a "noisy" fashion because the plates could be jumbled together with overlaps;
- a layer of words, represented by the letter sequences.

The program illustrated the need to use ontologies at different levels of abstraction processed concurrently, using a mixture of top-down, bottom-up and middle-out processing, where lower levels are not *parts* of the higher levels, though each ontological layer has part-whole hierarchies. Going from one ontological layer to another is not a matter of grouping parts into a whole, but *mapping* from one sort of structure to another, for instance, interpreting configurations of dots as representing configurations lines, and interpreting configurations of lines as representing overlapping 2-D plates.

Text represented using several ontological layers may be regarded as a very contrived example, but similar comments about ontological layers can be made when a working machine is perceived, such as the internals of an old-fashioned clock. There will be sensory layers concerned with changing patterns of varying complexity in the optic array. A perceiver will have to interpret those changing sensory patterns as representing 3-D surfaces and their relationships, some of which change over time. At a higher level of abstraction there are functional categories of objects, e.g. levers, gears, pulleys, axles, strings, and various more or less complex clusters of such objects, such as escapement mechanisms.

Many vision researchers appreciate the need for a vision system to move between a 2-D ontology and a 3-D ontology. For a recent survey see Breckon and Fisher (2005). The need for such layers will be evident to anyone who works on vision-based text-understanding. However it is rare to include as many ontological categories, in different layers, as I claim are needed by an intelligent human-like agent interacting with a 3-D environment, or to relate those layers to different processing layers in the central architecture as explained in Sloman (2001).

5.4 Seeing mental states

Perception of intelligent agents in the environment involves yet another level of abstraction, insofar as some perceived movements are interpreted as actions with purposes, for instance a hand moving towards a cup. If eyes and face are visible, humans will often see not just actions but also mental states, such as focus of attention in a certain direction, puzzlement, worry, relief, happiness, sadness, and so on. Insofar as these are all *seen* rather than inferred in some non-visual formalism, the percepts will be at least approximately in registration with the optic array. Happiness is seen in one face and not in another. The requirement for perceptual mechanisms to use an ontological layer that includes mental states raises many problems that will not be discussed here, for example the need to be able to cope with referential opacity. Representing something that is itself an information user requires meta-semantic competences. These subtleties are ignored by researchers who train computer programs to label pictures of faces using words such as "happy", "angry", and claim that their programs can recognise emotional states.

5.5 Perceiving 2-D processes in the optic array

2-D processes involving changes in the optic array are also important, as J. J. Gibson pointed out. As noted in Sloman (1982), apart from perception of static scenes, vision is also required for online control of continuous actions (visual servoing) which requires different forms of representation from those required for perception of structures to be described, remembered, used in planning actions, etc.

Sometimes a 2-D projection is more useful than a 3-D description for the control problem, as it may be simpler and quicker to compute, and can suffice for particular task, such as steering a vehicle through a gap. But it is a mistake to think that only continuously varying percepts are involved in online visual control of actions: there is also checking whether goals or sub-goals have been achieved, whether the conditions for future processes have not been violated, whether new obstacles or new opportunities have turned up, and so on. Those can involve checking discrete conditions.

Unfortunately research on ventral and dorsal streams of neural processing has led some researchers (e.g. Goodale & Milner, 1992) to assume that control of action is separate from cognition, or worse, that spatial perception ("where things are") is a completely separate function from categorisation ("what things are"), apparently ignoring the fact that what an object is may depend on where its parts are in relation to one another, or where it is located in a larger whole.

5.6 *Layered ontologies*

We have seen, in section 5.3, that in addition to part-whole decomposition, perception can use layered ontologies. For example, one sub-ontology might consist entirely of 2-D image structures and processes, whereas another includes 3-D spatial structures and processes, and another kinds of "stuff" of which objects are made and their properties (e.g. rigidity, elasticity, solubility, thermal conductivity, etc.), to which can be added mental states and processes, e.g. seeing a person as happy or sad, or as intently watching a crawling insect. The use of multiple ontologies is even more obvious when what is seen is text, or sheet music, perceived using different geometric, syntactic, and semantic ontologies.

From the combination of the two themes (a) the content of what is seen is often processes and process-related affordances, and (b) the content of what is seen involves both hierarchical structure and multiple ontologies, we can derive a set of requirements for a visual system that makes current working models seem very inadequate.

Many people are now working on how to cope with pervasive problems of noise and ambiguity in machine vision systems. This has led to a lot of research on mechanisms for representing and manipulating uncertainty. What is not always noticed is that humans have ways of seeing high level structures and processes whose descriptions are impervious to the low level uncertainties: you can see that there definitely is a person walking away from you on your side of the road and another walking in the opposite direction on the other side of the road, even though you cannot tell the precise locations, velocities, accelerations, sizes, orientations and other features of the people and their hands, feet, arms, legs, etc. The latter information may be totally irrelevant for your current purposes.

Some notes on this can be found in this discussion paper on predicting affordance changes, including both action affordances and epistemic affordances: http://www.cs.bham.ac.uk/research/projects/cosy/papers/#dp0702

5.7 *Seeing is prior to recognising*

Much research on visual perception considers only one of the functions of perception, namely recognition. It is often forgotten that there are many things we can see, and can act in relation to, that we cannot recognise or label, and indeed that is a precondition for learning to categorise things.

When you see a complex new object that you do not recognise you may see a great deal of 3-D structure, which includes recognising many types of *surface fragment*, including flat parts, curved parts, changes of curvature, bumps, ridges, grooves, holes, discontinuous curves where two curved parts meet, and many more. In addition to many surface fragments, many of their relationships are also seen, along with relationships between objects, and also between different parts of objects. When those relations change we get different *processes* occurring concurrently, some at the same level of abstractions, others at different levels. They can be seen without necessarily recognising the objects involved.

Biological considerations suggest that, for most animals, perception of processes must be the most important function, since perception is crucial to the control of action, in a dynamic, sometimes rapidly changing environment that can include mobile predators and mobile prey, and where different parts of the environment provide different nutrients, shelter, etc. So from this viewpoint perception of structures is just a special case of perception of processes – processes in which not much happens.

Unfortunately, not only has very little (as far as I know) been achieved in designing visual systems that can perceive a wide range of 3-D spatial structures, there is even less AI work on perception of *processes*, apart from things like online control of simple movements which involves sensing one or two changing values and sending out simple control signals, for instance "pole balancing" control systems. There seems also to be very little research in psychology and neuroscience on the forms of representations and mechanisms required for perception of processes involving moving or changing structures, apart from research that merely finds out who can do what under what conditions. Examples of the latter include Heider and Simmel (1944), Michotte (1962) and Johansson (1973). Finding out which bits of the brain are active does not answer the design questions.

Addressing those deficiencies, including, for instance, explaining how GLs for process representation could work, should be a major goal for future vision research, both in computational modelling but also in neuroscience. Some speculations about mechanisms are presented in section 6.

5.8 Seeing possible processes: proto-affordances

We have already noted that an important feature of process perception is the ability to consider different ways a process may continue, some of them conditional on other processes intervening, such as an obstacle being moved onto or off the path of a moving object. Many cases of predictive control include some element of uncertainty based on imprecise measurements of position, velocity or acceleration. This sort of uncertainty can be handled using fuzzy or probabilistic control devices which handle intervals instead of point values.

However there are cases where the issue is not uncertainty or inaccuracy of measurement but the existence of very different opportunities, such as getting past an obstacle by climbing over it, or going round it on the left or on the right. It may be very clear what the alternatives are, and what their advantages and disadvantages are. E.g one alternative may involve a climb that requires finding something to stand on, while another requires a heavy object to be pushed out of the way, and the third requires squeezing through a narrow gap.

The ability to notice and evaluate distinct possible futures is required not only when an animal is controlling its own actions but also when it perceives something else whose motion could continue in different ways. How the ability to detect such vicarious affordances is used may depend on whether the perceived mover is someone (or something) the perceiver is trying to help, or to eat, or to escape from.

In simple cases, prediction and evaluation of alternative futures can make use of a simulation mechanism. But the requirement to deal explicitly with alternative possibilities requires a more sophisticated simulation than is needed for prediction: a predictive simulation can simply be run to derive a result, whereas evaluation of alternatives requires the ability to start the simulation with different initial conditions so that it produces different results. It also requires some way of recording the different results so that they can be used later for evaluation or further processing.

The ability to cope with branching futures in a continuous spatial environment poses problems that do not arise in "toy" discrete grid-based environments. The agent has to be able to chunk continuous ranges of options into relatively small sets of alternatives in order to avoid dealing with explosively branching paths into the future. How to do this may be something learnt by exploring good ways to group options by representing situations and possible motions at a high level of abstraction.

Learning to see good ways of subdividing continuous spatial regions and continuous ranges of future actions involves developing a good descriptive ontology at a higher level of abstraction than sensor and motor signals inherently provide. The structure of the environment, not some feature of sensorimotor signals makes it sensible to distinguish the three cases: moving forward to one side of an obstacle, moving so as to make contact with the obstacle and moving so as to go to the other side.

In addition to "chunking" of possibilities on the basis of differences between opportunities for an animal or robot to move as a whole, there are ways of chunking them on the basis of articulation of the agent's body into independently movable parts. For example, if there are two hands available, and some task requires both hands to be used, one to pick an object up and the other to perform some action on it (e.g. removing its lid) then each hand can be considered for each task, producing four possible combinations. However if it is difficult or impossible for either hand to do both tasks, then detecting that difficulty in advance may make it clear that the set of futures should be pruned by requiring each hand to do only one task, leaving only two options.

In humans, and some other species, during the first few years of life a major function of play and exploration in an infant is providing opportunities for the discovery of many hundreds of concepts that are useful for chunking sets of possible states of affairs and possible process, and learning good ways to represent them so as to facilitate predicting high level consequences, which can then be used in rapid decision-making strategies.

The ability to perceive not just what is happening at any time but what the possible branching futures are – including, good futures, neutral futures, and bad futures from the point of view of the perceiver's goals and actions, is an aspect of J.J. Gibson's theory of perception as being primarily about *affordances for the perceiver* rather than acquisition of information about some objective and neutral environment. However, I don't think Gibson considered the need to be able to represent, compare and evaluate multi-step branching futures: that would have been incompatible with his adamant denial of any role for representations and computation.

6 Speculation about mechanisms required: new kinds of dynamical systems

Preceding sections have assembled many facts about animal and human vision that help to constrain both theories of how brains, or the virtual machines implemented on brains work, and computer-based models that are intended to replicate or explain human competences. One thing that has not been mentioned so far is the extraordinary speed with which animal vision operates. This is a requirement for fast moving animals whose environment can change rapidly (including animals that fly through tree-tops). An informal demonstration of the speed with which we can process a series of unrelated photographs and extract quite abstract information about them is available online (http://www.cs.bham.ac.uk/research/projects/cogaff/misc/multipic-challenge.pdf). No known mechanism comes anywhere near explaining how that is possible especially at the speed with which we do it.

6.1 Sketch of a possible mechanism

Perhaps we need a new kind of dynamical system. Some current researchers (e.g., Beer, 2000) investigate cognition based on dynamical systems composed of simple "brains" closely coupled with the environment through sensors and effectors. We need to extend those ideas to allow a multitude of interacting dynamical systems, some of which can run decoupled from the environment, for instance during planning and reasoning, as indicated crudely in Figure 14.4. During process perception, changing sensory information will drive a collection of linked processes at different levels of abstraction. Some of the same processes may occur when possible but non-existent processes are imagined in order to reason about their consequences.

Many dynamical systems are defined in terms of continuously changing variables and interactions defined by differential equations, whereas our previous discussion, e.g. in section 4.3, implies that we need mechanisms that can represent discontinuous as well as continuous changes, for example to cope with topological changes that occur as objects are moved, or goals become satisfied. Another piece of evidence for such a requirement is the sort of discrete "flip" that can occur when viewing well known ambiguous figures such as the Necker cube, the duck-rabbit, and the old-woman/young-woman picture. It is significant that such internal flips can occur without any change in sensory input.

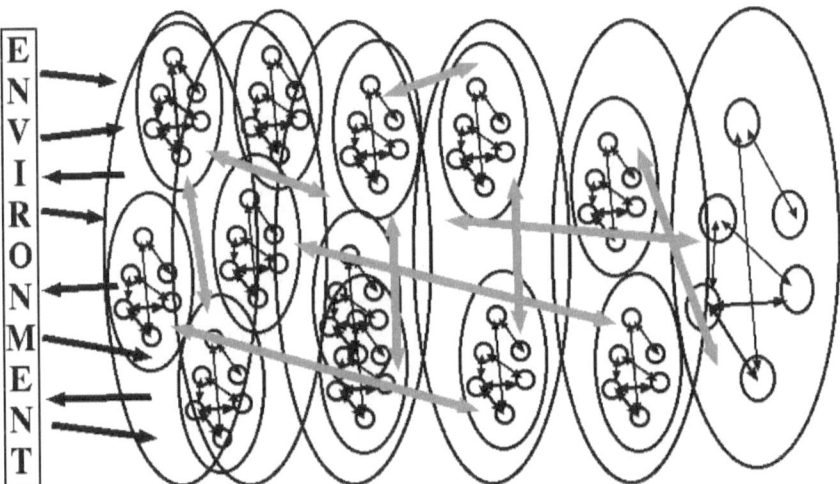

Figure 14.4 A crude impressionistic sketch indicating a collection of dynamical systems some closely coupled with the environment through sensors and effectors others more remote, with many causal linkages between different subsystems, many of which will be dormant at any time. Some of the larger dynamical systems are composed of smaller ones. The system does not all exist at birth but is grown, through a lengthy process of learning and development partly driven by the environment, as sketched in Chappell and Sloman (2007).

It is possible that adult human perception depends on the prior construction of a very large number of multi-stable dynamical systems each made of many components that are themselves made of "lower level" multi-stable dynamical systems. Many of the subsystems will be dormant at any time, but the mechanisms must support rapidly activating an organised, layered, collection of them partly under the control of current sensory input, partly under control of current goals, needs, or expectations, and partly under the control of a large collection of constraints and preferences linking the different dynamical systems.

On this model, each new perceived scene triggers the activation of a collection of dynamical systems driven by the low level contents of the optic array and these in turn trigger the activation of successively higher level dynamical systems corresponding to more and more complex ontologies, where the construction process is constrained simultaneously by general knowledge, the current data, and, in some cases, immediate contextual knowledge. Subsystems that are activated can also influence and help to constrain the activating subsystems, influencing grouping, thresholding, and removing ambiguities, as happened in the Popeye program described in section 5.3.

As processes occur in the scene or the perceiver moves, that will drive changes in some of the lower level subsystems which in turn will cause changes elsewhere, causing the perceived processes to be represented by internal processes at different levels of abstraction. Some of the same internal processing may be used to represent non-existent processes that could occur, when they are imagined in order to reason about what their consequences would be if they occurred.

On this view, a human-like visual system is a very complex multi-stable dynamical system:

- composed of multiple smaller multi-stable dynamical systems,
- that are grown over many years of learning,
- that may be (recursively?) composed of smaller collections of multi-stable dynamical systems that can be turned on and off as needed,
- some with only discrete attractors, others capable of changing continuously,
- many of them inert or disabled most of the time, but capable of being activated rapidly,
- each capable of being influenced by other subsystems or sensory input or changing current goals, i.e. turned on, then kicked into new (more stable) states bottom up, top down or sideways,
- constrained in parallel by many other multi-stable subsystems,
- with mechanisms for interpreting configurations of subsystem-states as representing scene structures and affordances, and interpreting changing configurations as representing processes,
- using different such representations at different levels of abstraction changing on different time scales,

- where the whole system is capable of growing new subsystems, permanent or temporary, some short-term (for the current environment) and some long term (when learning to perceive new things), e.g.

  - learning to read text
  - learning to sight read music
  - learning to play tennis expertly,
    etc.

That specification contrasts with "atomic-state dynamical systems", described in Sloman (1993) as dynamical systems (a) with a fixed number of variables that change continuously, (b) with one global state, (c) that can only be in one attractor at a time (d) with a fixed structure (e.g. a fixed size state vector).

The difficulties of implementing a dynamical system with the desired properties (including components in which spatial GLs are manipulated) should not be underestimated. The mechanisms used by brains for this purpose may turn out to be very different from mechanisms already discovered.

7 Concluding comments

In Sloman (1989) it was proposed that we need to replace "modular" architectures with "labyrinthine" architectures, reflecting both the variety of components required within a visual system and the varieties of interconnectivity between visual subsystems and other subsystems (e.g. action control subsystems, auditory subsystems, and various kinds of central systems).

One way to make progress may be to start by relating human vision to the many evolutionary precursors, including vision in other animals. If newer systems did not replace older ones, but built on them, that suggests that many research questions need to be rephrased to assume that many different kinds of visual processing are going on concurrently, especially when a process is perceived that involves different levels of abstraction perceived concurrently, e.g. continuous physical and geometric changes relating parts of visible surfaces and spaces at the lowest level, discrete changes, including topological and causal changes at a higher level, and in some cases intentional actions, successes, failures, near misses, etc. at a still more abstract level. The different levels use different ontologies, different forms of representation, and probably different mechanisms, yet they are all interconnected, and all in partial registration with the optic array (not with retinal images, since perceived processes survive saccades).

It is very important to take account of the fact that those ontologies are not to be defined only in terms of what is going on inside the organism (i.e. in the nervous system and the body) since a great deal of the information an organism needs is not about what is happening in it, but what is happening in the environment, though the environment is not some unique given (as implicitly assumed in Marr's theory of vision (1982), for example) but is different for different organisms, even when located in the same place. They have different niches.

As Ulric Neisser (1976) pointed out it is folly to study only minds and brains without studying the environments those minds and brains evolved to function in.

One of the major points emphasised here is that coping with our environment requires humans to be able to perceive, predict, plan, explain, reason about, and control processes of many kinds, and some of that ability is closely related to our ability to do mathematical reasoning about geometric and topological structures and processes. So perhaps trying to model the development of a mathematician able to do spatial reasoning will turn out to provide a major stepping stone to explaining how human vision works and producing convincing working models. Perhaps it will show that Immanuel Kant got something right about the nature of mathematical knowledge, all those years ago.

8 Acknowledgements

Many of the ideas reported in this chapter were developed as part of the requirements analysis activities in the EU funded CoSy project www.cognitivesystems.org. I am especially indebted to Jeremy Wyatt and other members of the Birmingham CoSy team. Mohan Sridharan kindly read and commented on a draft. Jackie Chappell has helped with biological evidence, the work on nature-nurture tradeoffs and the difference between Humean and Kantian causal understanding. Dima Damen helped me improve Figure 14.4. Shimon Edelman made several very useful comments. I thank Dietmar Heinke, the editor, for organising the workshop, driving the production of this volume and his patience with me.

9 References

Beer, R. (2000). Dynamical approaches to cognitive science. *Trends in Cognitive Sciences*, 4(3), 91–99. (http://vorlon.case.edu/beer/Papers/TICS.pdf)

Berthoz, A. (2000). *The brain's sense of movement*. London: Harvard University Press.

Breckon, T. P., & Fisher, R. B. (2005). Amodal volume completion: 3D visual completion. *Computer Vision and Image Understanding*, 99, 499–526. (doi:10.1016/j.cviu.2005.05.002)

Brooks, R. A. (1991). Intelligence without representation. *Artificial Intelligence*, 47, 139–159.

Fidler, S., & Leonardis, A. (2007). Towards scalable representations of object categories: Learning a hierarchy of parts. In *Proceedings conference on computer vision and pattern recognition* (pp. 1–8). Minneapolis: IEEE Computer Society. (http://vicos.fri.uni-lj.si/data/alesl/cvpr07fidler.pdf)

Fikes, R., & Nilsson, N. (1971). STRIPS: A new approach to the application of theorem proving to problem solving. *Artificial Intelligence*, 2, 189–208.

Funt, B. V. (1977). Whisper: A problem-solving system utilizing diagrams and a parallel processing retina. In *Ijcai* (pp. 459–464). Cambridge, MA: IJCAI'77.

Ghallab, M., Nau, D., & Traverso, P. (2004). *Automated planning, theory and practice.* San Francisco: Elsevier, Morgan Kaufmann.

Gibson, E. J., & Pick, A. D. (2000). *An ecological approach to perceptual learning and development.* New York: Oxford University Press.

Gibson, J. J. (1979). *The ecological approach to visual perception.* Boston, MA: Houghton Mifflin.

Glasgow, J., Narayanan, H., & Chandrasekaran, B. (Eds.). (1995). *Diagrammatic reasoning: Computational and cognitive perspectives.* Cambridge, MA: MIT Press.

Goodale, M., & Milner, A. (1992). Separate visual pathways for perception and action. *Trends in Neurosciences, 15*(1), 20–25.

Heider, F., & Simmel, M. (1944.). An experimental study of apparent behaviour. *American Journal of Psychology, 57*, 243–259.

Jablonka, E., & Lamb, M. J. (2005). *Evolution in four dimensions: Genetic, epigenetic, behavioral, and symbolic variation in the history of life.* Cambridge MA: MIT Press.

Jackson, A. (2002). The world of blind mathematicians. *Notices of the American Mathematical Society, 49*(10). (http://www.ams.org/notices/200210/comm-morin.pdf)

Jamnik, M., Bundy, A., & Green, I. (1999). On automating diagrammatic proofs of arithmetic arguments. *Journal of Logic, Language and Information, 8*(3), 297–321.

Johansson, G. (1973). Visual perception of biological motion and a model for its analysis. *Perception and Psychophysics, 14*, 201–211.

Kaneff, S. (Ed.). (1970). *Picture language machines.* New York: Academic Press.

Kant, I. (1781). *Critique of pure reason.* London: Macmillan. (Translated (1929) by Norman Kemp Smith)

Lakatos, I. (1976). *Proofs and refutations.* Cambridge, UK: Cambridge University Press.

Lakatos, I. (1980). The methodology of scientific research programmes. In J. Worrall & G. Currie (Eds.), *Philosophical papers, Vol I.* Cambridge: Cambridge University Press.

Lungarella, M., & Sporns, O. (2006). Mapping information flow in sensorimotor networks. *PLoS Computational Biolology, 2*(10:e144). (DOI: 10.1371/journal.pcbi.0020144)

Marr, D. (1982). *Vision.* San Francisco: W. H. Freeman.

Merrick, T. (2006). What Frege meant when he said: Kant is right about geometry. *Philosophia Mathematica, 14*(1), 44–75. (doi:10.1093/philmat/nkj013)

Michotte, A. (1962). *The perception of causality.* Andover, MA: Methuen.

Neisser, U. (1967). *Cognitive psychology.* New York: Appleton-Century-Crofts.

Neisser, U. (1976). *Cognition and reality.* San Francisco: W. H. Freeman.

Nelsen, R. B. (1993). *Proofs without words: Exercises in visual thinking.* Washingon, DC: Mathematical Association of America.

Poincaré, H. (1905). *Science and hypothesis.* London: W. Scott. (http://www.archive.org/details/scienceandhypoth00poinuoft)

Ryle, G. (1949). *The concept of mind.* London: Hutchinson.

Sauvy, J., & Suavy, S. (1974). *The child's discovery of space: From hopscotch to mazes – an introduction to intuitive topology.* Harmondsworth, UK: Penguin Education. (Translated from the French by Pam Wells)

Sloman, A. (1962). *Knowing and understanding: Relations between meaning and truth, meaning and necessary truth, meaning and synthetic necessary truth.* Unpublished doctoral dissertation, Oxford University, UK. (http://www.cs.bham.ac.uk/research/projects/cogaff/07.html#706)

Sloman, A. (1971). Interactions between philosophy and AI: The role of intuition and non-logical reasoning in intelligence. In *Proc 2nd ijcai* (pp. 209–226). London: William Kaufmann. (http://www.cs.bham.ac.uk/research/cogaff/04.html#200407)

Sloman, A. (1978). *The computer revolution in philosophy.* Hassocks, UK: Harvester Press (and Humanities Press). (http://www.cs.bham.ac.uk/research/cogaff/crp)

Sloman, A. (1979). The primacy of non-communicative language. In M. MacCafferty & K. Gray (Eds.), *The analysis of meaning: Informatics 5 Proceedings ASLIB/BCS conference, Oxford, March 1979* (pp. 1–15). London: Aslib. (http://www.cs.bham.ac.uk/research/projects/cogaff/81–95.html#43)

Sloman, A. (1982). Image interpretation: The way ahead? In O. Braddick & A. Sleigh (Eds.), *Physical and biological processing of images (Proceedings of an international symposium organised by The Rank Prize Funds, London, 1982)* (pp. 380–401). Berlin: Springer-Verlag. (www.cs.bham.ac.uk/research/projects/cogaff/06.html#0604)

Sloman, A. (1984). The structure of the space of possible minds. In S. Torrance (Ed.), *The mind and the machine: Philosophical aspects of artificial intelligence.* Chichester, UK: Ellis Horwood.

Sloman, A. (1989). On designing a visual system (towards a Gibsonian computational model of vision). *Journal of Experimental and Theoretical AI, 1*(4), 289–337. (http://www.cs.bham.ac.uk/research/projects/cogaff/81–95.html#7)

Sloman, A. (1993). The mind as a control system. In C. Hookway & D. Peterson (Eds.), *Philosophy and the cognitive sciences* (pp. 69–110). Cambridge, UK: Cambridge University Press. (www.cs.bham.ac.uk/research/projects/cogaff/81-95.html#18)

Sloman, A. (1994). Explorations in design space. In A. Cohn (Ed.), *Proceedings 11th European conference on AI, Amsterdam, August 1994* (pp. 578–582). Chichester: John Wiley.

Sloman, A. (2000). Interacting trajectories in design space and niche space: A philosopher speculates about evolution. In M. Schoenauer et al. (Eds.), *Parallel problem solving from nature – ppsn vi* (pp. 3–16). Berlin: Springer-Verlag.

Sloman, A. (2001). Evolvable biologically plausible visual architectures. In T. Cootes & C. Taylor (Eds.), *Proceedings of British machine vision conference* (pp. 313–322). Manchester: BMVA.

Sloman, A. (2008, March). *Kantian philosophy of mathematics and young Robots* (To appear in proceedings MKM08 No. COSY-TR-0802). School of Computer Science, University of Birmingham, UK. (http://www.cs.bham.ac.uk/research/projects/cosy/papers#tr0802)

Sloman, A., & Chappell, J. (2007). Computational cognitive epigenetics (Commentary on Jablonka & Lamb, 2005). *Behavioral and Brain Sciences, 30*(4), 375–376. (http://www.cs.bham.ac.uk/research/projects/cosy/papers/#tr0703)

Warneken, F., & Tomasello, M. (2006, 3 March). Altruistic helping in human infants and young chimpanzees. *Science,* 1301–1303. (DOI:10.1126/science.1121448)

Weir, A. A. S., Chappell, J., & Kacelnik, A. (2002). Shaping of hooks in New Caledonian crows. *Science, 297,* 981.

Winterstein, D. (2005). *Using diagrammatic reasoning for theorem proving in a continuous domain.* Unpublished doctoral dissertation, University of Edinburgh, School of Informatics, UK. (http://www.era.lib.ed.ac.uk/handle/1842/642)

15 Computational modelling in behavioural neuroscience

Methodologies and approaches (minutes of discussions at the workshop in Birmingham, UK, in May 2007)

Dietmar Heinke

This chapter reports on discussions at the international workshop "Closing the Gap between Neurophysiology and Behaviour: A Computational Modelling Approach" held in Birmingham, UK, May–June 2007. The aim of this chapter is not to set out a coherent scientific framework for computational modelling in behavioural neuroscience. As will become clearer in the course of the chapter, this would be an impossible task since the field employs a broad range of approaches and no agreed procedure exists. Instead, the report is an attempt to summarize the opinions of the workshop's participants as truthfully as possible, and, in so doing, to highlight the different approaches and methods utilized in the field. Thus, this chapter provides a kind of summary of the methodologies represented in this volume. In summarizing these approaches, though, the intention is to take at least a small step on the path of developing a coherent framework for computational modelling of human behaviour.

To begin with, there was general agreement between participants that the aim of computational modelling in behavioural neuroscience can be defined as "understanding how human/animal behaviour is generated by the brain". By the term "understanding", we mean "discovering the underlying computational principles behind human/animal behaviour". These computational principles should be grounded in some way in the neurophysiology of the organism. This definition also implies that computational modelling represents an explanatory approach to understanding human behaviour and that explanations are based on the neural substrate. It is important to point out that this type of computational modelling contrasts with other mathematical methods employed in behavioural neuroscience that are merely descriptive, insofar as they only provide equations that fit observed data, without explaining how the behaviours are generated. The participants at the workshop discussed numerous questions arising from this aim of modelling. Which is the best methodology to achieve this objective? Which is the correct mathematical model of the neural substrate? In other words, which is the correct abstraction level? How much shall data (behavioural, electro-physiological, etc.) constrain the modelling effort?

1 Methodologies

To report on the discussions about the methodology in computational mod-elling, it is useful to structure opinions into three approaches: the Marrian approach, the "physics" approach and the "engineering" approach. The Marrian approach follows the framework suggested by David Marr (1982). He proposed three levels of analysis: computational theory; representation and algorithm; hardware implementation. Computational theory specifies the goal, the reason and the strategy for a given computation. The second level (representation and algorithm) defines the input/output representations and the algorithm of how these two representations are linked. This level is actually oblivious to the form of implementation at the neural level. Charac-teristics of the neural substrate only become important at the third level, the level of hardware implementation. The levels are only loosely related, and there is no clearly defined way of moving from one level to the next. How-ever, as will be reported later, Kevin Gurney argued for a generalization of the Marrian approach in which the three levels are applied to different levels of abstraction from the neural substrate (see Churchland & Sejnowski, 1992; see also Gurney, chapter 6, this volume).

The "physics" approach begins with a theory of a certain human behaviour (equivalent to Marr's computational level) and considers compu-tational models as a method of testing the hypotheses of the theory. In behavioural neuroscience, the model is based on general neurophysiological principles (e.g., certain types of excitatory and inhibitory connections) and explores the conditions (e.g., parameter settings) under which the model produces the target behaviour (e.g., certain types of attractor behaviour). At the same time, these conditions provide an explanation for evidence found in behavioural experiments – for example, a certain synaptic connectivity proves to be responsible for a certain psychological law (for an example, see Deco & Rolls, chapter 3, this volume). The physics approach also stresses that models should be "minimal", as pointed out by Li Zhaoping. In such a minimal model, the removal of just one element should lead to a breakdown of the model's target behaviours ("Occam's razor"; William of Ockham, 1285–1347/49). However, it was also noted in discussions that in the evolution of the brain, new structures were incorporated into old structures, therefore not necessarily following the principle of minimality.

Finally, the "engineering" approach also bases the design of the model on general principles derived from neurophysiology. However, in contrast to the physics approach, this approach goes through an extensive design phase using the neurophysiological principles as building blocks to implement the model's target behaviours. During this design process, the engineering approach may employ mechanisms that are only loosely related to evidence from neurophysiological research. For instance, the model may contain a topology of connections that is motivated only by the objective of implementing the target behaviour and not directly by neurophysiological evidence (for an example, see Heinke et al., chapter 5, this volume). This framework allows more abstraction from the neural substrate than the physics approach and, contrary to the Marrian approach, is less concerned with the representations utilized in the model.

As the cross-references in the definitions of the different approaches show, the distinctions among the three approaches are not as clear-cut as the labels may suggest. Instead, they mainly differ in the emphasis the approaches make. However, it has to be said that the Marrian approach leaves more room for abstractions from the neural substrate with its second level. But then again, the level of abstraction is a matter of debate in any case, as the next section will show.

2 Level of abstractions

Every approach has to decide which aspects of the neural substrate have to be taken into account. There was general agreement between discussants that, at the current stage of play, it is not possible to take into account all aspects of the neural substrate. It was even argued that it is not really sensible to consider all details of neurophysiology, as the objective is to unearth the general principles behind human information processing. However, there was disagreement over which aspects of the neural substrate are crucial properties of human information processing. The discussions at the meeting mainly revolved around the issue of rate-coded neurons versus spiking neurons. Some discussants considered rate-coded neurons as the suitable level of abstraction from processes in the neural substrate; others suggested that the spiking level includes important mechanisms for understanding human behaviour. In addition, Fred Hamker pointed out that the organisation of the brain in functionally specialised regions represents a crucial feature of the neural substrate.

A few people suggested that moving between levels of abstractions is an important part of the modelling enterprise. Gustavo Deco suggested using the mean-field level (rate-coded neurons) to explore the relationship between a model's behaviour (attractors) and parameters at first and, then, using a veridical transformation of the model to the spiking level to simulate the real behaviour. On this new level, additional crucial features would be included, such as noise, different types of synaptic transmitters, and so on (see Deco & Rolls, chapter 3, this volume). However, Thomas Trappenberg pointed out that noise and different types of synaptic transmitters can also be taken into account at the rate-coded level in a similar way. On the other hand, Kevin Gurney advocated a generalized Marrian framework for moving between levels of abstractions from the neural substrate (see Gurney, chapter 6, this volume) whereby, to each level of abstraction, the three Marrian levels of analysis can be applied, and the links between the levels of abstractions are given by the first level and the third level in Marr's hierarchy. For instance, a dendritic tree can be described as a general-purpose digital signal processor (hardware implementation). An analysis of this tree may show that, under certain circumstances, this general-purpose processor performs multiplication and addition of incoming rate codes (algorithmic level). Hence, moving up in the levels of abstractions, a neuron's input processing can be modelled by simple additions and multiplications, which, in turn, can be seen as hardware implementation on the rate-coded level.

Other criteria for choosing abstraction levels, apart from the characteristics of the neural substrate, were also suggested in discussions. Interestingly, a point raised by several discussants was that the abstraction level can also be determined by the target audience for a model. For instance, spiking neurons facilitate the communication between modellers and neurophysiologists; or, models that are linked to brain structures are of interest to neuropsychologists. Such an "adaptation" of a model's vocabulary can be particularly important for its success in an interdisciplinary science such as computational modelling. Moreover, it was pointed out that even within the research community, the choice of the neuron model is crucial for the success of a particular model. For instance, a model based on competitive and cooperative processes with rate-coded neurons will have a larger impact on the area of computational modelling, as there is a large community of researchers utilizing this type of approach. This argument led some discussants to make a general point that the term "understanding" in the definition of the field strongly depends on the audience anyway. (Perhaps this point should have been elaborated more in discussions.)

Workshop participants also pointed out that the relevant experimental data determine the appropriate abstraction level. For instance, data from fMRI and ERP studies may best be modelled with rate-coded neuron models (Karl Friston). On the other hand, to model behavioural data from stroke patients where there is evidence that strokes affect certain neurotransmitters, a spiking model that includes such neurotransmitters can be the appropriate choice (see Humphreys et al., chapter 4, this volume).

Finally, the pragmatic problem of computer time was raised. Obviously, some approaches (e.g., spiking models) are computationally more demanding than others. However, as Deco pointed out, his approach of combining the rate-coded and spiking approaches can be a solution to this problem, as the time-consuming issue of determining parameters can be executed on the rate-coded (mean-field) level and then the real behaviour can be simulated on the spiking level (see Deco & Rolls, chapter 3, this volume).

3 Experimental data

Participants also discussed the role of data (experimental, behavioural, fMRI, physiological, etc.) in evaluating models. Again, opinions can be split into three factions: "oblivious", "qualitative", "quantitative".

The first faction argued that simply implementing a working solution would be enough to show that one has achieved understanding of human information processing – for example, implementing an object recognition system with spiking neurons. The model would be indicative of general processing principles behind object recognition.

The second faction argued that the data still tell us something about the human process. However, at the current stage of play, the exact quantitative outcome of an experiment is influenced by several factors that may not (or cannot) be covered by the model, and a quantitative fit would lead to a spurious fit, overplaying the model's hand. For instance, reaction times in visual search experiments exhibit roughly two types of outcomes: a highly efficient (parallel) search and a highly inefficient (serial) search. However, the exact search efficiency is influenced by numerous factors, such as similarity between items, search display density, and so on. Of these factors, a model would normally cover only a subset. Hence, an exact fit between data and model would insinuate a false correctness of the model (for an example, see Heinke et al., chapter 5, this volume).

The third faction, however, argued that the qualitative approach would make it difficult to refute the model. Instead, they suggested that if the problem is constrained enough (e.g., a simple psychophysical experiment), the numerous constraints from the neural substrate (e.g., biological plausibility of parameters and topologies) lead to a well-constrained problem. Therefore, the degrees of freedom in a quantitative fit are sufficiently reduced.

Nevertheless, the three fractions agreed that, in general, computational modelling should feed back to experimenters and raise new questions for experimentalists by generating novel predictions. However, some participants also pointed out that a model does not necessarily lead to predictions for absolutely new data, but can also capture data that it was not initially designed for. Such a generalization of the model's behaviour could also constitute a validation of the model, comparable to a prediction of novel data.

4 Final remark

As pointed out at the beginning, this chapter aims to contribute towards the development of a coherent science of computational modelling by reporting on discussions held at the Birmingham workshop. At this junction, it is also worth looking at the way science progresses in general. Here, as Aaron Sloman pointed out, a look at the Lakatosian philosophy of science can be helpful. According to Lakatos (Lavor, 1998), the progress of science is best conceptualized in terms of a "research programme". A research programme is a collection of different theories sharing a theoretical "core", heuristics and a protective belt. The "core" of a research programme is a set of theoretical assumptions, on which it is based – for example, the three Newtonian laws of motion. The heuristics are methodological prescriptions of how theories can be constructed from the core. These heuristics are particularly important when the core is confronted with "anomalous" experimental results that seemingly or not seemingly refute the core assumptions. In such situations, the heuristics lead to the construction of the "protective belt".

Now, rather than charting the scientific progress by looking at the individual theories, Lakatos suggested to gauge a research programme as either progressive or degenerative. A progressive research programme generates interesting research and experimental questions, whereas a degenerative research programme merely adapts to new data by extending the protective belt, but generates no new thoughts. A classical example for his theory is the development of Copernican theory. According to Lakatos, at its onset Copernican theory was of a similar quality, or even less, compared to the competing Ptolemaic world view, especially since ellipses had not been invented and, therefore, no parsimonious descriptions of the observed planet trajectories were available. However, the Copernican view sparked a new, thought-provoking train of research, making the Copernican world view a progressive research programme and, through the developments of the mathematical description of ellipses and Newton's gravitation theory, a successful theory of the solar system. However, this can only be said with hindsight – at the advent of Copernican theory, this was not at all clear.

The Lakatosian philosophy of science applied to computational modelling takes on the following shape. The "core" of computational modelling can be seen as the assumption that human behaviour is best described by computational principles. The heuristics are the Marrian approach, the physics approach and the engineering approach, and the various models represent the protective belt. Interestingly, in the protective belt lies a warning from Lakatosian philosophy. If this protective belt grows without generating new ideas – for example, models becoming more and more detailed with every piece of new evidence and leading to an increasing number of types of models – the scientific research programme may turn into a degenerative exercise.

On the other hand, the Lakatosian view suggests that scientific progress lies with the research programmes as such, and not so much with the individual theories. Hence, the diversity of approaches and methods reported in this chapter may even be appreciated, and the aim of coherence set out at the beginning of it may not be useful. Finally, it would have been interesting to compare computational modelling with other, competing research programmes such as qualitative modelling or mathematical descriptive approaches. However, this would have led too far beyond the remit of the workshop, but it might be the topic of a future workshop.

Anyway, the Lakatosian view is that only time will tell. Until then I would like to leave the reader with the exciting and interesting chapters in this book, which are certainly evidence for a progressive scientific research programme.

5 Acknowledgement

I would like to thank Glyn Humphreys, Aaron Sloman, Karl Friston and Thomas Trappenberg for comments on an earlier draft of this chapter.

6 References

Churchland, P. S., & Sejnowski, T. J. (1992). *The computational brain*. Cambridge, MA: MIT Press.

Lavor, B. (1998). *Lakatos: An introduction*. London: Routledge.

Marr, D. (1982). *Vision*. New York: W. H. Freeman.

Author index

Subject index